CONDITION: VG+
Edition: 1st

1970 289 I

Prtg: 1P (252511)

DJ/NO

COMMENTS: Volume 1 ONLY! Brown cloth wrs with white [illegible] lig + red+white sc denutur. Cover b fecs sly bd, vmd 1FC. OW ③ Text block + rest of bks in FC

KEYWORDS: (HLW1500), Mathematics, Applied Mathematics, Text books

19.4

— 0 42⁵⁰

INTRODUCTION TO APPLIED MATHEMATICS
Volume 1

INTRODUCTION TO APPLIED MATHEMATICS
Volume 1

JOSEPH GENIN
Purdue University

JOHN S. MAYBEE
University of Colorado

Holt, Rinehart and Winston, Inc.
*New York, Chicago, San Francisco, Atlanta, Dall
Montreal, Toronto, London, Sydney*

Copyright © 1970 by Holt, Rinehart and Winston, Inc.
All rights reserved
Library of Congress Catalog Number: 78-98512
SBN: 03-083112-1
Printed in the United States of America
1 2 3 4 5 6 7 8 9

PREFACE

This text is the outgrowth of our strong conviction that applied mathematics is usually presented in an unattractive form. Texts at the advanced undergraduate level often consist of a collection of "topics," with little or no attempt made to relate the topics. In this book we have attempted to present the material in a unified way, motivated by more or less realistic problems drawn from a variety of fields. Moreover, we have tried to give a clear description of the nature and extent of the practical information that can be obtained with the techniques presented.

In order to succeed in achieving our aim, we have laid a heavy burden upon the instructor and, to a lesser degree, upon the student. For example, our efforts to describe significant applications in physics and engineering require both the student and instructor to have more background than is required to study other applied mathematics texts. It is our hope that this additional factor will be more a challenge than a deterrent to the study of the book.

There are a variety of topics normally included in comparable texts at the same level which we have omitted. We are currently preparing a second volume written in the same style. In it we will, presumably, repair some of these omissions.

In developing this text we have constantly kept in mind the recommendations of the Committee on the Undergraduate Program in Mathematics (CUPM) with regard to applied mathematics curricula. The extent to which we have lived up to these recommendations is a matter upon which others will have to pass judgment. We have also been influenced by the very excellent pioneering work of Ben Noble in his book "Applications of Undergraduate Mathematics in Engineering" (The Mathematical Association of America and The Macmillan Company, New York, 1967); a work which itself was directed by CUPM. In fact, there are problems in this text based upon examples treated by Professor Noble.

A source written in the spirit of our text is material assembled at Cornell by Block, Cranch, Hilton, and Walker (*Engineering Mathematics*, Vols. 1 and 2, Cornell University, Ithaca, New York). We saw this subject matter as we were finishing up the present book. In our opinion it provides an excellent introduction and preparation at the sophomore level for the material we present.

A few remarks are in order regarding specific ways in which the material can be used. Our aim has been primarily to provide a skeleton, a fundamental selection of topics upon which the instructor can build in a variety of ways and some of which can be omitted. For example, we have built into Chapter 2 an introduction to a variety of topics in linear algebra. Many students nowadays will have learned some linear algebra during their first two years of mathematics courses and therefore, for such students, this material can be treated very cursorily or even partly omitted. On the other hand, the topic treated in the last two sections of the chapter is somewhat specialized in that special properties of commuting matrices are used. Thus, although the material represents a particularly elegant chapter in applied mathematics, it can be omitted. Again, in Chapter 3 we use a diffusion problem as a vehicle for presenting some basic material on Fourier series. Many instructors will wish to do more with this subject than we have done here. Many possibilities suggest themselves. An example is the very interesting use of Fourier series methods and linear algebra to find the boundaries of the regions of instability for the Mathieu equation. This is a current, real problem in applied mathematics.

Since we hope and anticipate that our text will be used by various instructors with great flexibility, it is somewhat difficult to provide any precise guidelines for using the book for a specific type of course. There is, perhaps, enough material in the text for a full year course if it is suitably expanded and taught immediately after the student's first two years of mathematics. On the other hand, there is probably not much more than a semester's material for students who have already had a good advanced calculus course. We have found that the first four chapters require $1\frac{1}{2}$ semesters for the first type of student even when several sections are omitted from Chapters 2 and 3.

We have included a wide variety of problems interspersed at the end of various sections. They form an essential part of the entire presentation and every instructor is urged to read them carefully so that he may assign them wisely. A few problems are so essential to later developments that we have marked them with an asterisk. All such problems should be assigned as homework or worked in the classroom.

Finally we would like to thank a number of people who have been encouraging or helpful in one way or another. First and foremost we would like to thank the people at Holt, Rinehart and Winston, Inc. who took the unusual step of persuading us to write this book and provided aid and encouragement at every stage. We wish to thank E. L. Infante and Jane Scanlan for providing exceptionally thoughtful reviews and for their encouragement. Finally thanks are due to Mrs. Linda Bell, Mrs. Janis Smith, and Mrs. Marie Kindgren, who typed various portions of the manuscript at one stage or another, for their patience and diligence in dealing with two difficult authors.

West Lafayette, Indiana J. G.
Boulder, Colorado J. S. M.

January 1970

CONTENTS

Preface, v

1. **ORDINARY DIFFERENTIAL EQUATIONS**   1

    1.1. The RLC Electric Circuit, 1
    1.2. The Simplest Leakage Problem, 4
    1.3. The Wronski Determinant, 12
    1.4. General Solution of the RLC Circuit, 15
    1.5. Uniqueness and Stability, 20
    1.6. The Energy Method, 24
    1.7. A More General Uniqueness Theorem, 33
    1.8. The Simple Pendulum, 36
    1.9. The Linear Pendulum, 44

2. **ORDINARY DIFFERENTIAL EQUATIONS—SIMPLE SYSTEMS**   51

    2.1. Introduction, 51
    2.2. The Linear Double Pendulum, 52
    2.3. A System of First-Order Differential Equations, 65
    2.4. Solution of the Homogeneous Problems of Cascaded Loops, 70
    2.5. Eigenvalues and Eigenvectors of the Problem of Cascaded Loops, 82
    2.6. The Nonhomogeneous Problem of Cascaded Loops, 87

2.7. Remarks about Linear First-Order Systems, 95
2.8. Eigenvalues and Eigenvectors of a Class of Damped Dynamic Systems of Equations, 100
2.9. Whirling Motion of a Rotating Shaft, 106

3. PARTIAL DIFFERENTIAL EQUATIONS 115

3.1. Heat Conduction in a Rod, 115
3.2. Some Properties of the Heat Equation, 122
3.3. Remarks on the Fourier Method, 131
3.4. A Diffusion Problem, 139
3.5. The Maximum-Value Principle, 147
3.6. Uniqueness in the Heat-Conduction Problem, 149
3.7. The Nonhomogeneous Heat Equation, 154
3.8. Asymptotic Behavior of Solutions, 162
3.9. A Nonlinear Problem, 171

4. OTHER TIME-DEPENDENT PARTIAL DIFFERENTIAL EQUATIONS 181

4.1. Longitudinal Vibration of a Prismatic Rod, 181
4.2. The Cauchy Problem for the Wave Equation, 187
4.3. The Energy Method for the Wave Equation: Uniqueness, 193
4.4. The Method of Separation of Variables, 201
4.5. Transverse Vibrations of a Rod, 210
4.6. A Nonconservative Problem, 215
4.7. The Follower Problem, 224
4.8. The Timoshenko Beam Equation, 231

5. THE VARIOUS ASPECTS OF PHYSICAL PROBLEMS 237

5.1. Introduction, 237
5.2. The Energy Integral, 239
5.3. Separation of Variables, 240
5.4. A Difference Approximation, 246
5.5. Some Properties of the Eigenvalues, 252
5.6. A Discrete System, 256
5.7. The Influence Function, 259
5.8. Remarks on the Integro-Differential Equation, 264
5.9. Some Properties of the Symmetric Integral Equation, 270
5.10. Solutions of the Integral Equation, 277

Index, 285

INTRODUCTION TO APPLIED MATHEMATICS
Volume 1

Chapter 1

ORDINARY DIFFERENTIAL EQUATIONS

1.1 THE RLC ELECTRIC CIRCUIT

We begin the investigation of ordinary differential equations by studying one of the standard problems in electric circuit theory. Consider the circuit shown in Figure 1.1 which consists of a resistance R, an inductance L, and a capacitance C, all in series with an electromotive source $e = e(t)$. The circuit has two switches K_1 and K_2 as shown. We wish to study the variation in the charge q on the capacitor as a function of time. The charge is defined as

$$\frac{dq}{dt} = i(t), \tag{1.1}$$

where $i = i(t)$ is the current passing through the element in question and is a function of time as indicated.

To formulate the problem we use Kirchhoff's first law of circuit theory, which states that in an electric network the sum of the voltage drops taken in a specified direction around any closed loop equals the sum of the voltage rises in that direction.

Let us start the analysis by closing switch K_1 while leaving switch K_2 open. Thus there exists but one loop. The voltage drops across the circuit elements R, L, and C in the loop are determined by the well-known relationships

$$e_R = iR, \quad e_L = L\frac{di}{dt}, \quad \text{and} \quad e_C = \frac{1}{C}\int i\, dt, \tag{1.2}$$

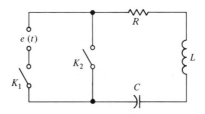

FIGURE 1.1 The RLC circuit.

respectively. The numbers R, L, and C are physical constants for the circuit elements which we may assume to be measurable, hence known quantities.

Summing the voltage drops in the circuit, we obtain

$$e(t) - iR - L\frac{di}{dt} - \frac{1}{C}\int i\,dt = 0 \qquad (1.3)$$

as the expression of Kirchhoff's law for the circuit with K_2 open and K_1 closed. In (1.3) we presume as known the input electromotive force e as a function of t so that the equation is an *integro-differential equation* for the determination of the current in the loop as a function of time.

Note that the last of Equations (1.2) gives e_C as an indefinite integral of the function i. It immediately follows that e_C can be determined only to within an arbitrary constant. Of course, knowledge of the charge stored in the capacitor at any fixed time t_0 will enable us to determine this arbitrary constant. Henceforth, such information about a system which is contingent on a fixed time will be referred to as an *initial condition*.

Let us assume, without loss of generality, that the initial condition is $q(t_0) = q_0$. Thus, integrating Equation (1.1) from time t_0 to any later time t, we obtain

$$q(t) - q(t_0) = \int_{t_0}^{t} i(s)\,ds,$$

where s is a dummy variable. On substituting q_0 for $q(t_0)$, we obtain

$$q(t) = q_0 + \int_{t_0}^{t} i(s)\,ds. \qquad (1.4)$$

Placing Equation (1.4) into (1.3) and rearranging terms yields

$$\frac{di}{dt} + \frac{R}{L}i + \frac{1}{LC}\int_{t_0}^{t} i(s)\,ds = -\frac{q_0}{L} + \frac{e(t)}{L}, \quad \text{for } t > t_0. \qquad (1.5)$$

Note that in formulating (1.5) we were required to select a particular indefinite integral of i. Further, since (1.5) is an equation involving the first derivative of i another initial condition is required if a complete solution is to be found. Again, without loss of generality, let us say that the required initial condition is

THE *RLC* ELECTRIC CIRCUIT   3

$$\frac{dq}{dt}\bigg|_{t=t_0} = i(t_0) = \dot{q}_0. \tag{1.6}$$

Equation (1.5) may now be solved for a function i subject to the initial condition (1.6). It is customary in most engineering practice to convert this integro-differential equation to a differential equation. This may be done easily because we have additional information concerning the behavior of the dependent variable $i(t)$ in (1.3), specifically (1.1). Thus we may substitute (1.1) into (1.3) to obtain the differential equation

$$L\frac{d^2q}{dt^2} + R\frac{dq}{dt} + \frac{1}{C}q = e(t), \tag{1.7}$$

where q is now the dependent variable.

We will now proceed to formulate and solve a number of problems of both mathematical and engineering interest, thereby providing the reader with insight to both aspects of Equation (1.7) and, more important, the physical system which it represents. To facilitate notation let us use a dot above a quantity to represent its time derivative. Therefore

$$\dot{q} = \frac{dq}{dt}, \ddot{q} = \frac{d^2q}{dt^2}, \text{ and so on.}$$

Our problem now is to determine the charge q on the condenser in the electric circuit of Figure 1.1 for all times $t \geq t_0$ when K_1 is closed and K_2 is open.

Thus we are given the differential equation

$$\ddot{q} + \frac{R}{L}\dot{q} + \frac{1}{LC}q = \frac{e(t)}{L}, \quad \text{for all} \quad t > t_0 \tag{1.8}$$

subject to the initial conditions

$$q(t_0) = q_0 \quad \text{and} \quad \dot{q}(t_0) = \dot{q}_0. \tag{1.9}$$

Equation (1.8), subject to conditions (1.9), is referred to as an *initial-value problem*.

Definition 1.1

When we say that we seek a solution to this problem what is meant is that we wish to find a function $q(t)$ defined and continuous for all $t \geq t_0$ having a continuous second derivative for $t > t_0$ and satisfying (1.8) for $t > t_0$ and (1.9) at $t = t_0$. Thus q must have a first derivative from the right at $t = t_0$.

With this understanding we shall only admit functions $e(t)$ which are continuous for $t > t_0$. Of course, if $\ddot{q}$ is continuous for $t > t_0$, then q and $\dot{q}$

are also continuous for $t > t_0$. Thus, our restrictions insure the continuity for each term in Equation (1.8).

The differential Equation (1.8) is called a *second-order differential equation* because its highest derivative of q is the second derivative. The variable q is usually called the *dependent variable* in the equation and t the *independent variable*. This is because we are required to find q as a function of t in order to find a solution of the equation.

Definition 1.2

The order of a differential equation is the order of the highest derivative of the dependent variable appearing in the equation.

1.2 THE SIMPLEST LEAKAGE PROBLEM

It will turn out that the solution to the equation

$$\ddot{q} + \frac{R}{L}\dot{q} + \frac{1}{LC}q = 0, \quad t > t_0 \tag{1.10}$$

will be one part of the solution to (1.8). In addition (1.10) is, in general, easier to solve than (1.8) and, more important, represents another mathematical model which can be formulated representing the circuit in Figure 1.1. Mathematically, it is the simplest problem we can formulate for the RLC circuit.

To formulate (1.10) let us suppose that an electromotive force $e(t)$ has been applied to the circuit of Figure 1.1 for a period of time $t = 0$ to $t = t_0$, with the result that at time t_0 the condensor is charged. That is, $q(t_0) = q_0$ and $\dot{q}(t_0) = \dot{q}_0$. Then at time t_0, K_1 is opened and simultaneously K_2 is closed. We wish to determine what happens to the charge on C for all time $t > t_0$. We shall refer to this problem as the *leakage problem* for the RLC circuit. Intuitively, one expects that since the external voltage has been removed the charge on C will dissipate with time, hence the term *leakage*.

We now start the solution by showing how to find a family of functions satisfying (1.10). This is done by seeking solutions having the form

$$q(t) = e^{\lambda t} \tag{1.11}$$

where λ is to be determined so as to make (1.11) a solution to (1.10). Thus, placing (1.11) into (1.10) we obtain

$$\left(\lambda^2 + \frac{R}{L}\lambda + \frac{1}{LC}\right)e^{\lambda t} = 0.$$

It will be essential for us to know that $e^{\lambda t} \neq 0$ for all real or complex values of λ and all real values of t. If μ is real $e^{\mu t} > 0$ for all real t so that it is only necessary to discuss the case where λ is complex.

Suppose that $\lambda = \mu + i\nu$ where μ and ν are real. Then

$$e^{\lambda t} = e^{\mu t}e^{i\nu t} = e^{\mu t}(\cos \nu t + i \sin \nu t).$$

Here we have made use of Euler's formula to write $e^{\lambda t}$ as a complex number of the form $z = x + iy$, with

$$x = e^{\mu t} \cos \nu t \quad \text{and} \quad y = e^{\mu t} \sin \nu t.$$

The absolute value of the complex number z is defined to be

$$|z| = (x^2 + y^2)^{1/2},$$

and it is clear that $|z| = 0$ if and only if $z = 0$. We then have

$$|e^{\lambda t}| = e^{\mu t}(\cos^2 \nu t + \sin^2 \nu t)^{1/2} = e^{\mu t} > 0$$

for all real μ and t. This shows that $e^{\lambda t} \neq 0$.

Since $e^{\lambda t} \neq 0$ it follows that the equality above can be valid, if and only if

$$\lambda^2 + \frac{R}{L}\lambda + \frac{1}{LC} = 0. \tag{1.12}$$

Equation (1.12) is a quadratic algebraic equation and thus there exist two values for λ. They are

$$\lambda_1 = -\frac{R}{2L} + \frac{1}{2}\sqrt{\left(\frac{R}{L}\right)^2 - \frac{4}{LC}},$$

and $\tag{1.13}$

$$\lambda_2 = -\frac{R}{2L} - \frac{1}{2}\sqrt{\left(\frac{R}{L}\right)^2 - \frac{4}{LC}}.$$

Note that the λ's are constants.

Anticipating future terminology we refer to Equation (1.12) as the *characteristic equation* of (1.10) and to λ_1 and λ_2 as the *eigenvalues* of (1.10). An inspection of (1.13) reveals that three possibilities arise depending upon whether $(R/L)^2 - 4/LC$ is positive, negative, or zero. Let us study these possibilities.

Case 1

$$\left(\frac{R}{L}\right)^2 - \frac{4}{LC} > 0$$

Therefore λ_1 and λ_2 are real and distinct. In this case we obtain as solutions to (1.10)

$$q_1(t) = e^{\lambda_1 t} \quad \text{and} \quad q_2(t) = e^{\lambda_2 t}. \tag{1.14}$$

At this point the reader may be slightly confused, for in (1.11) we appeared to be looking for "a" solution. To clarify this let us note the following proposition.

6 ORDINARY DIFFERENTIAL EQUATIONS

Proposition 1.1

If q_1 and q_2 are any two solutions of the differential equation (1.10) and a_1 and a_2 any two real or complex numbers, then $q = a_1 q_1 + a_2 q_2$ is also a solution of Equation (1.10).

The proof of this proposition consists in computing $\dot{q} = a_1 \dot{q}_1 + a_2 \dot{q}_2$, $\ddot{q} = a_1 \ddot{q}_1 + a_2 \ddot{q}_2$ and substituting these into (1.10), thus obtaining

$$a_1 \ddot{q}_1 + a_2 \ddot{q}_2 + \frac{R}{L}(a_1 \dot{q}_1 + a_2 \dot{q}_2) + \frac{1}{LC}(a_1 q_1 + a_2 q_2)$$

$$= a_1 \left(\ddot{q}_1 + \frac{R}{L} \dot{q}_1 + \frac{1}{LC} q_1 \right) + a_2 \left(\ddot{q}_2 + \frac{R}{L} \dot{q}_2 + \frac{1}{LC} q_2 \right)$$

$$= a_1 \cdot 0 + a_2 \cdot 0 = 0.$$

With Proposition 1.1 at hand and the knowledge that the functions given in (1.14) are in fact solutions of (1.10), we can conclude that every function q having the form

$$q(t) = a_1 e^{\lambda_1 t} + a_2 e^{\lambda_2 t} \tag{1.15}$$

for arbitrary choice of the constants a_1 and a_2 is also a solution of Equation (1.10). The leakage problem may now be solved in Case 1 provided that we can successfully choose from among all of the solutions (1.15) one which also satisfies the initial conditions (1.9), henceforth to be referred to as (1.16). That is,

$$q(t_0) = q_0 \quad \text{and} \quad \dot{q}(t_0) = \dot{q}_0. \tag{1.16}$$

By direct substitution of the first of (1.16) into (1.15) we obtain

$$q_0 = q(t_0) = a_1 e^{\lambda_1 t_0} + a_2 e^{\lambda_2 t_0}, \tag{1.16a}$$

and by differentiation of (1.15), and substitution of the second of (1.16) into the resulting equation we obtain

$$\dot{q}_0 = \dot{q}(t_0) = a_1 \lambda_1 e^{\lambda_1 t_0} + a_2 \lambda_2 e^{\lambda_2 t_0}. \tag{1.16b}$$

In these equations all quantities represent given information except a_1 and a_2 which still remain to be determined. We attempt to determine them so that (1.16a) and (1.16b) are simultaneously satisfied. Setting up the appropriate determinants in accordance with Cramer's rule we obtain

$$a_1 = \frac{\begin{vmatrix} q_0 & e^{\lambda_2 t_0} \\ \dot{q}_0 & \lambda_2 e^{\lambda_2 t_0} \end{vmatrix}}{\begin{vmatrix} e^{\lambda_1 t_0} & e^{\lambda_2 t_0} \\ \lambda_1 e^{\lambda_1 t_0} & \lambda_2 e^{\lambda_2 t_0} \end{vmatrix}} = \frac{\begin{vmatrix} q_0 & e^{\lambda_2 t_0} \\ \dot{q}_0 & \lambda_2 e^{\lambda_2 t_0} \end{vmatrix}}{W}$$

and
$$a_2 = \frac{\begin{vmatrix} e^{\lambda_1 t_0} & q_0 \\ \lambda_1 e^{\lambda_1 t_0} & \dot{q}_0 \end{vmatrix}}{W},$$

where
$$W = \begin{vmatrix} e^{\lambda_1 t_0} & e^{\lambda_2 t_0} \\ \lambda_1 e^{\lambda_1 t_0} & \lambda_2 e^{\lambda_2 t_0} \end{vmatrix} = e^{(\lambda_1+\lambda_2)t_0}(\lambda_2 - \lambda_1) \neq 0$$

because by the proposition of Case 1 $\lambda_2 \neq \lambda_1$ and clearly $e^{(\lambda_1+\lambda_2)t_0} \neq 0$. This means that we can determine the constants a_1 and a_2 in (1.15) in such a way that q will satisfy (1.10) and the initial conditions (1.16). Thus we have solved the leakage problem in Case 1. Note that the magnitudes of a_1 and a_2 are a function of both the initial conditions and the eigenvalues of the circuit.

Case 2

$$\left(\frac{R}{L}\right)^2 - \frac{4}{LC} < 0.$$

Therefore λ_1 and λ_2 are complex numbers, hence the conjugate complex numbers $\lambda_1 = \alpha + i\beta$ and $\lambda_2 = \alpha - i\beta$, where

$$\alpha = -\frac{R}{2L} \quad \text{and} \quad \beta = \frac{1}{2}\sqrt{\frac{4}{LC} - \left(\frac{R}{L}\right)^2}.$$

In this case we have the complex solutions

$$q_1(t) = e^{(\alpha+i\beta)t} \quad \text{and} \quad q_2(t) = e^{(\alpha-i\beta)t}. \tag{1.17}$$

Using Euler's formula we have $e^{i\beta t} = \cos \beta t + i \sin \beta t$. Therefore

$$e^{(\alpha+i\beta)t} = e^{\alpha t} e^{i\beta t} = e^{\alpha t}(\cos \beta t + i \sin \beta t).$$

Writing the latter as two terms we see that the complex solution $q_1(t)$ may be written as a linear combination of the real functions

$$\tilde{q}_1 = e^{\alpha t} \cos \beta t \quad \text{and} \quad \tilde{q}_2 = e^{\alpha t} \sin \beta t$$

with complex coefficients. That is, $q_1 = \tilde{q}_1 + i\tilde{q}_2$. We have a similar representation for q_2 in terms of $\tilde{q}_1$ and $\tilde{q}_2$, namely $q_2 = \tilde{q}_1 - i\tilde{q}_2$. Proposition 1.1 suggests the possibility that $\tilde{q}_1$ and $\tilde{q}_2$ are in fact a solution of Equation (1.10). To see that this is indeed so we observe that $\tilde{q}_1 = \frac{1}{2}(q_1 + q_2)$ and that $\tilde{q}_2 = 1/2i\,(q_1 - q_2)$. Consequently $\tilde{q}_1$ and $\tilde{q}_2$ are solutions of (1.10).

Again using Proposition 1.1 we set

$$q(t) = a_1\tilde{q}_1(t) + a_2\tilde{q}_2(t) \tag{1.18}$$

and seek to determine the constants a_1 and a_2 so as to satisfy the initial conditions (1.16). Following the procedure of Case 1 we obtain

$$q_0 = a_1 e^{\alpha t_0} \cos \beta t_0 + a_2 e^{\alpha t_0} \sin \beta t_0,$$

and

$$\dot{q}_0 = a_1(\alpha \cos \beta t_0 - \beta \sin \beta t_0)e^{\alpha t_0} + a_2(\alpha \sin \beta t_0 + \beta \cos \beta t_0)e^{\alpha t_0}$$

for the determination of a_1 and a_2. Using Cramer's rule to solve for a_1 and a_2 gives

$$a_1 = \frac{1}{W} \begin{vmatrix} q_0 & e^{\alpha t_0} \sin \beta t_0 \\ \dot{q}_0 & e^{\alpha t_0}(\alpha \sin \beta t_0 + \beta \cos \beta t_0) \end{vmatrix},$$

and

$$a_2 = \frac{1}{W} \begin{vmatrix} e^{\alpha t_0} \cos \beta t_0 & q_0 \\ (\alpha \cos \beta t_0 - \beta \sin \beta t_0)e^{\alpha t_0} & \dot{q}_0 \end{vmatrix},$$

where

$$W = \begin{vmatrix} e^{\alpha t_0} \cos \beta t_0 & e^{\alpha t_0} \sin \beta t_0 \\ (\alpha \cos \beta t_0 - \beta \sin \beta t_0)e^{\alpha t_0} & (\alpha \sin \beta t_0 + \beta \cos \beta t_0)e^{\alpha t_0} \end{vmatrix}$$
$$= \beta e^{2\alpha t_0} \neq 0,$$

since by the hypothesis of Case 2 $\beta > 0$ and $e^{2\alpha t_0} \neq 0$. In this way we determine a_1 and a_2 so that (1.18) is a solution to the leakage problem in Case 2.

Case 3

$$\left(\frac{R}{L}\right)^2 - \frac{4}{LC} = 0.$$

Therefore $\lambda_1 = \lambda_2 = -R/2L$. In this case the solutions $q_1(t) = e^{\lambda_1 t}$ and $q_2(t) = e^{\lambda_2 t}$ are identical. The experience gained in Cases 1 and 2 indicates that we should seek a second solution so that we can once again form a linear combination in order to satisfy the initial conditions at $t = t_0$. To this end we shall use the method of *variation of parameters*. This important tool will be of continued use to us in our subsequent work.

The method consists in seeking a second solution q_2' in the form

$$q_2'(t) = u(t)e^{\lambda_1 t}. \tag{1.19}$$

We readily find that

$$\dot{q}_2'(t) = \dot{u}e^{\lambda_1 t} + \lambda_1 u e^{\lambda_1 t},$$
$$\ddot{q}_2'(t) = \ddot{u}e^{\lambda_1 t} + 2\lambda_1 \dot{u}e^{\lambda_1 t} + \lambda_1^2 u e^{\lambda_1 t}.$$

Hence satisfying (1.10) we obtain

$$\ddot{q}_2' + \frac{R}{L}\dot{q}_2' + \frac{1}{LC}q_2' = \left(\ddot{u} + 2\lambda_1\dot{u} + \lambda_1^2 u + \frac{R}{L}\dot{u} + \lambda_1\frac{R}{L}u + \frac{1}{LC}u\right)e^{\lambda_1 t}$$

$$= \ddot{u}e^{\lambda_1 t} + ue^{\lambda_1 t}\left(\lambda_1^2 + \frac{R}{L}\lambda_1 + \frac{1}{LC}\right) + \dot{u}e^{\lambda_1 t}\left(2\lambda_1 + \frac{R}{L}\right)$$

and upon substituting for λ_1 and $\lambda_2\ (=-R/2L)$, we have

$$= \ddot{u}e^{\lambda_1 t} = 0.$$

Therefore if q_2' is to be a solution of (1.10) u must satisfy $\ddot{u} = 0$. By integrating twice we obtain

$$u(t) = b_1 t + b_2.$$

If we also insist that $u(0) = 0$ and $\dot{u}(0) = 1$, we obtain $u(t) = t$ and

$$q_2'(t) = te^{\lambda_1 t}.$$

Let us now attempt to solve the leakage problem with the linear combination

$$q(t) = a_1 q_1(t) + a_2 q_2'(t). \tag{1.20}$$

Once again applying the initial conditions (1.16) we have the following results:

$$q_0 = a_1 e^{\lambda_1 t_0} + a_2 t e^{\lambda_1 t_0},$$
$$\dot{q}_0 = a_1 \lambda_1 e^{\lambda_1 t_0} + a_2(1 + \lambda_1 t_0)e^{\lambda_1 t_0},$$

$$a_1 = \frac{1}{W}\begin{vmatrix} q_0 & te^{\lambda_1 t_0} \\ \dot{q}_0 & (1+\lambda_1 t_0)e^{\lambda_1 t_0} \end{vmatrix}$$

and

$$a_2 = \frac{1}{W}\begin{vmatrix} e^{\lambda_1 t_0} & q_0 \\ \lambda_1 e^{\lambda_1 t_0} & \dot{q}_0 \end{vmatrix}$$

where

$$W = \begin{vmatrix} e^{\lambda_1 t_0} & te^{\lambda_1 t_0} \\ \lambda_1 e^{\lambda_1 t_0} & (1+\lambda_1 t_0)e^{\lambda_1 t_0} \end{vmatrix}$$

$$= e^{2\lambda_1 t_0} \neq 0.$$

It follows that we can again determine a_1 and a_2 and consequently are able to solve the leakage problem in Case 3.

PROBLEMS

1. Derive a differential equation for the current flowing in the RL circuit illustrated in Figure 1.2 when K_1 is closed and K_2 is open. [*Hint:* Make use of Equations (1.2) and Kirchhoff's voltage law.]
   State exactly what is meant by a solution to this problem.

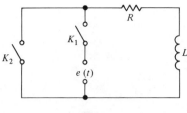

FIGURE 1.2

2. Let the current flowing in the RL circuit of Problem 1 at time t_0 be i_0. At time t_0, K_1 is opened and simultaneously K_2 is closed. Find the current in the circuit for $t \geq t_0$ by methods similar to those used in Section 1.2.
3. The differential equation for the RL circuit in Figure 1.2 with K_1 open and K_2 closed can be thought of as arising from the differential equation for the leakage problem of Section 1.2 by permitting c to tend to $+\infty$ and replacing dq/dt by i. Investigate the behavior of the solution of the leakage problem of Section 1.2 as $c \to \infty$.
4. Make a comparison of the results obtained for Problems 2 and 3.
5. Work Problem 1 with the inductance coil replaced by a capacitor.
6. Work Problem 1 with the resistor replaced by a capacitor.
7. Work Problem 2 for an RC circuit.
8. Work Problem 2 for a CL circuit. Add any hypotheses you may need to solve this problem.
9. In Section 1.1 we found an integro-differential equation for the current i flowing in the RLC circuit, namely Equation (1.3). For the leakage problem $[e(t) \equiv 0]$ investigate the possibility that this equation has exponential solutions $i(t) = e^{vt}$. Formulate appropriate initial conditions for the circuit problem which will enable you to find i directly by this method.
10. Investigate the behavior of the solutions to the leakage problem as $c \to \infty$. Relate the result to the solution of Problem 2.
11. Using Newton's second law derive the equation of motion for the system shown in Figure 1.3. Compare this equation to Equation (1.7). What conclusions do you draw from this comparison?

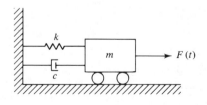

FIGURE 1.3

12. The differential equation for the RLC circuit is a special case of the general second-order differential equation with constant coefficients

$$\ddot{u} + a\dot{u} + bu = f(t), \quad a, b \text{ real constants.}$$

The equation where $f(t) \equiv 0$ is called *homogeneous;* hence the leakage problem involves a homogeneous equation. Show that the analysis of the leakage problem in Section 1.2 can be applied to the initial-value problem consisting of finding a function u satisfying the general homogeneous equation and satisfying the initial conditions

$$u(0) = u_0, \quad \dot{u}(0) = u_1.$$

13. Find a solution to each of the differential equations subject to the given initial conditions:

    (a) $\dfrac{dy}{dx} + \cos x = 0$ $y = 1$ when $x = 0$.

    (b) $\dfrac{d^2y}{dx^2} + \dfrac{dy}{dx} + y = 0$ $y = 0$ and $\dfrac{dx}{dy} = 1$ when $x = 0$.

    (c) $\dfrac{d^2y}{dt^2} + e^t = 0$ $y = \dfrac{dy}{dt} = 1$ when $t = 0$.

14. Verify that the functions given satisfy the differential equation

$$\frac{d^2y}{dx^2} + y = 0.$$

    (a) $y = \cos x$.   (c) $y = e^{\lambda x}$.
    (b) $y = \sin x$.   (d) $y = a_1 \cos x + a_2 \sin x$
                                (a_1 and a_2 arbitrary constants).

15. How do you explain the existence of four seemingly different solutions to the differential equation in Problem 14?

16. Prove Cramer's rule for the linear system

$$a_{11}x_1 + a_{12}x_2 = b_1$$
$$a_{21}x_1 + a_{22}x_2 = b_2.$$

    That is, prove that if b_1 and b_2 are not both zero the system has a solution for x_1 and x_2 if

$$\begin{vmatrix} a_{11} & a_{12} \\ a_{21} & a_{22} \end{vmatrix} = a_{11}a_{22} - a_{12}a_{21} \neq 0.$$

17. Prove that if the condition of Problem 16 is satisfied then the solution is unique, that is, there exists but one solution of the system.

18. Consider the homogeneous system obtained from the linear system of Problem 16. That is, set $b_1 = b_2 = 0$. A solution to this system will clearly be $x_1 = x_2 = 0$. This solution is called the *trivial solution.* All

other solutions are called *nontrivial*. Prove that there exists a nontrivial solution to the homogeneous system if and only if the determinant of the system $D = \begin{vmatrix} a_{11} & a_{12} \\ a_{21} & a_{22} \end{vmatrix}$, has the value zero. How many different solutions will there be to the problem if the condition is satisfied?

1.3 THE WRONSKI DETERMINANT

In this section we wish to reflect on the results of the previous two sections, analyze some of the features of the problem treated therein, and discuss the important properties of the solutions we have obtained.

The three most obvious properties developed are:

a. Each of the various functions obtained as a solution to the leakage problem is actually defined for all real values of t and satisfies the differential equation (1.10) for all real values of t, not just for $t > t_0$.

b. In each of the three cases of Section 1.2 we found two functions satisfying (1.10) and constructed the solution of the initial-value problem by forming a linear combination of these two functions.

c. In each case the two functions and their derivatives had to satisfy the condition that a certain determinant $W \neq 0$. Note that this fact is in no way influenced by the initial conditions.

One of our principal objectives in this book is to learn how to analyze difficult problems in applied mathematics. To this end we first study simple problems such as the leakage problem of Section 1.2 in order to gain insights that we may apply to the more difficult problems. We shall begin the process here by generalizing somewhat from a, b, and c above. To do this we will require knowledge of the following terms:

(i) The symbol $\equiv$ means identically equal.

(ii) The symbol (t_1, t_2) for $t_1 < t_2$ denotes the set of t, such that $t < t < t_2$, and this set is called an *open interval*.

(iii) A *real-valued function* is one whose range is a subset of the real line.

(iv) A point t is called an *interior point* of a set S if an open interval with t as its midpoint is contained in S. In other words, T is an interior point of S if all of the points in $(t - h, t + h)$ are in S for some $h > 0$.

Definition 1.3

Let f and g be real-valued functions defined on the open interval (t_1, t_2). We say f and g are *linearly dependent* on (t_1, t_2) if there exist two constants α and β, not both zero, such that

$$\alpha f(t) + \beta g(t) = 0 \quad \text{in } (t_1, t_2). \tag{1.21}$$

If the only solution of (1.21) is $\alpha = \beta = 0$ we say that f and g are *linearly independent* on (t_1,t_2).

To illustrate this definition let us show that the functions $e^{\lambda_1 t}$ and $e^{\lambda_2 t}$, where $\lambda_1 \neq \lambda_2$, are linearly independent on the entire real line. Consider the linear combination

$$\alpha e^{\lambda_1 t} + \beta e^{\lambda_2 t} \equiv 0, \quad \text{for all } t. \tag{1.22}$$

Divide this identity by the nonvanishing function $e^{\lambda_1 t}$ to obtain

$$\alpha + \beta e^{(\lambda_2 - \lambda_1) t} \equiv 0, \quad \text{for all } t.$$

Since this is an identity it must have a zero derivative for all values of t, that is,

$$\beta(\lambda_2 - \lambda_1) e^{(\lambda_2 - \lambda_1) t} = 0, \quad \text{for all } t.$$

Since $\lambda_2 \neq \lambda_1$ and $e^{(\lambda_2 - \lambda_1) t} \neq 0$, this last equation implies that $\beta = 0$. But this reduces (1.22) to

$$\alpha e^{\lambda_1 t} = 0$$

which implies that $\alpha = 0$. Hence $\alpha = \beta = 0$ is the only solution of (1.22) and we have proved $e^{\lambda_1 t}$, $e^{\lambda_2 t}$, $\lambda_1 \neq \lambda_2$, are linearly independent functions on the entire real line. A different proof of this is implied by Proposition 1.2 below.

Definition 1.4

Let f and g be real-valued functions defined and differentiable on (t_1,t_2). The function

$$W(f,g)(t) = \begin{vmatrix} f(t) & g(t) \\ \dot{f}(t) & \dot{g}(t) \end{vmatrix}$$

defined on (t_1,t_2) is called the *Wronskian of f and g* or the Wronski determinant of f and g.

Theorem 1.1

Let f and g be real-valued functions defined and differentiable on (t_1,t_2). Then if there exists a point t_0 in (t_1,t_2) such that $W(f,g)(t_0) \neq 0$, f and g are linearly independent.

PROOF. We will prove this theorem by contradiction. Suppose that f and g are linearly dependent on (t_1,t_2). Then we can find constants α and β both different from zero such that

$$\alpha f(t) + \beta g(t) \equiv 0 \quad \text{on } (t_1,t_2).$$

Differentiating this identity we must also have

$$\alpha \dot{f}(t) + \beta \dot{g}(t) \equiv 0 \quad \text{on } (t_1, t_2).$$

In particular, at $t = t_0$ we have

$$\alpha f(t_0) + \beta g(t_0) = 0$$
$$\alpha \dot{f}(t_0) + \beta \dot{g}(t_0) = 0.$$

We now have a system of two linear homogeneous equations for the determination of α and β. A necessary and sufficient condition for this system to have a nontrivial solution pair (α, β) at any point t_0 is that

$$\begin{vmatrix} f(t_0) & g(t_0) \\ \dot{f}(t_0) & \dot{g}(t_0) \end{vmatrix} = W(f,g)(t_0) = 0$$

(see Problem 18, Section 1.2). But the determinant of the system is, by hypothesis, $W(f,g)(t_0) \neq 0$. We have thus reached the desired contradiction, and it follows that f and g must be linearly independent.

Let us now tie these ideas together in the study of the differential equation (1.10). Suppose q_1 and q_2 are solutions of (1.10) so that we have

$$\ddot{q}_1 + \frac{R}{L} \dot{q}_1 + \frac{1}{LC} q_1 = 0,$$

$$\ddot{q}_2 + \frac{R}{L} \dot{q}_2 + \frac{1}{LC} q_2 = 0.$$

Multiply the first equation by q_2 and the second by q_1 and subtract to obtain

$$-q_1 \ddot{q}_2 + q_2 \ddot{q}_1 + \frac{R}{L}(-q_1 \dot{q}_2 + q_2 \dot{q}_1) = 0. \tag{1.23}$$

Now

$$\frac{d}{dt}(-\dot{q}_1 q_2 + \dot{q}_2 q_1) = -\ddot{q}_1 q_2 - \dot{q}_1 \dot{q}_2 + \ddot{q}_2 q_1 + \dot{q}_2 \dot{q}_1$$
$$= -\ddot{q}_1 q_2 + \ddot{q}_2 q_1,$$

so that (1.23) becomes

$$\frac{d}{dt} W(q_1, q_2) + \frac{R}{L} W(q_1, q_2) = 0. \tag{1.24}$$

The left-hand side of (1.24) can also be written as

$$\frac{d}{dt}[W(q_1, q_2) e^{(R/L)t}] = 0, \tag{1.25}$$

since $e^{(R/L)t} \neq 0$. Integrating (1.25) from t_0 to t yields

GENERAL SOLUTION OF THE RLC CIRCUIT

$$W(q_1,q_2)(t)e^{(R/L)t} = W(q_1,q_2)(t_0)e^{(R/L)t_0}$$

or

$$W(q_1,q_2)(t) = W(q_1,q_2)(t_0)e^{(R/L)(t_0-t)}. \qquad (1.26)$$

An inspection of (1.26) reveals that if $W(q_1,q_2)(t_0) \neq 0$ then $W(q_1,q_2)(t) \neq 0$ for all values of t because $e^{(R/L)(t_0-t)} \neq 0$. In other words, if q_1 and q_2 are solutions of (1.10) then their Wronskian is either identically zero for all real t or never equals zero. We can now formulate and prove:

Proposition 1.2

If q_1 and q_2 are solutions of Equation (1.10), then they are linearly independent functions on the entire real line if and only if at some point t_0

$$W(q_1,q_2)(t_0) \neq 0.$$

PROOF. The proof of the "only if" portion of the proposition—that is, the assertion that q_1 and q_2 are linearly independent implies $W(q_1,q_2)(t_0) \neq 0$ for some t_0—will not be pursued here.

To prove the "if" portion we note that if q_1 and q_2 were linearly dependent we could find constants α and β, such that $\alpha \neq 0$ and $\beta \neq 0$ satisfy

$$\alpha q_1(t) + \beta q_2(t) = 0 \quad \text{for all } t,$$

and

$$\alpha \dot{q}_1(t) + \beta \dot{q}_2(t) = 0 \quad \text{for all } t.$$

In particular we could find a t_1 such that $W(q_1,q_2)(t_1) = 0$. By (1.26) it follows that $W(q_1,q_2) \equiv 0$ since q_1 and q_2 are solutions of (1.10). This contradicts the fact that $W(q_1,q_2)(t_0) \neq 0$, so we conclude that q_1 and q_2 are linearly independent.

In retrospect we can now summarize the material presented in Section 1.2 by saying that in each case we found a pair of linearly independent solutions to (1.10) and represented the solution of the leakage problem as a linear combination of these solutions.

1.4 GENERAL SOLUTION OF THE RLC CIRCUIT

Let us now return to the problem of Section 1.1, where the switch K_1 of the RLC circuit of Figure 1.1 is closed and the switch K_2 is open, our objective being to find a general solution $q(t)$ to that problem. The technique to be used will be to build upon the solution obtained in Section 1.2 for the leakage problem. To this end we observe that any solution to the latter problem is simply a solution of the original problem with $e(t) \equiv 0$.

Suppose we can find a function q_p such that q_p is a solution of the differential equation

$$\ddot{q}_p + \frac{R}{L}\dot{q}_p + \frac{1}{LC}q_p = \frac{e(t)}{L}, \tag{i}$$

subject to the very special initial conditions

$$q_p(t_0) = 0 \quad \text{and} \quad \dot{q}_p(t_0) = 0. \tag{ii}$$

Denoting by q_l the solution of the leakage problem we can formulate the function

$$q(t) = q_l(t) + q_p(t), \tag{1.27}$$

where, as indicated above, $q(t)$ is a general solution of the problem of Section 1.1. From our previous work we possess q_l, consequently our task is to find the function q_p.

For this purpose we use the technique developed in Section 1.2. Namely, since

$$q_l(t) = a_1 q_1(t) + a_2 q_2(t)$$

is the solution of the leakage problem, we draw upon this knowledge and as an informed guess try to find q_p in the form

$$q_p(t) = a_1(t)q_1(t) + a_2(t)q_2(t). \tag{1.28}$$

We have therefore to determine the functions a_1 and a_2 given that q_1 and q_2 are solutions of (1.10).

Differentiating (1.28) once yields

$$\dot{q}_p(t) = \dot{a}_1 q_1 + \dot{a}_2 q_2 + a_1 \dot{q}_1 + a_2 \dot{q}_2.$$

Simplify this expression by imposing the condition

$$\dot{a}_1 q_1 + \dot{a}_2 q_2 = 0, \tag{1.29}$$

so that

$$\dot{q}_p = a_1 \dot{q}_1 + a_2 \dot{q}_2.$$

Differentiating again yields

$$\ddot{q}_p = \dot{a}_1 \dot{q}_1 + \dot{a}_2 \dot{q}_2 + a_1 \ddot{q}_1 + a_2 \ddot{q}_2.$$

The requirement that q_p shall be a function satisfying (i) gives us

$$\dot{a}_1 \dot{q}_1 + \dot{a}_2 \dot{q}_2 + a_1 \ddot{q}_1 + a_2 \ddot{q}_2 + \frac{R}{L} a_1 \dot{q}_1 + \frac{R}{L} a_2 \dot{q}_2 + \frac{1}{LC} a_1 q_1 + \frac{1}{LC} a_2 q_2$$

$$= \dot{a}_1 \dot{q}_1 + \dot{a}_2 \dot{q}_2 + a_1 \left(\ddot{q}_1 + \frac{R}{L} \dot{q}_1 + \frac{1}{LC} q_1 \right) + a_2 \left(\ddot{q}_2 + \frac{R}{L} \dot{q}_2 + \frac{1}{LC} q_2 \right)$$

$$= \dot{a}_1 \dot{q}_1 + \dot{a}_2 \dot{q}_2 = \frac{e}{L}.$$

We thus have two linear equations for $\dot{a}_1$ and $\dot{a}_2$, namely (1.29) and

$$\dot{a}_1 \dot{q}_1 + \dot{a}_2 \dot{q}_2 = \frac{e}{L}. \tag{1.30}$$

The determinant of this system is precisely the Wronskian of q_1 and q_2 which is everywhere different from zero, as was proved in Section 1.3. We may therefore solve for $\dot{a}_1$ and $\dot{a}_2$ using Cramer's rule to obtain

$$\dot{a}_1 = \frac{1}{W(q_1,q_2)} \begin{vmatrix} 0 & q_2 \\ \frac{e}{L} & \dot{q}_2 \end{vmatrix} = -\frac{1}{W(q_1,q_2)} \frac{eq_2}{L},$$

$$\dot{a}_2 = \frac{1}{W(q_1,q_2)} \begin{vmatrix} q_1 & 0 \\ \dot{q}_1 & \frac{e}{L} \end{vmatrix} = \frac{1}{W(q_1,q_2)} \frac{eq_1}{L}.$$

Let us next integrate each of these equations from t_0 to t. This gives

$$a_1(t) = -\frac{1}{L} \int_{t_0}^{t} \frac{e(s)q_2(s)}{W(q_1,q_2)(s)} \, ds,$$

and

$$a_2(t) = \frac{1}{L} \int_{t_0}^{t} \frac{q_1(s)e(s)}{W(q_1,q_2)(s)} \, ds.$$

Placing these integrals into (1.28) yields

$$q_p(t) = \frac{1}{L} \int_{t_0}^{t} \frac{q_2(t)q_1(s) - q_1(t)q_2(s)}{W(q_1,q_2)(s)} e(s) \, ds. \tag{1.31}$$

The function defined by the integral (1.31) is by its construction a solution of the differential equation (i). We must still verify that it satisfies the initial conditions (ii).

That $q_p(t_0) = 0$ is an immediate consequence of a fundamental property of the definite integral. To verify that $\dot{q}_p(t_0) = 0$ we must differentiate with respect to t. Since t occurs in the upper limit of the integral, we obtain by Leibnitz' rule (see Problem 11 below)

$$\dot{q}_p(t) = \frac{1}{L} \frac{q_2(t)q_1(t) - q_1(t)q_2(t)}{W(q_1,q_2)(t)} e(t) + \frac{1}{L} \int_{t_0}^{t} \frac{\dot{q}_2(t)q_1(s) - \dot{q}_1(t)q_2(s)}{W(q_1,q_2)(s)} e(s) \, ds$$

$$= \frac{1}{L} \int_{t_0}^{t} \frac{\dot{q}_2(t)q_1(s) - \dot{q}_1(t)q_2(s)}{W(q_1,q_2)(s)} e(s) \, ds.$$

The last integral clearly vanishes when $t = t_0$ and we have verified that $\dot{q}_p(t_0) = 0$.

To complete this section, let us compute the integral (1.31) for Case 1 of Section 1.2, namely for the case where $q_1(t) = e^{\lambda_1 t}$, $q_2(t) = e^{\lambda_2 t}$, and $\lambda_1 \neq \lambda_2$. Recalling that

18 ORDINARY DIFFERENTIAL EQUATIONS

$$W(q_1,q_2)(s) = e^{(\lambda_1+\lambda_2)s}(\lambda_2 - \lambda_1) \quad [\text{Equation (1.17)}]$$

we obtain

$$q_p(t) = \frac{1}{L}\int_{t_0}^{t} \frac{e^{\lambda_2(t-s)} - e^{\lambda_1(t-s)}}{\lambda_2 - \lambda_1} e(s)\, ds.$$

PROBLEMS

1. (a) What is the difference between the equal sign $=$ and the sign $\equiv$?
   (b) What is a real-valued function?
   (c) What is meant by the term linearly independent functions?
2. Show that the functions $e^{\lambda_1 t}$, $e^{\lambda_2 t}$, and $e^{\lambda_3 t}$, where $\lambda_3 > \lambda_2 > \lambda_1$ are linearly independent on the entire real line. [*Hint:* Refer to Equation (1.22) for a starting point.]
3. Determine which of the following sets of functions are linearly dependent on their entire interval of definition:

   (a) $\sin \lambda$, $\cos \lambda$, $\sin 2\lambda$.
   (b) $\cos 2x$, $\cos^2 x$, $\sin^2 x$.
   (c) e^q, qe^q, $\sinh q$.
   (d) $\sinh x$, e^x, e^{-x}.
   (e) $1 + q$, $1 + 2q$, q^2.
   (f) $t^2 - t + 1$, $3t^2 - t - 1$, $t^2 - 1$.
   (g) y, y^2, y^3.
   (h) e^θ, $e^{-\theta}$.

4. Prove that the functions x and x^3 are linearly independent on any interval. State whether they can be solutions of the same differential equation on an interval containing the point $x = 0$ if the equation they must satisfy has the form

$$\frac{d^2u}{dx^2} + a(x)\frac{du}{dx} + b(x)u = 0$$

where a and b are continuous functions.

5. Prove that the functions x, $|x|$ are linearly dependent on any interval not including $x = 0$ and that they are linearly independent on any interval containing $x = 0$ as an interior point.
6. Prove that if u_1 and u_2 are solutions of the differential equation

$$\ddot{u} + a\dot{u} + bu = 0,$$

where a and b are constants, then $W(u_1,u_2)(t)$ is either identically zero or never zero.

7. Given that u_1 and u_2 are solutions of

$$\ddot{u} + a(t)\dot{u} + b(t)u = 0$$

on (α,β) where $\beta > \alpha$ and a and b are functions of time in the interval (α,β). Show that $W(u_1,u_2)(t)$ is either identically zero or never zero on (α,β).

8. Use the method of variation of parameters to find the general solution of the following equations:

   (a) $\ddot{y} + 4y = \sec 2t$.
   (b) $\ddot{x} + x = \cot t$.
   (c) $y'' - y = e^x$.
   (d) $\ddot{q} - q = \log(t+1)$.
   (e) $u''' - 6u'' + 11u' - 6u = e^{4q}$.
   (f) $\ddot{s} + 2A\dot{s} + \omega^2 s = B \sin \lambda t$
       $(A > 0, \lambda > 0)$.

9. Use the method of variation of parameters to solve the nonhomogeneous differential equation
$$\ddot{u} + a\dot{u} + bu = f(t)$$
subject to the initial conditions $u(0) = \dot{u}(0) = 0$.

10. Suppose that u_1 and u_2 are linearly independent solutions of
$$\ddot{u} + a(t)\dot{u} + b(t)u = 0$$
on $[0, \infty)$. The symbol $[0, \infty)$ reads $\{0 \leq t < \infty\}$. The functions a and b are defined on $[0, \infty)$. Find the solution of the equation
$$\ddot{u} + a(t)\dot{u} + b(t)u = f(t)$$
satisfying the initial condition $u(0) = \dot{u}(0) = 0$. Comment on the hypotheses required for the function f in order that the solution be mathematically valid.

11. Show that if f is continuous and u and v are differentiable then
$$\frac{d}{dt} \int_{u(t)}^{v(t)} f(s)\, ds = \dot{v}(t) f(v(t)) - \dot{u}(t) f(u(t)).$$
Where must f be defined for the result to be valid? (This result is called *Liebnitz' rule*.)

12. Let the functions $a(t)$, $b(t)$, and $c(t)$ be defined in the interval (α, β) and suppose $a(t)$ is differentiable there. Show that the differential equation
$$a(t)\ddot{u} + b(t)\dot{u} + c(t)u = 0$$
can be transformed by suitable changes of variables into the differential equation
$$\frac{d}{dt}\left(p(t) \frac{dV}{dt}\right) + r(t) V = 0$$
defined on the same interval.

13. Find the general solution to the equation of motion derived in Problem 11 of Section 1.2.

14. Determine the general solution for the current in the RLC circuit of Figure 1.1 with K_2 open and K_1 closed when $E(t) = E_0 \sin at$.

15. Physicists normally think in terms of displacements, velocities, and accelerations when formulating dynamics problems. Mathematically it is sometimes desirable to obtain even higher derivatives of the variables describing the motion of a system in order to facilitate solving the problem. As an example consider the system shown in Figure 1.4.

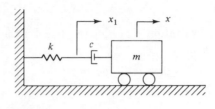

FIGURE 1.4

Using Newton's second law, formulate the two equations involving x and x_1 necessary in the setup of the equation of motion of the third-order system. Obtain the third-order equation. What is the form of the general solution of this equation?

16. Work Problem 18 of Section 1.2 for the case of the homogeneous linear system in three variables, namely,

$$a_{11}x_1 + a_{12}x_2 + a_{13}x_3 = 0,$$
$$a_{21}x_1 + a_{22}x_2 + a_{23}x_3 = 0,$$
$$a_{31}x_1 + a_{32}x_2 + a_{33}x_3 = 0.$$

17. Prove that if f_1, f_2, and f_3 are twice differentiable on an interval (a,b) and their Wronskian

$$W(f_1, f_2, f_3)(t) \equiv \begin{vmatrix} f_1 & f_2 & f_3 \\ f_1' & f_2' & f_3' \\ f_1'' & f_2'' & f_3'' \end{vmatrix}$$

is different from zero at a point t_0 in (a,b), then f_1, f_2, and f_3 are linearly independent on (a,b).

1.5 UNIQUENESS AND STABILITY

The results of Section 1.2 were used in Section 1.4 to show that there exists at least one solution to the problem of Section 1.1. We may now reasonably ask if other solutions exist. This is the *problem of uniqueness*.

To investigate the uniqueness problem let us suppose that there exists a second solution $p(t)$. Thus

$$p(t_0) = q_0, \; \dot{p}(t_0) = \dot{q}_0,$$
$$\ddot{p} + \frac{R}{L}\dot{p} + \frac{1}{LC}p = \frac{e}{L},$$

and q, of course, also satisfies these conditions. Let $r(t)$ equal $p(t) - q(t)$. It immediately follows that

$$r(t_0) = 0, \dot{r}(t_0) = 0,$$

and (1.32)

$$\ddot{r} + \frac{R}{L}\dot{r} + \frac{1}{LC}r = 0.$$

That is, r is a solution of the leakage problem with zero initial conditions at t_0. This problem will be called the *homogeneous problem* for the RLC circuit and, more specifically, the differential equation (1.10) with zero right-hand side will be called a *homogeneous differential equation*.

Suppose we can show that the only solution of the homogeneous problem is $r(t) \equiv 0$, for all t. This will imply that $p(t) \equiv q(t)$, for all t, thus proving the uniqueness of the solution of our problem.

We will present the proof outlined above by a simple yet extremely powerful method, an energy-like method. Let us multiply (1.32) by $\dot{r}$ to obtain

$$\dot{r}\ddot{r} + \frac{R}{L}\dot{r}^2 + \frac{1}{LC}\dot{r}r = 0. \tag{1.33}$$

Note the following identities which can be formulated from the terms in (1.33).

$$\dot{r}\ddot{r} = \frac{1}{2}\frac{d}{dt}(\dot{r}^2) \quad \text{and} \quad \dot{r}r = \frac{1}{2}\frac{d}{dt}r^2.$$

Using these identities, (1.33) may be rewritten as

$$\frac{d}{dt}\left(\dot{r}^2 + \frac{1}{LC}r^2\right) = -2\frac{R}{L}\dot{r}^2. \tag{1.34}$$

Next integrate (1.34) from t_0 to t, obtaining

$$\dot{r}^2(t) + \frac{1}{LC}r^2(t) = -2\frac{R}{L}\int_{t_0}^{t}\dot{r}^2(s)\,ds + \dot{r}^2(t_0) + \frac{1}{LC}r^2(t_0). \tag{1.35}$$

In (1.32) we showed that $r(t_0) = \dot{r}(t_0) = 0$. These reduce (1.35) to

$$\dot{r}^2(t) + \frac{1}{LC}r^2(t) = -2\frac{R}{L}\int_{t_0}^{t}\dot{r}^2(s)\,ds. \tag{1.36}$$

The integral on the right-hand side of (1.36) is nonnegative since it is the integral of a nonnegative function, and R and L are positive constants. Here we are making use of the following result.

Lemma 1.1

Let f be defined, nonnegative and integrable on the interval (a,b). Then
$$\int_a^b f(x)\, dx \geq 0.$$

Note that the function $\dot{r}^2$ is integrable on the interval (t_0, t) because r is the solution of the differential equation (1.32) and hence has continuous first and second derivatives. Thus $\dot{r}$ is continuous and so is $\dot{r}^2$. Finally a continuous function is integrable. From this information we deduce the inequality
$$-2\frac{R}{L}\int_{t_0}^t \dot{r}^2(s)\, ds \leq 0.$$

On the other hand, inspecting the left-hand side of (1.36) we see that
$$\dot{r}^2(t) \geq 0,$$
which implies that
$$\dot{r}^2(t) + \frac{1}{LC} r^2(t) \geq \frac{1}{LC} r^2(t) \geq 0.$$

We therefore deduce from (1.36) that
$$0 \leq \frac{1}{LC} r^2(t) \leq 0, \quad \text{for all } t. \tag{1.37}$$

Since L and C are positive constants, (1.37) implies that $r(t) = 0$ as we set out to prove. Let us formalize this result as:

Proposition 1.3

There exists at most one solution to the RLC circuit problem.

The reader should note that the proof we have just given establishing the uniqueness of the solution to our problem also shows that the leakage problem, which is a special case, has a unique solution. Let us now return to the leakage problem and using some of the arguments presented in this section extract additional information from its solution.

We have, by (1.13), the values
$$\lambda_1 = -\frac{R}{2L} + \frac{1}{2}\sqrt{\left(\frac{R}{L}\right)^2 - \frac{4}{LC}},$$
$$\lambda_2 = -\frac{R}{2L} - \frac{1}{2}\sqrt{\left(\frac{R}{L}\right)^2 - \frac{4}{LC}}.$$

In Case 1 of Section 1.2, noting that L and C are positive constants, we have

$$0 < \left(\frac{R}{L}\right)^2 - \frac{4}{LC} < \left(\frac{R}{L}\right)^2,$$

or more specifically

$$\frac{1}{2}\sqrt{\left(\frac{R}{L}\right)^2 - \frac{4}{LC}} < \frac{R}{2L},$$

from which it follows that $\lambda_1 < 0$ and $\lambda_2 < 0$. Consequently the solutions $q_1(t) = e^{\lambda_1 t}$ and $q_2(t) = e^{\lambda_2 t}$ decrease as t increases. Moreover,

$$\lim_{t \to +\infty} q_i(t) = 0, \quad i = 1, 2,$$

or

$$\lim_{t \to +\infty} [a_1 q_1(t) + a_2 q_2(t)] = 0 \tag{1.38}$$

whatever the constants a_1 and a_2. It follows that in Case 1 the charge eventually dissipates, or more descriptively leaks off the condenser.

Similarly in Case 2 we have the solutions $e^{-(R/2L)t} \cos \beta t$, $e^{-(R/2L)t} \sin \beta t$, where

$$\beta = \frac{1}{2}\sqrt{\frac{4}{LC} - \left(\frac{R}{L}\right)^2}.$$

Since $|\sin \beta t| \leq 1$ and $|\cos \beta t| \leq 1$ for all t, we have

$$\lim_{t \to +\infty} e^{-(R/2L)t} \cos \beta t = \lim_{t \to +\infty} e^{-(R/2L)t} \sin \beta t = 0.$$

The limit (1.38) is therefore again valid in Case 2.

Finally in Case 3 we have

$$\lim_{t \to +\infty} e^{-(R/2L)t} = \lim_{t \to +\infty} t e^{-(R/2L)t} = 0$$

and (1.38) is also valid for this case.

Therefore we have proved:

Proposition 1.4

Every solution $q(t)$ of the leakage problem for the RLC circuit satisfies $\lim_{t \to \infty} q(t) = 0$.

The property just established is a very special case of a property of considerable importance in all such problems. Therefore, within reason, let us generalize the results of this section.

24 ORDINARY DIFFERENTIAL EQUATIONS

We remind the reader that a real-valued function f is *bounded* if there exists a constant $M > 0$ such that $|f(t)| \leq M$ for all t in the domain of definition of f. In particular, the results obtained above imply that all solutions of the leakage problem are bounded for $t \geq t_0$. This fact in turn is usually expressed by saying that the differential equation (1.10) is *stable* in the classical sense. Here is a reasonably general formulation of this concept for problems similar to the one we have been considering.

Definition 1.5

We say that the differential equation

$$\ddot{u} + f(t)\dot{u} + g(t)u = 0 \tag{1.39}$$

is stable in the sense of Laplace (or in the classical sense) if every solution of the initial-value problem remains bounded as $t \to \infty$.

The initial-value problem for Equation (1.39) consists in finding a function $u = u(t)$ satisfying (1.39) for all $t > 0$ such that $u(0) = u_0$, $\dot{u}(0) = \dot{u}_0$. Of course our definition will only be of interest for those problems which admit solutions defined for all positive values of t.

Actually for the RLC-circuit problem we have been studying, we were able to show that every solution $q(t)$ satisfies $q(t) \to 0$ as $t \to \infty$. Mathematically this situation is described by saying that the zero solution, $q(t) \equiv 0$, of (1.10) is *asymptotically stable*. This concept of stability is usually more subtle than the above concept, and we shall not attempt to study it systematically in this text.

1.6 THE ENERGY METHOD

The detailed analysis made thus far of the circuit problem was possible largely because of the inherent simplicity of the system represented by Figure 1.1. Let us see what happens when the problem becomes more complex. For example, suppose that the inductance L is a function of time, say,

$$L = L_0(1 + \alpha t), \quad \text{where } \alpha > 0, \tag{1.40}$$

with L defined for all $t \geq 0$. With this the differential equation (1.10) for the leakage problem becomes

$$\ddot{q} + \frac{R}{L_0(1 + \alpha t)} \dot{q} + \frac{1}{CL_0(1 + \alpha t)} q = 0, \quad t \geq 0, \tag{1.41}$$

with the initial conditions remaining

$$q(0) = q_0 \quad \text{and} \quad \dot{q}(0) = \dot{q}_0. \tag{1.42}$$

If we try, as in Section 1.2, to find exponential solutions to Equation (1.41) having the form

$$q(t) = e^{\lambda t}$$

we obtain the characteristic equation:

$$\lambda^2 + \frac{R}{L_0(1 + \alpha t)} \lambda + \frac{1}{CL_0(1 + \alpha t)} = 0$$

which shows that, in the case of a nonconstant coefficient, λ is not independent of time as was found in the earlier problem. Therefore, we suspect that the entire technique built up in Section 1.2 is not applicable to the present problem. The question now arises of how to proceed. If we cannot readily exhibit an explicit solution, what information, if any, can we obtain about the behavior of the system described by (1.41)? The answer is that quite a lot of information can be obtained about the solutions of Equation (1.41) without actually having them at hand. Thus physical insight may be obtained about the system without having to, or being able to, find its solutions. Moreover, in many situations it turns out that the information available is what the engineer or scientist really wishes to know about the system. If he wants the actual solutions in terms of numbers they are usually obtainable by numerical methods.

For the present we wish to illustrate how to extract useful information about the solutions of the leakage problem described by Equations (1.41) and (1.42) without actually computing the solution. Our starting point is the tacit assumption which analysts make about all problems whose mathematical model was deduced from a physical system; that is, that solutions exist for all $t \geq 0$. The method to be used is the energy technique introduced in Section 1.5 to prove the uniqueness theorem. Multiply Equation (1.41) by $\dot{q}$:

$$\dot{q}\ddot{q} + \frac{R}{L_0(1 + \alpha t)} \dot{q}^2 + \frac{1}{CL_0(1 + \alpha t)} q\dot{q} = 0, \quad \text{for all } t \geq 0.$$

Again using the identities

$$\dot{q}\ddot{q} = \frac{1}{2}\frac{d}{dt}(\dot{q}^2) \quad \text{and} \quad q\dot{q} = \frac{1}{2}\frac{d}{dt}(q^2), \tag{1.43}$$

we obtain

$$\frac{1}{2}\frac{d}{dt}(\dot{q}^2) + \frac{R}{L_0(1 + \alpha t)} \dot{q}^2 + \frac{1}{CL_0(1 + \alpha t)} \frac{1}{2}\frac{d}{dt}(q^2) = 0.$$

Note the following identity with respect to the last term of the above equation

$$\frac{1}{CL_0(1+\alpha t)}\frac{d}{dt}(q^2) = \frac{d}{dt}\left(\frac{q^2}{CL_0(1+\alpha t)}\right) + \frac{\alpha q^2}{CL_0(1+\alpha t)^2},$$

and substitute this into the previous equation to obtain

$$\frac{d}{dt}\left[\frac{\dot{q}^2}{2} + \frac{q^2}{2CL_0(1+\alpha t)}\right] = -\frac{R\dot{q}^2}{L_0(1+\alpha t)} - \frac{\alpha q^2}{2CL_0(1+\alpha t)^2}.$$

Finally, integrate the resulting equation from 0 to t, obtaining

$$\dot{q}^2(t) + \frac{q^2(t)}{CL_0(1+\alpha t)} = \dot{q}^2(0) + \frac{q^2(0)}{CL_0}$$
$$- 2\int_0^t \left\{\frac{R\dot{q}^2(s)}{L_0(1+\alpha s)} + \frac{\alpha q^2(s)}{2CL_0(1+\alpha s)^2}\right\}ds. \quad (1.44)$$

We shall refer to Equation (1.44) as the fundamental energy identity for Equation (1.41). Let us examine (1.44) to see what information can be extracted from it.

The first thing to be observed is that

$$\frac{R\dot{q}^2(s)}{L_0(1+\alpha s)} + \frac{\alpha q^2(s)}{2CL_0(1+\alpha s)} \geq 0 \quad \text{for all } s \geq 0.$$

Hence, by virtue of Lemma 1.1,

$$\int_0^t \left\{\frac{R\dot{q}^2(s)}{L_0(1+\alpha s)} + \frac{\alpha q^2(s)}{2CL_0(1+\alpha s)^2}\right\}ds \geq 0. \quad (1.45)$$

The inequality (1.45) permits us to deduce from (1.44) that

$$\dot{q}^2(t) + \frac{q^2(t)}{CL_0(1+\alpha t)} \leq \dot{q}^2(0) + \frac{q^2(0)}{CL_0}, \quad t \geq t_0. \quad (1.46)$$

If we define the function $\mathcal{E}$ by

$$\mathcal{E}(t) = \dot{q}^2(t) + \frac{q^2(t)}{CL_0(1+\alpha t)} \equiv i^2(t) + \frac{q^2(t)}{CL_0(1+\alpha t)} \quad (1.47)$$

then (1.46) may be written more compactly as

$$\mathcal{E}(t) \leq \mathcal{E}(0), \quad t \geq t_0. \quad (1.48)$$

We may refer to the number $\mathcal{E}(t)$ as the total electrical energy in the circuit at time t, or $\mathcal{E}$ is the total energy function for the circuit. This terminology will be more evident later in this chapter when we consider an analogous mechanical system.

The inequality (1.46) [or (1.48)] shows that the electrical energy in the circuit at any time $t > 0$ is bounded by the electrical energy in the circuit at time $t = 0$. Such a condition is referred to as a *dissipative circuit*.

Returning to Equation (1.44) let us define the function

$$\tilde{\mathcal{E}}(t) = i^2(t) + \frac{q^2(t)}{\frac{2CR}{\alpha}(1+\alpha t)}. \tag{1.49}$$

Observe now that $\tilde{\mathcal{E}}(t) = 0$ for a fixed value of $t \geq 0$ if and only if $\mathcal{E}(t) = 0$. This is because $\tilde{\mathcal{E}}(t) = 0$ if and only if both $i(t)$ and $q(t)$ are identically zero. The same is true of $\mathcal{E}(t)$.

Using (1.49) we may write (1.44) in the form

$$\mathcal{E}(t) = \mathcal{E}(0) - 2\int_0^t \frac{R}{L_0(1+\alpha s)} \tilde{\mathcal{E}}(s)\,ds, \quad t \geq 0. \tag{1.50}$$

Moreover, the time $t = 0$ can be replaced by any later time t_0 so that (1.50) can also be written as

$$\mathcal{E}(t) = \mathcal{E}(t_0) - 2\int_{t_0}^t \frac{R}{L(1+\alpha s)} \tilde{\mathcal{E}}(s)\,ds, \quad t \geq t_0 \tag{1.51}$$

and (1.46) becomes

$$\mathcal{E}(t) \leq \mathcal{E}(t_0) \quad t \geq t_0. \tag{1.52}$$

Now suppose $\mathcal{E}(t) = 0$ at some time $t_0 \geq 0$. Then (1.52) says that $\mathcal{E}(t) \leq 0$ for $t \geq t_0$. But

$$\mathcal{E}(t) = i^2(t) + \frac{q^2(t)}{CL_0(1+\alpha t)} \geq \frac{q^2(t)}{CL_0(1+\alpha t)} \geq 0,$$

and since $1/CL(1+\alpha t) \neq 0$, we must have $q(t) = 0$ for all $t \geq t$. It follows by exactly the same reasoning that $i(t) = 0$ for $t \geq t_0$. Hence $\mathcal{E}(t) = 0$ for $t \geq t_0$. Consequently we have proved:

Proposition 1.5

The total energy $\mathcal{E}$ of the leakage problem (1.41), (1.42) is either different from zero for all $t \geq 0$ or identically zero from some time t_0 on, the time t_0 being the first time after $t = 0$ at which $\mathcal{E}(t) = 0$.

It is clear that the function $\tilde{\mathcal{E}}(t)$ could be substituted for $\mathcal{E}(t)$ in Proposition 1.5.

We now know that every solution of our problem either diminishes to zero after some finite time or has positive total energy for all finite times. We can say more about the latter case. If $\mathcal{E}$ remains different from zero so does $\tilde{\mathcal{E}}$ which means the integrand of

$$I(t) = \int_0^t \frac{R}{L_0(1+\alpha s)} \tilde{\mathcal{E}}(s)\,ds$$

is positive.

Let us observe that if the function f is defined on the interval (a,b) and if $f(x_1) \leq f(x_2)$ whenever $a \leq x_1 < x_2 \leq b$, f is said to be *nondecreasing* on (a,b). If f satisfies $f(x_1) < f(x_2)$ for all $a \leq x_1 < x_2 \leq b$, then f is called *strictly increasing* on (a,b). The same concepts can obviously be defined relative to the *closed interval* $[a,b] = \{a \leq x \leq b\}$ and the *half-open intervals* $(a,b]$ and $[a,b)$ as well as to semi-infinite or infinite intervals. Clearly we may also introduce the analogous concepts of a decreasing or a strictly decreasing function.

Next we note that Lemma 1.1 can be strengthened for the case of a function f which is positive on (a,b). In this case we can assert that

$$\int_a^b f(x)\,dx > 0.$$

Applying these concepts to $I(t)$ we have for $t_2 > t_1$

$$I(t_2) = \int_0^{t_2} \frac{R}{L_0(1+\alpha s)} \tilde{\mathcal{E}}(s)\,ds = \int_0^{t_1} \frac{R}{L_0(1+\alpha s)} \tilde{\mathcal{E}}(s)\,ds + \int_{t_1}^{t_2} \frac{R}{L_0(1+\alpha s)} \tilde{\mathcal{E}}(s)\,ds.$$

Hence

$$I(t_2) = I(t_1) + \int_{t_1}^{t_2} \frac{R}{L_0(1+\alpha s)} \tilde{\mathcal{E}}(s)\,ds > I(t_1)$$

because $\tilde{\mathcal{E}}$ is positive. It follows that on the semi-infinite interval $(0, \infty)$ I is a strictly increasing function of t.

Returning to (1.50) which may now be written in the form

$$\mathcal{E}(t) = \mathcal{E}(0) - 2I(t)$$

we see that $\mathcal{E}(t)$ is strictly decreasing. This result is certainly sharper than inequality (1.48) which only implies that $\mathcal{E}(t)$ is bounded by $\mathcal{E}(0)$. We now have

$$\mathcal{E}(t_1) < \mathcal{E}(t_0) \quad \text{if} \quad t_1 > t_0; \tag{1.53}$$

that is, the electrical energy is dissipated.

We shall now make use of the following theorem of advanced calculus.

Theorem 1.2

Let f be a function defined and bounded on the semi-infinite interval (x_0, ∞) and suppose f is nondecreasing or nonincreasing on (x_0, ∞). Then $\lim_{x \to \infty} f(x)$ exists.

In this connection, note that if f is nondecreasing on (x_0, ∞) it is bounded below by $f(x_0)$. Hence in order to apply the theorem in this case it is sufficient to know that f is bounded above. A similar remark applies if f is nonincreasing. Since a strictly decreasing function is a special case of a nonincreasing function, the theorem applies to such functions as well.

The foregoing arguments apply to any initial-value problem for Equation (1.40). Since we always have $\mathcal{E}(t) \geq 0$ for all $t \geq 0$, we have now proved:

Proposition 1.6

The total energy function $\mathcal{E}(t)$ for the leakage problem (1.41), (1.42) is a strictly decreasing function of time and

$$\lim_{t \to \infty} \mathcal{E}(t) = \text{const.} \geq 0 \tag{1.54}$$

showing that the limit of the energy function exists as $t \to \infty$.

These results may also be used to estimate the rate of growth of q and i. Since

$$\mathcal{E}(t) \geq \frac{q^2(t)}{CL_0(1 + \alpha t)},$$

we have, by applying (1.53) for $t_1 = t$, $t_0 = 0$, the estimate

$$\frac{q^2(t)}{CL_0(1 + \alpha t)} < i^2(0) + \frac{q^2(0)}{CL_0}$$

hence

$$q^2(t) < CL_0(1 + \alpha t)\mathcal{E}(0).$$

This inequality implies in turn that

$$|q(t)| < \sqrt{CL_0 \mathcal{E}(0)(1 + \alpha t)}. \tag{1.55}$$

Thus, for sufficiently large t, $|q|$ grows more slowly than $\sqrt{t}$. The same sort of argument applies to $i(t)$. Namely

$$\mathcal{E}(t) \geq i^2(t),$$

hence

$$i^2(t) < i^2(0) + \frac{q^2(0)}{CL_0} = \mathcal{E}(0),$$

or

$$|i(t)| < \sqrt{\mathcal{E}(0)}, \quad \text{for all } t \geq 0. \tag{1.56}$$

The inequalities (1.55) and (1.56) show that for the leakage problem (1.41), (1.42) the absolute value of the charge, $|q|$, can be unbounded, while the absolute value of its derivative, $|i|$, is always bounded. The reason for the disparity between this result and the bounded behavior (stability) of the basic leakage problem of Section 1.2 is attributable to the unbounded behavior of the function L. However, remember that L was a "made up" function. It is more physically reasonable that L will grow linearly for a while, say to some time t_1, after which it asymptotically approaches some finite limiting value, say L_1. This type of function is pictured graphically in Figure 1.5, where $L = L_0(1 + \alpha t)$ from 0 to t_1. For such a function L we will have $L_0 \leq L(t) \leq L_1$, for all $t \geq 0$, and $0 < \dot{L}(t) \leq \alpha$, for all $t > 0$. Equation (1.41) becomes

$$\ddot{q} + \frac{R}{L(t)} \dot{q} + \frac{1}{CL(t)} q = 0, \quad t \geq 0. \tag{1.57}$$

Repeating the energy analysis for (1.41) in the case of (1.57) yields the basic energy identity

$$\mathcal{E}(t) = \mathcal{E}_0(t) - 2 \int_0^t \mathcal{E}^*(s) \, ds, \tag{1.58}$$

where

$$\mathcal{E}(t) = i^2(t) + \frac{q^2(t)}{CL(t)}, \quad L(0) = L_0,$$

and (1.59)

$$\mathcal{E}^*(t) = \frac{Ri^2(t)}{L(t)} + \frac{\dot{L}(t) q^2(t)}{2CL^2(t)}.$$

Since L and $\dot{L}$ are positive functions we have $\mathcal{E}^*(t) \geq 0$ for all $t \geq 0$ and it follows that (1.48) is valid for the present problem. Moreover, because of the boundedness of $L(t)$, we find in place of (1.55) the estimate

$$|q(t)| \leq \sqrt{\mathcal{E}(0) L(t)} < \sqrt{\mathcal{E}(0) L_1}, \quad \text{for all } t \geq 0. \tag{1.60}$$

The estimate (1.56) continues to hold for $|i(t)|$. Thus we are fully justified

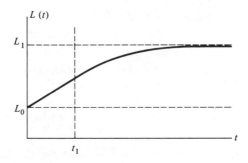

FIGURE 1.5

in attributing the possible unbounded growth of $|q|$ to the unbounded character of L.

The important thing is, of course, that we were able to deduce a great deal of useful information about the initial-value problem without ever having to, or being able to, solve the problem explicitly. The foregoing example should enable the reader to begin to appreciate the strength of the energy method.

PROBLEMS

1. Find a general solution for the equation
$$\ddot{u} + a\dot{u} + bu = 0,$$
where a and b are constants, subject to the initial conditions
$$u(0) = u_0 \quad \text{and} \quad \dot{u}(0) = \dot{u}_0.$$

2. Determine which of the following equations has a stable solution in the sense of Laplace when $a > b > 0$.
   (a) $\ddot{u} - a\dot{u} + bu = 0$.
   (b) $\ddot{u} + a\dot{u} - bu = 0$.
   (c) $\ddot{u} - a\dot{u} - bu = 0$.
   (d) $\ddot{u} - bu = 0$.
   (e) $\ddot{u} + bu = 0$.
   (f) $\ddot{u} - a\dot{u} = 0$.
   (g) $\ddot{u} + a\dot{u} = 0$.
   (h) $\ddot{u} = 0$.

3. Physically, explain what is taking place when the equations of Problem 2 describe
   (a) an electrical system,
   (b) a mechanical system.

4. Determine which of the following equations has a stable solution in the sense of Laplace when $a > b > 0$.
   (a) $\ddot{u} + a\dot{u} + bu = Ae^{\lambda t}$; for $\lambda > 0, \lambda = 0, \lambda < 0; A > 0$.
   (b) $\ddot{u} + a\dot{u} - bu = Ae^{\lambda t}$; for $\lambda > 0, \lambda = 0, \lambda < 0; A > 0$.
   (c) $\ddot{u} - a\dot{u} + bu = Ae^{\lambda t}$; for $\lambda > 0, \lambda = 0, \lambda < 0; A > 0$.
   (d) $\ddot{u} - a\dot{u} - bu = A \cos \lambda t; A > 0$.
   (e) $\ddot{u} + a\dot{u} + bu = At + B \cos \lambda t; A, B > 0$.

5. Physically, explain what is taking place when the equations of Problem 4 describe
   (a) an electrical system,
   (b) a mechanical system.

6. Derive an energy identity for the initial-value problem given in Problem 1 by multiplying the differential equation by $\dot{u}$ and suitable manipulations.

7. State the conditions on the constants a and b in the differential equation of Problem 1, which make all solutions remain at a constant-energy level for all time t (called *conservation of energy*). State the conditions which make all solutions dissipative.
8. Formulate and prove the uniqueness theorem for the differential equation of Problem 1 under the dissipative (or conservative) conditions you have obtained in Problem 7.
9. Find dissipative conditions for the differential equation
$$\ddot{u} + a(t)\dot{u} + b(t)u = 0,$$
subject to the initial conditions $u(0) = u_0$ and $\dot{u}(0) = \dot{u}_0$. Assume a and b to be differentiable or satisfy other conditions as needed. Are there any conservative conditions for this problem?
10. Using the fact that
$$\frac{d}{dt}\int_0^t i(s)\,ds = i(t),$$
derive an energy identity for the integro-differential equation (1.5).
11. We have shown that every solution of the leakage problem decreases to zero as $t \to \infty$. For what choices of the constants a and b of the differential equation of Problem 1 will there be solutions such that $|u|$ increases with increasing t?
12. In a nonlinear RLC circuit the resistance R is a function of the current having the form
$$R = R_0(1 + \alpha i^2), \quad \alpha > 0, \quad R_0 > 0.$$
Prove that if $q(0) = q_0$ and $\dot{q}(0) = i_0$, the solution to the leakage problem for this circuit is such that
$$R \leq R_0\left[1 + \alpha\left(i_0^2 + \frac{1}{LC}q_0^2\right)\right].$$
*13. Consider the initial-value problem of finding u defined for $t \geq t_0$ such that
$$\ddot{u} + a(t)\dot{u} + \beta(t)u = f(t), \quad t > t_0,$$
with $u(t_0) = u_0$ and $\dot{u}(t_0) = u_1$. Show that the solution of this problem can be reduced to finding the solution of the problem
$$\ddot{v} + \tilde{\alpha}(t)\dot{v} + \tilde{\beta}(t)v = \tilde{f}(t), \quad t > 0,$$
where $v(0) = \tilde{v}_0$ and $\dot{v}(0) = \tilde{v}_1$, by an appropriate change of variable. On the basis of the foregoing can we assume that all initial-value problems may be formulated with initial data at $t = 0$?
14. Prove that for the problem defined by (1.41) and (1.42),
$$\lim_{t \to \infty} \ddot{q}(t) = 0.$$

15. Show that the function defined by

$$L(t) = L_0 \frac{1 + \alpha t}{1 + \beta t}, \quad \alpha > \beta > 0,$$

satisfies the conditions

$$L_0 \leq L(t) \leq L_1, \quad \text{for all } t \geq 0,$$

and

$$0 < \dot{L}(t) \leq c, \quad c = \text{const.}, \quad \text{for all } t \geq 0.$$

Compute

$$\lim_{t \to \infty} L(t).$$

1.7 A MORE GENERAL UNIQUENESS THEOREM

In Section 1.5 we proved that the solution of the RLC circuit problem is unique by using the energy method. This technique is very well suited to such a problem but does not work for more general problems that the student will meet. Accordingly in this section we will present a uniqueness theorem for an equation which is in some ways more general than any discussed thus far. The equation will also provide us with a further illustration of the fact brought out in Section 1.6 that significant information can be extracted from differential equations by the use of a little mathematical insight. The latter is most important in those cases where solutions are not attainable or the functions involved are not explicitly known, as is the case below.

First we need the following result about the definite integral.

Lemma 1.2

Let the function g be defined and continuous on the interval (a,b), then

$$\left| \int_a^q g(t) \, dt \right| \leq \int_a^b |g(t)| \, dt. \tag{1.61}$$

With this at hand we can formulate the result we seek as:

Theorem 1.3

Let the function f be bounded and continuous and g be continuous on $(0, \infty)$. Then there is at most one solution of the differential equation

$$\ddot{u} + f(t)u = g(t) \tag{1.62}$$

on $[0, \infty)$ satisfying the initial conditions

$$u(0) = u_0, \quad \dot{u}(0) = \dot{u}_0. \tag{1.63}$$

PROOF. Assume that u_1 and u_2 are both solutions of the initial-value problem (1.62) and (1.63). Then from (1.62) we have

$$\ddot{u}_1 - \ddot{u}_2 + f(t)(u_1 - u_2) = 0.$$

We integrate this equation from 0 to t and use the fact that $\dot{u}_1(0) = \dot{u}_2(0) = \dot{u}_0$, to obtain

$$\dot{u}_1 - \dot{u}_2 + \int_0^t f(s)(u_1 - u_2)\, ds = 0,$$

where

$$|\dot{u}_1 - \dot{u}_2| = \left| \int_0^t f(s)(u_1 - u_2)\, ds \right|.$$

Using (1.61) we obtain

$$|\dot{u}_1(t) - \dot{u}_2(t)| \leq \int_0^t |f(s)|\, |u_1 - u_2|\, ds. \tag{1.64}$$

By hypothesis, f is bounded, hence let $M > 0$ be such that $|f(t)| \leq M$ for all $t > 0$. Then from (1.64) we obtain the inequality

$$|\dot{u}_1(t) - \dot{u}_2(t)| \leq M \int_0^t |u_1(s) - u_2(s)|\, ds. \tag{1.65}$$

Next let us set

$$\dot{u}_1 = v_1, \quad \dot{u}_2 = v_2$$

so that

$$\dot{u}_1 - \dot{u}_2 = v_1 - v_2 \tag{1.66}$$

and (1.65) becomes

$$|v_1(t) - v_2(t)| \leq M \int_0^t |u_1(s) - u_2(s)|\, ds. \tag{1.67}$$

Now integrate (1.66) from 0 to t and use the fact that $u_1(0) = u_2(0) = u_0$ to obtain

$$u_1(t) - u_2(t) = \int_0^t (v_1(s) - v_2(s))\, ds.$$

Again we make use of (1.61) to derive the inequality

$$|u_1(t) - u_2(t)| \leq \int_0^t |v_1(s) - v_2(s)|\, ds. \tag{1.68}$$

Adding inequalities (1.67) and (1.68) yields

$$|u_1(t) - u_2(t)| + |v_1(t) - v_2(t)|$$
$$\leq \int_0^t \{M|u_1(s) - u_2(s)| + |v_1(s) - v_2(s)|\}\, ds$$
$$\leq (1 + M) \int_0^t [|u_1(s) - u_2(s)| + |v_1(s) - v_2(s)|]\, ds.$$

With this preparation the remainder of the proof is quite simple. Set

$$\sigma(t) = |u_1(t) - u_2(t)| + |v_1(t) - v_2(t)| \tag{1.69}$$

so that the last inequality becomes

$$\sigma(t) \leq (1 + M) \int_0^t \sigma(s)\, ds. \tag{1.70}$$

Note that $\sigma(t) \geq 0$ for all t and $\sigma(t_0) = 0$ if and only if $u_1(t_0) = u_2(t_0)$ and $v_1(t_0) = v_2(t_0)$. Let us also set

$$r(t) = \int_0^t \sigma(s)\, ds \tag{1.71}$$

and observe that $r(0) = 0$ and $\dot{r}(t) = \sigma(t)$. In terms of the function r, inequality (1.70) becomes

$$\dot{r} - (1 + M)r \leq 0.$$

Multiply this equation by $e^{-(1+M)t}$ to obtain

$$e^{-(1+M)t}(\dot{r} - (1 + M)r) = \frac{d}{dt}\,(re^{-(1+M)t}) \leq 0.$$

Integrating this equation and using the fact that $r(0) = 0$ yields

$$e^{-(1+M)t}r(t) \leq 0.$$

Now noting that $r(t) \geq 0$ and $e^{-(1+M)t} > 0$ for all t it follows that $r(t) \equiv 0$ for all $t \geq 0$. Since (1.70) can be written as

$$\sigma(t) \leq (1 + M)r(t),$$

it follows that $\sigma(t) \equiv 0$ for all $t \geq 0$. But this means that

$$u_1(t) \equiv u_2(t), \quad t \geq 0,$$

which completes the proof of the theorem.

The proof of Theorem 1.3 remains valid if f and g are only defined on a finite interval, say on the interval (a,b). Let us formulate this case explicitly.

Theorem 1.4

Let the function f be bounded and continuous and g be continuous on (a,b). Then there is at most one solution of the differential equation

$$\ddot{u} + f(t)u = g(t)$$

on $[a,b)$ satisfying the initial conditions

$$u(a) = u_a, \quad \dot{u}(a) = \dot{u}_a.$$

PROBLEMS

1. Prove that if f and g satisfy the conditions of Theorem 1.3, then there is at most one solution of the differential equation
$$\dot{u} + f(t)u = g(t) \tag{1.72}$$
on $[0, \infty)$ satisfying the initial condition
$$u(0) = u_0. \tag{1.73}$$

2. Solve the initial-value problem (1.72), (1.73).
   [*Hint:* Observe that
   $$\frac{d}{dt}\left(e^{\int f(s)\,ds}u\right) = g(t)e^{\int f(s)\,ds},$$
   where $\int f(s)\,ds$ is any indefinite integral of f.]

3. Prove Theorem 1.4.

*4. Prove that if f is bounded and continuous and g is continuous on (a,b), there is at most one solution of the differential equation (1.62) on $(a,b]$ satisfying the initial conditions
$$u(b) = u_b, \quad \dot{u}(b) = \dot{u}_b.$$

5. Suppose that the differential equation (1.62) is changed to
$$\ddot{u} + h(t)\dot{u} + f(t)u = g(t) \tag{1.74}$$
with h bounded and continuous on $(0, \infty)$. Prove that under the hypotheses of Theorem 1.3 there exists at most one solution to the initial-value problem for this differential equation on $[0, \infty)$.

1.8 THE SIMPLE PENDULUM

All of the RLC-circuit problems considered thus far have been linear, in the sense the term was defined in Proposition 1.1. That is, if q_1 and q_2 are solutions to a differential equation and α_1 and α_2 are arbitrary real or complex numbers then $\alpha_1 q_1 + \alpha_2 q_2$ is also a solution. Most of the problems treated in this book will be, by intent, linear and the student will see again and again the importance of forming linear combinations of functions to obtain general solutions to differential equations.

All physical systems are nonlinear. Nevertheless, we find that linear problems are extremely important for a variety of reasons, the most important being that they often enable one to answer the question "What allows us to linearize an equation?" Given a physical system, one formulates a mathematical model from which one can obtain the equations of motion of the system. Rarely can these equations be solved. Thus the investigator makes assumptions which allow him to solve the equations—assumptions such as linearization. What good is such a solution?

To answer this question the investigator goes to the laboratory and experiments. If his experimental results and theoretical results agree he has justified his assumptions. But he still has a gimmick, not a method, in his hands, for his linearization solved a very special problem. If his method of linearization can be applied to a class of problems he then can say that he has a method of solution. Equally important, one can often obtain a great deal of insight from the solution of linear problems to enable one to attack sensibly the problem's nonlinear analog, especially in those cases where the linear solution is not sufficiently delicate to describe the behavior of the physical system. The discussion which follows on the simple pendulum is intended to illustrate, in a small way, both situations discussed above.

A simple pendulum consists of a heavy particle swinging in a vertical plane at the end of a rod or taut string of negligible weight. Let mg be the weight of the particle and $\theta = \theta(t)$ the angle which the rod of length l makes with the vertical as shown in Figure 1.6. Neglecting friction the forces on the particle are the weight mg and the tension of the rod. Since the tension in the rod acts normal to the path the resultant force along the tangent to the path is $mg \sin \theta$. From kinematics we write the acceleration vector of the particle as

$$\mathbf{a} = l\ddot{\theta}\mathbf{e}_t + l\dot{\theta}^2\mathbf{e}_n,$$

where $\mathbf{e}_t$ and $\mathbf{e}_n$ are unit vectors in the tangential and normal directions respectively, as indicated in Figure 1.6. Applying Newton's second law of motion in the tangential direction, we obtain

$$-mg \sin \theta = ml\ddot{\theta},$$

or (1.75)

$$\ddot{\theta} + \frac{g}{l} \sin \theta = 0.$$

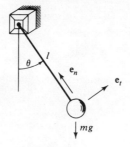

FIGURE 1.6

Physical intuition tells us that without an initial displacement or velocity the pendulum will remain motionless. Hence for a nontrivial solution we require the initial conditions

$$\theta(0) = \theta_0 \quad \text{and} \quad \dot{\theta}(0) = \omega_0. \tag{1.76}$$

We may now formulate a basic problem.

Find a function θ satisfying (1.75) for all $t \geq 0$ subject to the initial conditions (1.76) at time $t = 0$.

We begin the investigation of this problem by showing that it is not linear. We recall from Section 1.2 that a problem is linear if $c_1\theta_1 + c_2\theta_2$ is a solution whenever θ_1 and θ_2 are solutions and c_1, c_2 are constants. To do this suppose that there exist two solutions of Equation (1.75), θ_1 and θ_2. Therefore,

$$\ddot{\theta}_1 + \frac{g}{l} \sin \theta_1 = 0,$$

and

$$\ddot{\theta}_2 + \frac{g}{l} \sin \theta_2 = 0.$$

In accordance with Proposition 1.1 if (1.75) is linear the function $\theta = \theta_1 + \theta_2$ satisfies

$$\ddot{\theta} + \frac{g}{l} \sin \theta = 0.$$

But

$$\sin \theta_1 + \sin \theta_2 = 2 \sin \frac{\theta_1 + \theta_2}{2} \cos \frac{\theta_1 - \theta_2}{2} \neq \sin(\theta_1 + \theta_2)$$

in general. Thus $\theta_1 + \theta_2$ is not a solution of Equation (1.75) for all solutions θ_1 and θ_2. We may now say that the differential equation (1.75) is nonlinear.

Assuming, as seems physically reasonable, that this problem has a solution θ defined for all $t \geq 0$ for any choice of θ_0 and ω_0, what can we then say about this solution? In order to answer this question we shall first resort to a geometric analysis of Equation (1.75). To simplify the notation let $g/l = k^2$ so that (1.75) becomes

$$\ddot{\theta} + k^2 \sin \theta = 0.$$

We shall now replace this second-order equation by two first-order equations. This is carried out by setting

$$\dot{\theta} = \frac{d\theta}{dt} = \omega \tag{1.77}$$

so that (1.75) becomes

THE SIMPLE PENDULUM

$$\dot{\omega} = \frac{d\omega}{dt} = -k^2 \sin \theta. \tag{1.78}$$

Solving (1.77) and (1.78) for dt and equating the results we obtain

$$k^2 \sin \theta \, d\theta = -\omega \, d\omega.$$

Integrating this equation yields the fundamental equation

$$\frac{\omega^2}{2} - k^2 \cos \theta = C \tag{1.79}$$

where C is a constant of integration.

We shall gain considerable insight into the nature of the problem by graphing Equation (1.79) in the θ,ω-plane for different values of the constant C. The θ,ω-plane, or more generally the displacement-velocity plane, is called the *phase plane* for the differential equation (1.75) or for the system (1.77), (1.78). Solving (1.79) for ω yields

$$\omega = \pm [2(C + k^2 \cos \theta)]^{1/2}. \tag{1.80}$$

We see at once that if $C < -k^2$, Equation (1.79) has no real graph. If $C = -k^2$, the graph reduces to the points

$$\begin{aligned}\theta &= 2n\pi, \quad n = 0, \pm 1, \pm 2, \ldots \\ \omega &= 0.\end{aligned} \tag{1.81}$$

These points in the phase plane are called *singular solutions* of the system (1.77). We see in fact that if

$$\theta(t) \equiv 2n\pi, \quad \omega(t) \equiv 0$$

both (1.77) and (1.78) are satisfied [(1.75), of course, also has the solution $\theta(t) \equiv 2n\pi$]. To find all of the singular solutions of the system (1.77), (1.78), we seek all pairs of numbers θ, ω which make the right-hand sides of these equations zero. Thus we have from (1.77),

$$\omega = 0,$$

and from (1.78)

$$-k^2 \sin \theta = 0.$$

In this way we obtain, in addition to the singular solutions (1.81), the solutions

$$\begin{aligned}\theta &= n\pi, \quad n = 0, \pm 1, \pm 2, \ldots \\ \omega &= 0.\end{aligned} \tag{1.82}$$

The graph of each of these solutions in the phase plane is a single point. These points are called the *singular trajectories* of the system (1.77), (1.78).

Returning now to Equation (1.80), let us consider the graph obtained when $C > k^2$. We have two values of ω defined for every value of θ, one positive and one negative. The graphs in this case are the wavy lines in Figure 1.7. They are plotted for a particular value of $C > k^2$. The arrows indicate the direction in which these curves, called *nonsingular trajectories* (also open trajectories), are traced out with increasing time. The directions are seen at once from (1.77), for if $\omega > 0$, θ increases with t and if $\omega < 0$, θ decreases with t.

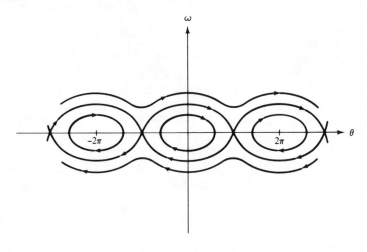

FIGURE 1.7

Here we are making use of the following facts. If f is defined and differentiable on the interval (a,b) and $f'(x) \geq 0$ there, then f is nondecreasing on (a,b). If $f'(x) > 0$ then f is strictly increasing. Similar results are valid if $f'(x) \leq 0$ or $f'(x) < 0$ on (a,b). These results are very easy to prove. Suppose for example that $f'(x) > 0$ on (a,b) and let $a < x_1 < x_2 < b$. We may apply the *mean-value theorem* on the closed interval $[x_1,x_2]$. It states that

$$f(x_2) - f(x_1) = f'(\bar{x})(x_2 - x_1)$$

where $x_1 < \bar{x} < x_2$. Since $f'(\bar{x}) > 0$ and $x_2 - x_1 > 0$, we have

$$f(x_2) - f(x_1) > 0.$$

It remains to discuss the nature of the trajectories when $-k^2 < C \leq k^2$. Let us first consider the case $-k^2 < C < k^2$. To be specific we may take as a typical example the case $C = k^2/2$ so that

$$\omega = \pm \sqrt{2}\, k[\tfrac{1}{2} + \cos \theta]^{1/2}.$$

Thus ω is only defined if

THE SIMPLE PENDULUM   41

$$\cos \theta \geq -\tfrac{1}{2},$$

that is, on the intervals

$$2n\pi - \frac{5\pi}{6} \leq \theta \leq 2n\pi + \frac{5\pi}{6}, \quad n = 0, \pm 1, \pm 2, \ldots.$$

These trajectories, called *closed trajectories* for obvious reasons, are also plotted in Figure 1.7.

Finally let us consider the case $C = k^2$. In this case ω is also defined for all values of θ and the corresponding curves are plotted in Figure 1.7 as the curves which pass through the points $\omega = 0$, $\theta = (2n+1)\pi$. These trajectories are called *separatrices*, and in the present case they separate the family of closed trajectories defined for $-k^2 < C < k^2$ from the family of open trajectories defined for $C > k^2$.

In order to see more clearly the physical meaning of the trajectories in the phase plane, let us revert to the energy method used in Section 1.6. Multiplying (1.75) by $\dot{\theta}$ gives

$$\dot{\theta}\ddot{\theta} + k^2\dot{\theta}\sin \theta = 0.$$

Applying the first of identities (1.43) to θ (in place of q) and noting that $(d/dt)\cos \theta = -\dot{\theta}\sin \theta$, we obtain

$$\frac{d}{dt}\left[\frac{\dot{\theta}^2}{2} - k^2\cos \theta\right] = 0.$$

Now integrate from 0 to t to get the identity

$$\frac{\dot{\theta}^2(t)}{2} - k^2 \cos \theta(t) = \frac{\dot{\theta}^2(0)}{2} - k^2 \cos \theta(0),$$

or

$$\frac{\omega^2}{2} - k^2 \cos \theta = \frac{\omega_0^2}{2} - k^2 \cos \theta_0. \qquad (1.83)$$

This identity is precisely the same as (1.79) if we identify

$$C = \frac{\omega_0^2}{2} - k^2 \cos \theta_0. \qquad (1.84)$$

The terms in (1.83) do not possess dimensionally correct units for energy. We can correct this by multiplying by ml^2 to obtain

$$\frac{ml^2\omega^2}{2} - mgl \cos \theta = \frac{ml^2\omega_0^2}{2} - mgl \cos \theta_0.$$

Defining the energy functions

$$T = \frac{ml^2\omega^2}{2} \quad \text{and} \quad V = -mgl \cos \theta + K, \qquad (1.85)$$

where K is an arbitrary constant, (1.83) can be written as

$$T + V = \text{const.} = ml^2 C. \tag{1.86}$$

Here T and V are the kinetic and potential energy functions, in the classical-mechanics sense, for the pendulum.

From the physical point of view the trajectories in the phase plane are now seen to represent specific energy levels of the pendulum which, in this case, is a conservative mechanical system. On the other hand, the trajectories also present a geometric picture of the solutions of the pendulum problem.

From Equation (1.84) we see that the restriction

$$C \geq -k^2$$

already noted from (1.80) is built into our problem, for the right-hand side of (1.84) is clearly bounded below by $-k^2$. The closed trajectories arise when

$$-k^2 < \frac{\omega_0^2}{2} - k^2 \cos \theta_0 < k^2.$$

These represent periodic motion of the pendulum and will arise whenever $\omega_0 = \omega$ if $|\theta_0| < \pi$.

It remains for us to examine the separatrices and the singular solutions. The separatrices indicate that if the proper amount of energy is supplied initially, namely

$$T + V = ml^2 k^2$$

we may lose uniqueness in the problem. The pendulum's bob may either rotate about the pivot following an open trajectory or it may oscillate back and forth periodically from the upward vertical position.

To examine the singular trajectories corresponding to $(\omega_0^2/2) - k^2 \cos \theta_0 = -k^2$, let us note that this relation implies that the initial conditions must be

$$\omega_0 = \theta_0 = 0.$$

Then from (1.83) we have

$$\frac{\omega^2}{2} - k^2 \cos \theta = -k^2.$$

But $\omega^2/2 \geq 0$, hence

$$-\cos \theta(t) \leq -1$$

or

$$\cos \theta(t) \geq 1 \tag{1.87}$$

for all $t \geq 0$. Obviously, the inequality in (1.87) is impossible by the definition of the cosine function, hence we must actually have

$$\cos \theta(t) = 1.$$

It follows immediately that $\theta(t) \equiv 0$. This explains the physical meaning of the singular trajectory $\theta = 0$, $\omega = 0$. The remaining singular trajectories $\theta = \pm 2n\pi$, $\omega = 0$ as well as the closed trajectories surrounding these points for $n \neq 0$ have no direct physical meaning. This is because for a "stationary motion" or oscillatory motion we need not assign any physical significance to values of θ such that $|\theta| > \pi$. This is not true for the open trajectories since for them these values count how many times the bob has rotated around the pivot.

It is customary to fix the constant K so that the potential energy is equal to zero for the position of equilibrium $\theta(t) = 0$ corresponding to the singular trajectory $\theta = 0$, $\omega = 0$. Thus

$$V = ml^2 k^2 (1 - \cos \theta) = mgl(1 - \cos \theta). \tag{1.88}$$

Before concluding our work on the simple pendulum, let us observe the following facts. We may write (1.79) in the form

$$\left(\frac{d\theta}{dt}\right)^2 - 2k^2 \cos \theta = \tilde{C}, \quad \tilde{C} = 2C.$$

Solving this equation for $d\theta/dt$ yields

$$\frac{d\theta}{dt} = \pm \sqrt{\tilde{C} + 2k^2 \cos \theta}$$

from which we deduce that

$$dt = \pm \frac{d\theta}{\sqrt{\tilde{C} + 2k^2 \cos \theta}}.$$

Integrating this equation from $t = 0$ to t, we obtain

$$t = \pm \int_{\theta_0}^{\theta} \frac{d\varphi}{\sqrt{\tilde{C} + 2k^2 \cos \varphi}}. \tag{1.89}$$

The integral occurring on the right-hand side of (1.89) can be shown to be an elliptic integral and hence cannot be evaluated by elementary methods. Nevertheless, we have exhibited a solution of the simple pendulum in the form

$$t = t(\theta)$$

where the function $t(\theta)$ appears in integral form. This type of solution may be useful for numerical computations, but it does not give us a great deal of additional insight. Instead a phase-plane diagram furnishes a far more informative analysis of the problem.

1.9 THE LINEAR PENDULUM

In the absence of any dissipative forces (friction) a pendulum initially displaced from rest through an angle θ_0 and then released will continue to oscillate between the maximum angles θ_0 and $-\theta_0$ if $\theta_0 < \pi$. This was shown geometrically in Section 1.8.

In Section 1.8 we also showed that the pendulum is in static equilibrium in the downward vertical position. By introducing a standard technique of analytical mechanics we will be able to study the pendulum's motion (vibration) in a neighborhood of this static equilibrium position. This consists of writing a Taylor expansion with remainder of the potential energy about $\theta = 0$ as follows:

$$V = mgl\left(1 - 1 + \frac{\theta^2}{2!} - \frac{\theta^4}{4!}\cos\theta^*\right) = mgl\left(\frac{\theta^2}{2!} - \frac{\theta^4}{4!}\cos\theta^*\right)$$

where θ^* lies between 0 and θ. We now discard the term $(\theta^4/4!)\cos\theta^*$ on the heuristic grounds that it makes a very small contribution to $dV/d\theta$ for small values of θ compared to the term $\theta^2/2$. Thus, for the small vibration analog of the problem we have

$$T(t) = \frac{ml\dot{\theta}^2(t)}{2}, \quad V(t) = \frac{mgl\theta^2(t)}{2}. \tag{1.90}$$

It is worth mentioning in passing that the process we are using here often leads to a linear system which can be used to furnish important information about the nonlinear problem at hand.

With the energy formulations at hand the equation of motion for the pendulum can be found with the aid of Lagrange's equations of motion. A derivation of Lagrange's equations of motion can be found in any advanced dynamics book. Since we are dealing with a conservative system, for our problem Lagrange's equation is simply

$$\frac{d}{dt}\frac{\partial T}{\partial \dot{\theta}} + \frac{\partial V}{\partial \theta} = 0. \tag{1.91}$$

Placing (1.90) into (1.91) we obtain

$$\ddot{\theta} + \frac{g}{l}\theta = 0 \tag{1.92}$$

as the equation of motion for the system. The initial conditions remain

$$\theta(0) = \theta_0 \quad \text{and} \quad \dot{\theta}(0) = \omega_0. \tag{1.93}$$

The static equilibrium position $\theta = 0$ is said to be a position of stable equilibrium if every solution of the initial-value problem (1.92), (1.93) is stable in the sense of Laplace. Let us begin by showing that this is a position of stable equilibrium.

The initial-value problem we now have to solve is mathematically equivalent to a special case of the leakage problem of Section 1.2. Note that if we set the constants $R = 0$, $L = l$, $g = 1/C$, $q_0 = \theta_0$ and $i_0 = \omega_0$ in all of the results of that section we will, in essence, transform the solution of the leakage problem into the solution to the pendulum problem. The characteristic equation for (1.92) is simply

$$\lambda^2 + \frac{g}{l} = 0,$$

which always has the conjugate complex solutions

$$\lambda_1 = i\sqrt{g/l}, \quad \lambda_2 = -i\sqrt{g/l}.$$

Thus only the second case treated in Section 1.2 needs to be discussed here and so we seek solutions of (1.92), (1.93) having the form

$$\theta(t) = \alpha \cos \sqrt{g/l}\, t + \beta \sin \sqrt{g/l}\, t. \tag{1.94}$$

Let us actually compute α and β for the pendulum problem. We have

$$W\left(\cos \sqrt{\frac{g}{l}}\, t, \sin \sqrt{\frac{g}{l}}\, t\right)(0) = \sqrt{\frac{g}{l}},$$

hence

$$\alpha = \frac{\begin{vmatrix} \theta_0 & 0 \\ \omega_0 & \sqrt{g/l} \end{vmatrix}}{\sqrt{g/l}} = \theta_0,$$

$$\beta = \frac{\begin{vmatrix} 1 & \theta_0 \\ 0 & \omega_0 \end{vmatrix}}{\sqrt{g/l}} = \frac{\omega_0}{\sqrt{g/l}}.$$

Thus

$$\theta(t) = \theta_0 \cos \sqrt{g/l}\, t + \frac{\omega_0}{\sqrt{g/l}} \sin \sqrt{g/l}\, t. \tag{1.95}$$

Two different representations of the solution (1.80) are of interest. The first one is

$$\theta(t) = \omega_0 G(t) + \theta_0 \frac{d}{dt} G(t), \quad G(t) = \frac{1}{\sqrt{g/l}} \sin \sqrt{g/l}\, t. \tag{1.96}$$

We shall see that this type of representation arises in a variety of problems. The function $G(t)$ is called the *Green's function* for the initial-value problem

(1.92), (1.93). Before proceeding to the second representation let us consider briefly the nonhomogeneous problem of solving $\ddot{\theta} + g_l\theta = f(t)$ with the initial conditions $\theta(0) = 0 = \dot{\theta}(0)$. For this purpose we make use of the integral (1.31). It yields

$$\theta(t) = \frac{1}{\sqrt{g/l}} \int_0^t (\sin\sqrt{g/l}\,t \cos\sqrt{g/l}\,s - \cos\sqrt{g/l}\,t \sin\sqrt{g/l}\,s)f(s)\,ds$$

when applied to the present equation. Using a trigonometric identity we have at once

$$\theta(t) = \int_0^t G(t-s)f(s)\,ds,$$

where

$$G(t-s) = \frac{1}{\sqrt{g/l}} \sin\sqrt{g/l}\,(t-s).$$

Thus the Green's function just used to solve the homogeneous problem solves the nonhomogeneous problem as well.

The second representation is obtained by multiplying and dividing (1.95) by the positive number

$$\sqrt{\theta_0^2 + \frac{l\omega_0^2}{g}} = \sqrt{2\mathcal{E}(0)}$$

which is the square root of twice the total energy of the system at time $t = 0$. We may then define an angle φ by setting

$$\sin\varphi = \frac{\theta_0}{\sqrt{2\mathcal{E}(0)}},$$

so that $\cos\varphi = \omega_0\sqrt{l/g}/\sqrt{2\mathcal{E}(0)}$. Thus φ may be visualized as an acute angle of a right triangle whose hypotenuse is $\sqrt{2\mathcal{E}(0)}$.

Placing the two equations for φ into (1.95) we obtain

$$\theta(t) = \sqrt{2\mathcal{E}(0)}\,[\sin\varphi \cos\sqrt{g/l}\,t + \cos\varphi \sin\sqrt{g/l}\,t],$$

which, using an appropriate trigonometric identity, can be written more compactly as

$$\theta(t) = \sqrt{2\mathcal{E}(0)} \sin(\sqrt{g/l}\,t + \varphi). \qquad (1.97)$$

The representation (1.97) is quite appealing in that one immediately sees that the vibrations of the linear pendulum are sinusoidal in character. Engineers and physicists often use the expression "simple harmonic motion" when expressing this fact. The number $\sqrt{2\mathcal{E}(0)}$ is called the amplitude of the vibration and is related as indicated to the total energy of the system at time $t = 0$ and hence can be controlled by the initial conditions as was foreshadowed in Section 1.8. The number $2\pi/\sqrt{g/l}$ is the period of the motion, since it is related to the period of the sine function; and φ is called the *phase angle*. It is clear that φ depends entirely upon the initial data θ_0 and ω_0.

From any of (1.95), (1.96), or (1.97) it is seen that all solutions of the initial value problem are bounded and from (1.97) in particular that $\sqrt{2\mathcal{E}(0)}$ is an upper bound. This observation establishes the stability of equilibrium of the solution $\theta \equiv 0$ of the pendulum problem.

It is also true that the linear pendulum satisfies the same conservation law satisfied by the nonlinear pendulum of Section 1.8. That is,

$$\mathcal{E}(t) = \mathcal{E}(0), \quad \text{for all } t \geq 0. \tag{1.98}$$

This may be verified by a direct calculation with the solution in the form (1.97) [or (1.95)] or by using the same energy-integral technique established in Section 1.6.

We wish to emphasize one point in closing. Namely that the position of stable equilibrium, $\theta(t) \equiv 0$ for all t, corresponds to an absolute minimum of the potential energy function $V(t)$. Indeed, a satisfactory method of defining stable equilibrium in mechanics is in terms of minima of the potential energy function.

PROBLEMS

1. Using Newton's second law, derive Equation (1.92).
2. Prove that if u_1 is a solution of $a\ddot{u} + b\dot{u} + cu = F_1(t)$ and u_2 is a solution of $a\ddot{u} + b\dot{u} + cu = F_2(t)$, then $u = u_1 + u_2$ is a solution of $a\ddot{u} + b\dot{u} + cu = F_1(t) + F_2(t)$. What conclusion can be deduced from the foregoing proof?
3. If x_1 is a solution to $a\ddot{x} + b\dot{x} + cx^3 = F_1(t)$ and x_2 is a solution to $a\ddot{x} + b\dot{x} + cx^3 = F_2(t)$, is $x = x_1 + x_2$ a solution to $a\ddot{x} + b\dot{x} + cx^3 = F_1(t) + F_2(t)$? Substantiate your answer. What conclusion can be deduced from the problem?
4. Would the method of linear superposition have worked in Problem 3 if the nonlinearity had occurred in the velocity or acceleration term rather than the displacement term?
5. Use the energy method where possible to determine which equations of Problem 2 of Section 1.6 are stable. Make appropriate assumptions where necessary.
6. Use the energy method where possible to determine which equations of Problem 4 of Section 1.6 are stable. Make appropriate assumptions where necessary.

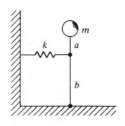

FIGURE 1.8

7. Use Newton's second law to derive the equation of motion for the system of Figure 1.8, which is in a vertical plane. Is the resulting equation linear or nonlinear? Substantiate your answer.
8. Write the potential and kinetic energy functions for the system of Figure 1.8 and take the time derivative of their sum. Compare the resulting equation to that obtained in Problem 7. What principle did you use here? (Here is an excellent example of a gimmick. It works for a one-degree-of-freedom conservative system only. Why?)
9. Determine a stability criterion for the system in Figure 1.8.
10. Use a Taylor expansion and linearize the energy functions you obtained in Problem 8. By the method of Problem 8 obtain the equation of motion.
11. Using small-angle assumptions, linearize the equation you obtained in Problem 7. Compare this equation to the result of Problem 10. What conclusion can you deduce from these results?
12. Work Problem 7 for the system of Figure 1.9.
13. Work Problem 8 for the system of Figure 1.9. Note that the gimmick no longer works and you have to resort to Lagrange's equations for nonconservative systems (or a direct application of Newton's second law)

$$\frac{d}{dt}\left(\frac{\partial T}{\partial \dot\theta}\right) - \frac{\partial T}{\partial \theta} + \frac{\partial D}{\partial \dot\theta} + \frac{\partial V}{\partial \theta} = 0.$$

14. Work Problem 9 for the system of Figure 1.9.
15. Work Problem 10 for the system of Figure 1.9.

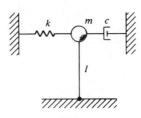

FIGURE 1.9

16. Work Problem 11 for the system of Figure 1.9.
17. Consider the solution to a linear, homogeneous system

$$u = A \cos \omega t + B \sin \omega t$$

where ω is the natural frequency of the system. How are A and B determined? Show that the solution can also be written in the form

$$u = C \sin (\omega t + \varphi)$$

where φ is called the *phase angle*. Give a physical interpretation of φ for a mechanical system and for an electrical system.

18. Make a plot of the trajectories in the phase plane of the linear pendulum equation (1.92) similar to that made in the text for the nonlinear pendulum.

19. An acceptable model for the motion of certain nonlinear spring-mass systems is furnished by the differential equation

$$\ddot{x} + \alpha x + \beta x^3 = 0, \quad \alpha, \beta \text{ constants,}$$

subject to given initial conditions at $t = 0$. That is, $x(0) = x_0, \dot{x}(0) = \dot{x}_0$ are given. This equation is called *Duffing's equation*. Make a plot of the trajectories in the phase plane for Duffing's equation when

(a) $\alpha > 0, \beta > 0,$ and
(b) $\alpha < 0, \beta > 0.$

20. A so-called "soft" nonlinear spring-mass system could be described by the differential equation

$$\ddot{x} + \alpha x - \beta x^{1/3} = 0, \quad \alpha > 0, \beta > 0, \qquad (1.99)$$

subject to given initial conditions at $t = 0$. Make a plot in the phase plane of the trajectories of this system.

21. Show that the equations for the trajectories of Duffing's equation and of Equation (1.99) can also be found by the energy method of Sections 1.6 and 1.7.

22. Find integral formulas similar to formula (1.88) for t as a function of x for the Duffing's equation and for Equation (1.99).

23. Can you make a phase-plane plot of the nonlinear RLC circuit problem given in Problem 12 of Section 1.6? That is, determine whether you can draw a phase-plane plot of

$$\ddot{q} + \frac{R_0(1 + \alpha \dot{q}^2)}{L} \dot{q} + \frac{1}{LC} q = 0, \quad q(0) = q_0, \, i(0) = i_0.$$

How does this problem differ from the simple pendulum, Duffing's equation, and Equation (1.99)?

24. We have seen that linear superposition is an acceptable way to solve

small (linear) vibrations problems. In engineering practice one usually drops the transient part of the solution (the solution of the homogeneous case) and focuses attention on the steady-state part of the solution (the solution of the nonhomogeneous case). Explain why this is done.

25. Considering the steady-state solution to a linear system acted upon by a harmonic forcing function, say $F_0 e^{i\omega t}$ where ω is the forcing frequency, the solution is often written in the form

$$u = A \cos \omega t + B \sin \omega t.$$

How are A and B determined? Show that the solution can also be written in the form

$$u = C \sin (\omega t + \varphi)$$

where φ is called the *phase angle*. Give a physical interpretation of φ for a mechanical system and for an electrical system.

Chapter 2

ORDINARY DIFFERENTIAL EQUATIONS— SIMPLE SYSTEMS

2.1 INTRODUCTION

Up to this point our attention has been directed toward physical problems which can be completely described by a single differential equation. Many problems require two or more simultaneous differential equations for their description. In our future work these will be referred to simply as *systems*.

In the next section we consider a classical example of such a problem, the double pendulum. Our purpose in doing this is to further illustrate the mechanical problems considered in Sections 1.8 and 1.9 and to indicate how rapidly these problems can become complex to analyze.

In Sections 2.2 through 2.6 we consider an electrical circuit problem which leads to a system of three first-order differential equations. In Section 2.7 we present a more general treatment of linear first-order differential equations. This provides the mechanism by which we introduce some vector-matrix notation. Further, we illustrate elementary operations with the vector-matrix concepts as they apply to differential equations.

The final two sections of this chapter are devoted to a discussion of the equations of motion for a whirling shaft. This problem, because of the far-reaching importance of such a system, has received much attention over the years. In Section 2.8 a linear symmetric rotating-shaft problem is explicitly solved, and in Section 2.9 we briefly discuss this solution, comparing the results with a simple nonlinear model of the same system.

Again, as in Chapter 1, we continue to stress the importance of giving some thought to the solution of a problem in the context of desired information rather than elaborating upon routine solution methods which may rapidly become useless when one attempts to solve new and more complex problems.

2.2 THE LINEAR DOUBLE PENDULUM

Consider the double pendulum illustrated in Figure 2.1. As before assume the masses m_1 and m_2 to be particles at the ends of rigid weightless rods of lengths l_1 and l_2, respectively. The motion is constrained to take place in a

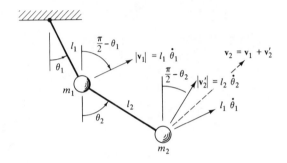

FIGURE 2.1

plane with the plane polar coordinates θ_1 and θ_2 serving as generalized coordinates for the system. The kinetic energy is, in general,

$$T = \tfrac{1}{2}(m_1 v_1^2 + m_2 v_2^2).$$

The required velocity vectors are shown in Figure 2.1. From the kinematics of the system we have

$$\mathbf{v}_1 = l_1 \dot{\theta}_1 \mathbf{e}_{t1}$$

where $\mathbf{e}_{t1}$ is a unit vector normal to l_1. By a simple vector addition we obtain

$$\mathbf{v}_2 = \mathbf{v}_1 + \mathbf{v}'_2$$

where

$$\mathbf{v}'_2 = l_2 \dot{\theta}_2 \mathbf{e}_{t2}.$$

Note that $\mathbf{v}'_2$ is the velocity of m_2 when the motion of m_1 is constrained. Hence $\mathbf{e}_{t2}$ is a unit vector normal to l_2.

With $\mathbf{v}_1$, $\mathbf{v}_2$ and some trigonometric manipulations we find

$$T = \frac{m_1}{2} l_1^2 \dot{\theta}_1^2 + \frac{m_2}{2} \{[l_1 \dot{\theta}_1 + l_2 \dot{\theta}_2 \cos(\theta_2 - \theta_1)]^2 + [l_1 \dot{\theta}_2 \sin(\theta_2 - \theta_1)]^2\}. \quad (2.1)$$

It should be noted that the kinetic and potential energy functions are scalar quantities. In writing the potential energy we will adjust the arbitrary constant by using the vertical position $\theta_1 = \theta_2 = 0$ as a position of equilibrium. Referring to Figure 2.2, we obtain the formula

$$V = m_1 g l_1(1 - \cos \theta_1) + m_2 g[l_1(1 - \cos \theta_1) + l_2(1 - \cos \theta_2)] \quad (2.2)$$

as the potential energy of the system.

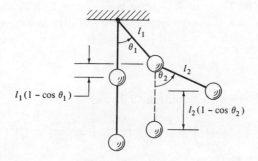

FIGURE 2.2

Since we are dealing with a conservative system, Lagrange's equations have the form

$$\frac{d}{dt}\left(\frac{\partial T}{\partial \dot{\theta}_i}\right) - \frac{\partial T}{\partial \theta_i} = -\frac{\partial V}{\partial \theta_i}, \quad i = 1, 2. \quad (2.3)$$

Performing the indicated operations in (2.3) with the aid of (2.1) and (2.2) we obtain the equations of motion for the double pendulum. These equations are nonlinear and very messy. We will leave it to the student to develop them.

Our attention here will be limited to the study of the small vibrations of the double pendulum about its equilibrium position. The method of finding these equations is similar to that used in Section 1.8 for the simple pendulum. We expand each of these functions T and V in a Taylor series in the variables θ_1 and θ_2 about the equilibrium position $\theta_1 = \theta_2 = 0$. Retaining only the nonzero terms of lowest degree in θ_1 and θ_2 yields the revised kinetic and potential energy functions for the small vibrations. Equations (2.3) are then applied to obtain the equations of motion.

In the functions T and V the variables θ_1 and θ_2 always appear as arguments of the trigonometric functions. This means we may use the series expansions

$$\cos x = 1 - \frac{x^2}{2!} + \frac{x^4}{4!} - \cdots,$$

and
$$\sin x = x - \frac{x^3}{3!} + \frac{x^5}{5!} - \cdots,$$

Placing these in (2.1) and retaining the first nonzero term only, gives
$$T = \frac{m_1}{2} l_1^2 \dot{\theta}_1^2 + \frac{m_2}{2} \{[l_1\dot{\theta}_1 + l_2\dot{\theta}_2]^2 + [l_2\dot{\theta}_2(\theta_2 - \theta_1)]^2\}.$$

Observe, however, that the last term is of degree 2 in θ_1 and θ_2 while the remaining terms are of degree zero in θ_1 and θ_2. Hence we may discard this term to obtain
$$T = \frac{m_1}{2} l_1^2 \dot{\theta}_1^2 + \frac{m_2}{2} [l_1\dot{\theta}_1 + l_2\dot{\theta}_2]^2. \tag{2.4}$$

Operating similarly with Equation (2.2) yields
$$V = m_1 g l_1 \left(1 - 1 + \frac{\theta_1^2}{2}\right) + m_2 g \left[l_1\left(1 - 1 + \frac{\theta_1^2}{2}\right) + l_2\left(1 - 1 + \frac{\theta_2^2}{2}\right)\right]$$
$$= \frac{m_1}{2} g l_1 \theta_1^2 + \frac{m_2}{2} g(l_1\theta_1^2 + l_2\theta_2^2),$$

or
$$V = \frac{m_1 + m_2}{2} g l_1 \theta_1^2 + \frac{m_2}{2} g l_2 \theta_2^2. \tag{2.5}$$

Lagrange's equations are now applied to T and V to obtain the equations of motion
$$\begin{aligned}(m_1 + m_2)l_1^2\ddot{\theta}_1 + m_2 l_1 l_2 \ddot{\theta}_2 + g l_1(m_1 + m_2)\theta_1 &= 0, \\ m_2 l_1 l_2 \ddot{\theta}_1 + m_2 l_2^2 \ddot{\theta}_2 + g l_2 m_2 \theta_2 &= 0.\end{aligned} \tag{2.6}$$

Prescribe the initial conditions for the problem as
$$\theta_1(0) = \varphi_1, \quad \theta_2(0) = \varphi_2, \quad \dot{\theta}_1(0) = \omega_1, \quad \dot{\theta}_2(0) = \omega_2. \tag{2.7}$$

We now have what can be termed a *standard initial-value problem*, namely that of finding functions θ_1, θ_2 satisfying (2.6) and (2.7).

The coefficients in Equations (2.6) are all constants. We might hope, therefore, that it will be possible to find exponential solutions having the form
$$\theta_1(t) = \alpha e^{\lambda t}, \quad \theta_2(t) = \beta e^{\lambda t} \tag{2.8}$$
for suitable constants α, β, and λ. Substituting (2.8) into (2.6) results in the equations
$$\{[(m_1 + m_2)l_1^2 \lambda^2 + g l_1(m_1 + m_2)]\alpha + m_2 l_1 l_2 \lambda^2 \beta\} e^{\lambda t} = 0,$$
and
$$\{m_2 l_1 l_2 \lambda^2 \alpha + (m_2 l_2^2 \lambda^2 + g l_2 m_2)\beta\} e^{\lambda t} = 0.$$

Since $e^{\lambda t} \neq 0$, the constants α and β must satisfy the linear equations

$$(m_1 + m_2)l_1(l_1\lambda^2 + g)\alpha + m_2 l_1 l_2 \lambda^2 \beta = 0,$$
$$m_2 l_1 l_2 \lambda^2 \alpha + m_2 l_2(l_2 \lambda^2 + g)\beta = 0. \qquad (2.9)$$

The system (2.9) is a homogeneous system and hence can have the trivial solution $\alpha = \beta = 0$. In view of Equations (2.8) the trivial solution is of no interest to us, therefore we must find nontrivial solutions, if possible. This will be possible if and only if the determinant of the system

$$\begin{vmatrix} (m_1 + m_2)l_1(l_1\lambda^2 + g) & m_2 l_1 l_2 \lambda^2 \\ m_2 l_1 l_2 \lambda^2 & m_2 l_2(l_2 \lambda^2 + g) \end{vmatrix} = 0. \qquad (2.10)$$

Expanding the determinant yields

$$m_1 m_2 l_1{}^2 l_2{}^2 \lambda^4 + g l_1 l_2 m_2 (m_1 + m_2)(l_1 + l_2)\lambda^2 + g^2 l_1 l_2 m_2 (m_1 + m_2) = 0, \qquad (2.11)$$

which is a polynomial of degree 4 in λ. Note, however, that the equation is actually biquadratic, that is, a quadratic equation in λ^2. The zeros of this biquadratic will be the values of λ for which we can determine solutions of (2.6) having the form (2.8).

By an application of the quadratic formula we obtain from (2.11)

$$\lambda^2 = -\frac{g(m_1 + m_2)(l_1 + l_2)}{2m_1 l_1 l_2} \pm \left\{ \frac{g^2(m_1 + m_2)^2(l_1 + l_2)^2}{4m_1{}^2 l_1{}^2 l_2{}^2} - \frac{g^2(m_1 + m_2)}{m_1 l_1 l_2} \right\}^{1/2}.$$

Set

$$M = \frac{g^2(m_1 + m_2)}{m_1 l_1 l_2}$$

and

$$L = \frac{l_1 + l_2}{2g},$$

then

$$\lambda^2 = -ML \pm \sqrt{M^2 L^2 - M}. \qquad (2.12)$$

Note that $M > 0$, $L > 0$, and $M^2 L^2 - M > 0$. The first two inequalities are obvious. The last inequality is not obvious. To show that it holds, consider

$$M^2 L^2 - M = \frac{g^2(m_1 + m_2)}{4m_1{}^2 l_1{}^2 l_2{}^2} [(m_1 + m_2)(l_1 + l_2)^2 - 4m_1 l_1 l_2]$$
$$= \frac{g^2(m_1 + m_2)}{4m_1{}^2 l_1{}^2 l_2{}^2} [m_2(l_1 + l_2)^2 + m_1(l_1 - l_2)^2] > 0.$$

These inequalities imply that

$$\sqrt{M^2 L^2 - M} < |ML|.$$

Hence both of the numbers

56 ORDINARY DIFFERENTIAL EQUATIONS—SIMPLE SYSTEMS

and
$$\lambda_1^2 = -ML + \sqrt{M^2L^2 - M},$$
$$\lambda_2^2 = -ML - \sqrt{M^2L^2 - M}$$

are negative. Let us set $-\mu_1 = \lambda_1^2$, and $-\mu_2 = \lambda_2^2$. Then the four zeros of the biquadratic equation (2.11) may be written as

$$i\sqrt{\mu_1}, \quad -i\sqrt{\mu_1}, \quad i\sqrt{\mu_2}, \quad \text{and} \quad -i\sqrt{\mu_2}.$$

Using each of these four values we can find nontrivial solutions to the homogeneous linear system (2.9).

At this point we shall assume that $m_1 = m_2 = m$ and $l_1 = l_2 = l$ in order to simplify the remainder of our calculations. Then the system (2.9) becomes

$$\begin{aligned} 2(l\lambda^2 + g)\alpha + l\lambda^2\beta &= 0, \\ l\lambda^2\alpha + (l\lambda^2 + g)\beta &= 0. \end{aligned} \tag{2.13}$$

The quantities M, L, μ_1, and μ_2 simplify as follows:

$$M = \frac{2g^2}{l^2} = 2\left(\frac{g}{l}\right)^2, \quad L = \frac{l}{g}, \quad ML = \frac{g}{l},$$

$$M^2L^2 - M = 4\left(\frac{g}{l}\right)^2 - 2\left(\frac{g}{l}\right)^2 = 2\left(\frac{g}{l}\right)^2, \quad \lambda_1^2 = -2\frac{g}{l} + \sqrt{2}\frac{g}{l},$$

$$\lambda_2^2 = -2\frac{g}{l} - \sqrt{2}\frac{g}{l},$$

and

$$\mu_1 = (2 - \sqrt{2})\frac{g}{l}, \quad \mu_2 = (2 + \sqrt{2})\frac{g}{l}. \tag{2.14}$$

Now let us set $\lambda = i\sqrt{\mu_1}$ in (2.13). We obtain, after some simplification, the equations

$$2(\sqrt{2} - 1)\alpha - (2 - \sqrt{2})\beta = 0,$$

and

$$-(2 - \sqrt{2})\alpha + (\sqrt{2} - 1)\beta = 0.$$

If we multiply the second of these equations by $\sqrt{2}$ we see it is the same as the first equation. Thus either equation may be solved for β in terms of α with the same result, namely

$$\beta = \frac{2 - \sqrt{2}}{\sqrt{2} - 1}\alpha = \sqrt{2}\,\alpha.$$

This result is rather surprising as it shows that α and β are not independent. They are related to within an arbitrary constant. It will be seen throughout this book that this will always be the case for systems of differential equations. It follows from (2.8) that the pair of functions

$$\theta_1(t) = \alpha e^{\sqrt{(2-\sqrt{2})\frac{g}{l}}\,t}, \quad \theta_2(t) = \sqrt{2}\,\alpha e^{i\sqrt{(2-\sqrt{2})\frac{g}{l}}\,t}$$

is a solution of (2.6) in the special case under consideration for an arbitrary choice of the constant α. Corresponding to the value $\lambda = -i\sqrt{(2-\sqrt{2})\frac{g}{l}}$ we will also find the solution pair

$$\theta_1^*(t) = \alpha e^{-i\sqrt{(2-\sqrt{2})\frac{g}{l}}\,t}, \quad \theta_2^*(t) = \sqrt{2}\,\alpha e^{-i\sqrt{(2-\sqrt{2})\frac{g}{l}}\,t}$$

for arbitrary α.

Let us now state a result which the reader can readily verify for himself by a direct computation.

Proposition 2.1

Suppose θ_1 and θ_2 are a solution pair for the system (2.6) and φ_1 and φ_2 a second solution pair. Then every linear combination

$$x_1 = a_1\theta_1 + b_1\varphi_1$$
$$x_2 = a_2\theta_2 + b_2\varphi_2$$

where a_1, a_2, b_1, b_2 are arbitrary real or complex numbers is also a solution pair.

This proposition then asserts the linearity of the system (2.6), as was shown in Section 1.8. Hence, as in the case of a single equation, we may revert to the real solution pairs

$$\theta_{11}(t) = \alpha \cos\sqrt{(2-\sqrt{2})\frac{g}{l}}\,t, \quad \theta_{12}(t) = \alpha\sqrt{2}\cos\sqrt{(2-\sqrt{2})\frac{g}{l}}\,t,$$

and

$$\theta_{21}(t) = \alpha \sin\sqrt{(2-\sqrt{2})\frac{g}{l}}\,t, \quad \theta_{22}(t) = \alpha\sqrt{2}\sin\sqrt{(2-\sqrt{2})\frac{g}{l}}\,t.$$

To see this we note that

$$\theta_{11} = \frac{\theta_1 + \theta_1^*}{2}, \quad \theta_{21} = \frac{\theta_1 - \theta_1^*}{2i},$$
$$\theta_{12} = \frac{\theta_2 + \theta_2^*}{2}, \quad \theta_{22} = \frac{\theta_2 - \theta_2^*}{2i}.$$

Solution pairs corresponding to the values $\lambda = i\sqrt{\mu_2}$ and $\lambda = -i\sqrt{\mu_2}$ can be similarly obtained. One obtains the pairs

$$\theta_{31} = \alpha \cos\sqrt{(2+\sqrt{2})\frac{g}{l}}\,t, \quad \theta_{32}(t) = -\alpha\sqrt{2}\cos\sqrt{(2+\sqrt{2})\frac{g}{l}}\,t,$$

and

$$\theta_{41}(t) = \alpha \sin \sqrt{(2+\sqrt{2})\frac{g}{l}}\, t, \quad \theta_{42}(t) = -\alpha \sqrt{2} \sin \sqrt{(2+\sqrt{2})\frac{g}{l}}\, t.$$

In order to complete the solution of the problem of finding functions Θ_1, Θ_2 for the special case we are considering we shall form linear combinations in the same manner as was done in the case of a single equation. Since we have shown that α and β are related, let us introduce the arbitrary real numbers $\alpha_1, \alpha_2, \alpha_3, \alpha_4$ and form the pair of functions

$$\begin{aligned}\Theta_1(t) &= \alpha_1 \cos \sqrt{\mu_1}\, t + \alpha_2 \sin \sqrt{\mu_1}\, t + \alpha_3 \cos \sqrt{\mu_2}\, t + \alpha_4 \sin \sqrt{\mu_2}\, t, \\ \Theta_2(t) &= \sqrt{2}\,[\alpha_1 \cos \sqrt{\mu_1}\, t + \alpha_2 \sin \sqrt{\mu_1}\, t] \\ &\quad - \sqrt{2}\,[\alpha_3 \cos \sqrt{\mu_2}\, t + \alpha_4 \sin \sqrt{\mu_2}\, t].\end{aligned} \quad (2.15)$$

Using the first two of the initial conditions given in Equations (2.7), we have

$$\begin{aligned}\varphi_1 &= \alpha_1 + \alpha_3, \\ \varphi_2 &= \sqrt{2}\,\alpha_1 - \sqrt{2}\,\alpha_3.\end{aligned}$$

These two linear equations can be solved uniquely for α_1 and α_3 in terms of the initial data φ_1 and φ_2 because the determinant of the system

$$\begin{vmatrix} 1 & 1 \\ \sqrt{2} & -\sqrt{2} \end{vmatrix} = -2\sqrt{2} \neq 0.$$

Similarly, using the last two of Equations (2.7), we have

$$\begin{aligned}\omega_1 &= \sqrt{\mu_1}\,\alpha_2 + \sqrt{\mu_2}\,\alpha_4, \\ \omega_2 &= \sqrt{2\mu_1}\,\alpha_2 - \sqrt{2\mu_2}\,\alpha_4.\end{aligned}$$

Here we have

$$\begin{vmatrix} \sqrt{\mu_1} & \sqrt{\mu_2} \\ \sqrt{2\mu_1} & -\sqrt{2\mu_2} \end{vmatrix} = -2\sqrt{2\mu_1\mu_2} \neq 0,$$

so that we may determine α_2 and α_4 uniquely. Once the values α_1 through α_4 have been determined in this way, we have a solution of the problem of finding functions θ_1, θ_2 satisfying (2.6) and (2.7).

In summary, the most significant observation about the linear problem treated in this section is that its solution proved to be remarkably complicated when compared to the solution of the leakage problem of Section 1.2. Mathematically the latter problem is the same as the linear single pendulum. Yet the formulation of the double-pendulum problem does not seem so much more complicated, for we needed only one more differential equation and only two additional initial conditions.

It is customary to refer to the leakage problem as a one-degree-of-freedom system and the double-pendulum system as a two-degree-of-freedom system. Hence, there are as many degrees of freedom as there are dependent variables.

Let us attempt to identify features of the solution methods which are similar in the two problems. In the first place we were able to obtain solutions of exponential type for both systems of equations. Secondly, by constructing linear combinations of solutions in each case we were able to solve the initial-value problem. Thus the general procedure was the same in each case. The difference in the difficulty of solving a system of two equations as compared with problems involving a single differential equation is completely tied up in the amount of algebraic manipulations required.

We may expect, moreover, that systems involving many degrees of freedom, hence many differential equations, will lead us into even more complicated algebraic problems, but the method of solution remains the same. This furnishes us with a compelling reason for studying some linear algebra before we attempt to construct any general theory of systems of differential equations. It will be seen that linear algebra will aid us in formulating a compact method of notation, and also furnish us with a very strong tool for manipulation purposes.

PROBLEMS

1. Consider the system of second-order differential equations

$$m_{11}\ddot{x} + m_{12}\ddot{y} + s_{11}x - s_{12}y = 0,$$
$$m_{21}\ddot{x} + m_{22}\ddot{y} - s_{21}x + s_{22}y = 0.$$

   Assume all of the coefficients in this system are positive and that

$$m_{11}m_{22} - m_{12}m_{21} > 0, \quad s_{11}s_{22} - s_{12}s_{21} > 0.$$

   Show that if one seeks solutions having the form

$$x = \alpha e^{\lambda t}, \quad y = \beta e^{\lambda t},$$

   all of the eigenvalues λ will have the form

$$\lambda = i\mu$$

   where μ is real.

2. For the system of Problem 1 set $M = m_{11}m_{22} - m_{12}m_{21}$ so that $M > 0$ when the conditions $m_{11}m_{22} - m_{12}m_{21} > 0$ and $s_{11}s_{22} - s_{12}s_{21} > 0$ are valid. Show that the system can be written in the form

$$M\ddot{x} + (m_{22}s_{11} + m_{12}s_{21})x - (s_{12}m_{22} + m_{12}s_{22})y = 0,$$
$$M\ddot{y} - (m_{21}s_{11} + m_{11}s_{21})x + (m_{11}s_{22} + m_{21}s_{12})y = 0,$$

   without making any change of variables.

3. Show that by a proper choice of generalized coordinates for the double pendulum problem, the system (2.6) can be replaced by a system of the form

$$\ddot{\varphi}_1 + [\nu_1 + M(\nu_1 + \nu_2)]\varphi_1 - M\nu_1\varphi_2 = 0$$
$$\ddot{\varphi}_2 - \mu\nu_2\varphi_1 + \nu_2\varphi_2 = 0$$

where $M = m_2/m_1$, $\nu_1 = g/l_1$, $\nu_2 = g/l_2$, and $\mu = l_1/l_2$.

4. Show that the number
$$[\nu_1 + M(\nu_1 + \nu_2)]\nu_2 - M\mu\nu_1\nu_2 > 0$$
where the numbers M, ν_1, ν_2, and μ are those defined in Problem 3.

5. In the system of Problem 3 we wish to make the change of variables
$$\varphi_1 = a_{11}\xi + a_{12}\eta$$
$$\varphi_2 = a_{21}\xi + a_{22}\eta$$
where a_{11}, a_{12}, a_{21}, a_{22} are constants. Show that these constants may be so chosen that the system can be reduced to the form
$$\ddot{\xi} + \omega_1\xi = 0$$
$$\ddot{\eta} + \omega_2\eta = 0.$$

6. The solution of the linear double-pendulum problem can be written in the form
$$\theta_1(t) = \sqrt{\mathcal{E}_1}\sin(\sqrt{\mu_1}\,t + \delta_1) + \sqrt{\mathcal{E}_2}\sin(\sqrt{\mu_2}\,t + \delta_2),$$
$$\theta_2(t) = \sqrt{2\mathcal{E}_1}\sin(\sqrt{\mu_1}\,t + \delta_1) - \sqrt{2\mathcal{E}_2}\sin(\sqrt{\mu_2}\,t + \delta_2).$$

(a) Find the constants $\mathcal{E}_1$, $\mathcal{E}_2$, δ_1, δ_2.
(b) Show that in the amplitude representation
$$\theta_1(t) = A\left\{\frac{\sqrt{\mathcal{E}_1}}{\sqrt{\mathcal{E}_1 + \mathcal{E}_2}}\sin(\sqrt{\mu_1}\,t + \delta) + \frac{\sqrt{\mathcal{E}_2}}{\sqrt{\mathcal{E}_1 + \mathcal{E}_2}}\sin(\sqrt{\mu_2}\,t + \delta_2)\right\},$$
$$\theta_2(t) = \sqrt{2}A\left\{\frac{\sqrt{\mathcal{E}_1}}{\sqrt{\mathcal{E}_1 + \mathcal{E}_2}}\sin(\sqrt{\mu_1}\,t + \delta)\right.$$
$$\left. - \frac{\sqrt{\mathcal{E}_2}}{\sqrt{\mathcal{E}_1 + \mathcal{E}_2}}\sin(\sqrt{\mu_2}\,t + \delta_2)\right\},$$
the amplitude
$$A = \sqrt{\mathcal{E}_1 + \mathcal{E}_2} = \sqrt{\frac{1}{2mgl}\mathcal{E}(0)},$$
where $\mathcal{E}(0)$ is the total energy in the system at time $t = 0$.

7. With the aid of the energy representation of Problem 6(b) prove that the total energy of the linear double pendulum, $\mathcal{E}(t) = T(t) + V(t)$, at any time t equals $\mathcal{E}(0)$. That is, $\mathcal{E}(t) \equiv \mathcal{E}(0)$.

8. For the solution of the linear double pendulum, find functions $g_1(t)$ and $g_2(t)$ such that
$$\theta_1(t) = \omega_1 g_1(t) + \varphi_1\dot{g}_1(t) + \omega_2 g_2(t) + \varphi_2\dot{g}_2(t)$$

and functions $g_3(t)$ and $g_4(t)$ such that
$$\Theta_2(t) = \omega_1 g_3(t) + \varphi_1 \dot{g}_3(t) + \omega_2 g_4(t) + \varphi_2 \dot{g}_4(t).$$

9. The system
$$m_{11}\ddot{x} + m_{12}\ddot{y} + d_{11}\dot{x} + d_{12}\dot{y} + s_{11}x - s_{12}y = 0$$
$$m_{21}\ddot{x} + m_{22}\ddot{y} + d_{21}\dot{x} + d_{22}\dot{y} + s_{21}x + s_{22}y = 0.$$

is called a *damped system*. Prove that this system can be reduced to the form
$$\ddot{x} + \tilde{d}_{11}\dot{x} + \tilde{d}_{12}\dot{y} + \tilde{s}_{11}x - \tilde{s}_{12}y = 0$$
$$\ddot{y} + \tilde{d}_{21}\dot{x} + \tilde{d}_{22}\dot{y} - \tilde{s}_{21}x + \tilde{s}_{22}y = 0$$

when $M = m_{11}m_{22} - m_{12}m_{21} \neq 0$.

10. The energy method can be applied to the damped system of Problem 9 by multiplying the first equation by $\dot{x}$, the second by $\dot{y}$, adding the resulting equations and finally integrating from 0 to t. Assume that $s_{12} = s_{21}$ and $m_{12} = m_{21}$. Identify the kinetic and potential energy terms in the resulting expression.

11. Use your solution to Problem 10 to show that the system of Problem 9 is conservative if and only if
$$\int_0^t [d_{11}\dot{x}^2 + (d_{12} + d_{22})\dot{x}\dot{y} + d_{22}\dot{y}^2]d\tau = 0$$

for every t. What does this mean physically?

12. For the system of Problem 9 if m_{11}, m_{12} and m_{22} are positive and
$$M = m_{11}m_{22} - m_{12}^2 > 0,$$

and if
$$M_2 = \frac{m_{11} + m_{22}}{2} + \frac{1}{2}\sqrt{(m_{11} + m_{22})^2 - 4M},$$
$$M_1 = \frac{m_{11} + m_{22}}{2} - \frac{1}{2}\sqrt{(m_{11} + m_{22})^2 - 4M},$$

show that
$$M_1 > 0, \quad M_2 > M_1,$$

and
$$M_1(\dot{x}^2 + \dot{y}^2) \leq m_{11}\dot{x}^2 + 2m_{12}\dot{x}\dot{y} + m_{22}\dot{y}^2 \leq M_2(\dot{x}^2 + \dot{y}^2).$$

13. Suppose that s_{11}, s_{12}, s_{22} of Problem 9 are arbitrary real numbers. Set $s = s_{11}s_{22} - s_{12}^2$ and show that the numbers
$$s_1 = \frac{s_{11} + s_{22}}{2} + \frac{1}{2}\sqrt{(s_{11} + s_{22})^2 - 4s},$$
$$s_2 = \frac{s_{11} + s_{22}}{2} - \frac{1}{2}\sqrt{(s_{11} + s_{22})^2 - 4s},$$

are real, and

$$s_1(x^2 + y^2) \le s_{11}x^2 - 2s_{12}xy + s_{22}y^2 \le s_2(x^2 + y^2).$$

14. Consider the system of Problem 9 under the conditions $m_{12} = m_{21}$, $d_{12} = d_{21}$, $s_{12} = s_{21}$. Associate with this system the numbers M, S, M_1, M_2, s_1, s_2 as given in Problems 12 and 13 and similarly the numbers

$$D = d_{11}d_{22} - d_{12}^2,$$

D_1 and D_2. Using the energy identity derived in Problem 10 prove that if $M_1 \ge 0$, $D_1 \ge 0$, and $s_1 > 0$, then

$$x^2(t) + y^2(t) \le \frac{1}{s_1}[M_2(\dot{x}_0^2 + \dot{y}_0^2) + s_2(x_0^2 + y_0^2)], \quad t > 0,$$

where $x_0, y_0, \dot{x}_0, \dot{y}_0$ is the initial data for the system. What does this result say about the stability of the system?

15. What are the weakest conditions on the coefficients of the system of Problem 14 which will guarantee that $M_1 \ge 0$, $D_1 \ge 0$ and $s_1 > 0$?

16. The system of springs and masses shown in Figure 2.3 is constrained to move in a horizontal plane by frictionless glides. Derive the equations

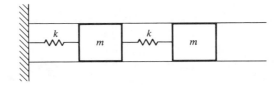

FIGURE 2.3

for the small vibrations of this system about an equilibrium position. Solve the resulting equations for general initial conditions.

17. Derive the equations for the small vibrations of the spring mass system of Figure 2.4 in which the masses are constrained to move horizontally.

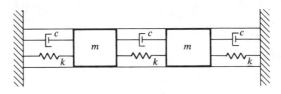

FIGURE 2.4

18. Consider the system of Figure 2.5. The force $F_0 \sin \omega t$ acts upon the mass M. Show that the equations of motion for the system can be put into the form

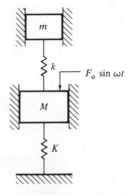

FIGURE 2.5

$$M\ddot{x} + (K+k)x - ky = F_0 \sin \omega t$$
$$m\ddot{x} - kx + ky = 0$$

and solve this system with zero initial conditions.

19. Use the energy method to show that the solution to the nonhomogeneous system

$$m_1\ddot{x} + s_{11}x - s_{12}y = f_1(t)$$
$$m_2\ddot{y} - s_{12}x + s_{22}y = f_2(t)$$

is unique if

$$m_1 \geq 0, \quad m_2 \geq 0, \quad s_{11} > 0, \quad s_{22} > 0,$$

and

$$s_{11}s_{22} - s_{12}^2 > 0.$$

20. Consider the electrical network of Figure 2.6. Obtain the integro-differential equations for the currents i_1 and i_2 flowing in this network if the capacitors are initially uncharged.

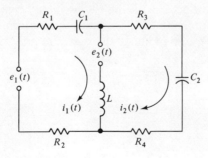

FIGURE 2.6

21. For the system of Problem 20 with e_1 and e_2 removed from the system establish the energy identity

$$\frac{1}{\sqrt{2}}(i_1 - i_2)^2 + \int_0^t [(R_1 + R_2)i_1{}^2 + (R_3 + R_4)i_2{}^2]\, dt$$
$$+ \frac{1}{2c_1}\left\{\left(\int_0^t i_1(\tau)\, d\tau\right)^2 + \left(\int_0^t i_2(\tau)\, d\tau\right)^2\right\} = c,$$

where

$$c = \frac{1}{\sqrt{2}}[i_1(0) - i_2(0)]^2.$$

With the help of this identity show that $c = 0$ implies $i_1 = i_2 = 0$ for all $t \geq 0$.

22. For the circuit of Figure 2.7 write the equations for the charge on the capacitors. Here M is the mutual inductance so that, for example, the

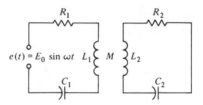

FIGURE 2.7

voltage drop in L has the two components $L_1(di_1/dt)$ and $M(di_2/dt)$; that is, the total voltage drop across L_1 is

$$-L_1 \frac{di_1}{dt} - M \frac{di_2}{dt}.$$

23. The model in Figure 2.8 is a very simplified representation of the motion of an automobile chassis with the engine idling. Here m is the mass of the chassis, G represents the position of the center of gravity and let r be the radius of gyration about the center of gravity. Show that with

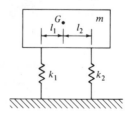

FIGURE 2.8

suitably chosen generalized coordinates the equations of small vibration have the form

$$m\ddot{x} + (k_1 + k_2)x + (k_1 l_1 - k_2 l_2)\omega = 0,$$
$$mr\ddot{\omega} + (k_1 l_1 - k_2 l_2)x + (k_1 l_1^2 + k_2 l_2^2)\omega = 0.$$

24. Derive the equation of motion for small vibrations for the system of Figure 2.9. Assume nonzero initial conditions and obtain a general solution for the system.

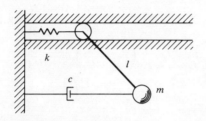

FIGURE 2.9

2.3 A SYSTEM OF FIRST-ORDER DIFFERENTIAL EQUATIONS

In Section 2.2 we considered a physical problem whose mathematical model was represented by a system of second-order differential equations. In this section we return to the analysis of electrical networks in order to motivate a problem leading to a system of first-order differential equations. Such systems arise in a wide variety of problems.

Consider the network of Figure 2.10, consisting of three loops in sequence (referred to as *cascaded loops*). The switch K is closed at $t = t_0 = 0$. Assume

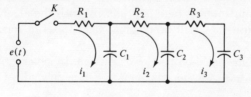

FIGURE 2.10

the loop currents to be i_1, i_2, and i_3 flowing as indicated. Applying Kirchhoff's voltage law to the loops, as was done in Section 1.1, yields the system of equations

$$i_1 R_1 + \frac{1}{C_1} \int (i_1 - i_2)\, dt = e(t)$$

$$i_2 R_2 + \frac{1}{C_1} \int (i_2 - i_1)\, dt + \frac{1}{C_2} \int (i_2 - i_3)\, dt = 0 \qquad (2.16)$$

and

$$i_3 R_3 + \frac{1}{C_2} \int (i_3 - i_2)\, dt + \frac{1}{C_3} \int i_3\, dt = 0.$$

In preference to these integral equations we may develop analogous differential equations by either differentiating with respect to time or by replacing i_k by dq_k/dt, $k = 1, 2$, or 3, where the q_k are the charges on the C_k. For our purposes we choose to differentiate each equation with respect to time. After some manipulations this leads to the system of equations

$$\frac{di_1}{dt} + \frac{1}{R_1 C_1} i_1 - \frac{1}{R_1 C_1} i_2 = \dot{e}(t),$$
$$\frac{di_2}{dt} - \frac{1}{R_2 C_1} i_1 + \frac{1}{R_2}\left(\frac{1}{C_1} + \frac{1}{C_2}\right) i_2 - \frac{1}{R_2 C_2} i_3 = 0, \quad (2.17)$$
$$\frac{di_3}{dt} - \frac{1}{R_3 C_3} i_2 + \frac{1}{R_3}\left(\frac{1}{C_2} + \frac{1}{C_3}\right) i_3 = 0.$$

Here, of course, the voltage $e(t)$ and hence its derivative $\dot{e}(t)$ is assumed known as a function of time for all $t \geq 0$. In addition to the system of differential equations (2.17) we prescribe the initial conditions

$$i_1(0) = a_1, \quad i_2(0) = a_2, \quad i_3(0) = a_3. \quad (2.18)$$

Before calling upon our previous experience to attempt to find a solution for this system we wish to write the problem in a more convenient form. For this purpose we introduce the concept of a vector as an ordered triple of real numbers written as

$$\mathbf{a} = (a_1, a_2, a_3),$$

where the numbers a_1, a_2, a_3 are called the *components* of $\mathbf{a}$. Geometrically a vector may be identified with a point in R^3, the space of ordinary analytic geometry. But our interest here is primarily with the algebraic properties of vectors because they will be of great help to us in formulating many problems.

Let us note some basic properties of vectors. In general, if $\mathbf{a} = (a_1, a_2, a_3)$ and $\mathbf{b} = (b_1, b_2, b_3)$ are vectors, the sum $\mathbf{a} + \mathbf{b}$ is defined by

$$\mathbf{a} + \mathbf{b} = (a_1 + b_1, a_2 + b_2, a_3 + b_3). \quad (2.19)$$

The product of a vector by a scalar, that is, by a real number, is defined by

$$\alpha \mathbf{a} = (\alpha a_1, \alpha a_2, \alpha a_3). \quad (2.20)$$

The vector

$$\mathbf{0} = (0, 0, 0)$$

has the important property that if $\mathbf{a}$ is any vector,

$$\mathbf{a} + \mathbf{0} = \mathbf{a}. \quad (2.21)$$

Moreover, this vector is unique. That is, the only solution of the equation

$$\mathbf{a} = \mathbf{0}$$

is $a_1 = a_2 = a_3 = 0$.

We define the scalar product of the vectors $\mathbf{a}$ and $\mathbf{b}$, written $\mathbf{a} \cdot \mathbf{b}$, by

$$\mathbf{a} \cdot \mathbf{b} = a_1 b_1 + a_2 b_2 + a_3 b_3. \tag{2.22}$$

It is easy to see that if $\mathbf{a}$, $\mathbf{b}$, and $\mathbf{c}$ are vectors

$$\mathbf{a} \cdot \mathbf{b} = \mathbf{b} \cdot \mathbf{a}$$
$$(\mathbf{a} + \mathbf{c}) \cdot \mathbf{b} = \mathbf{a} \cdot \mathbf{b} + \mathbf{c} \cdot \mathbf{b},$$
$$a(\mathbf{a} \cdot \mathbf{b}) = (a\mathbf{a}) \cdot \mathbf{b} = \mathbf{a} \cdot (a\mathbf{b}),$$

and

$$\mathbf{a} \cdot (\mathbf{b} + \mathbf{c}) = \mathbf{a} \cdot \mathbf{b} + \mathbf{a} \cdot \mathbf{c}.$$

With the aid of the scalar product we can define the length of the vector $\mathbf{a}$ by the formula

$$|\mathbf{a}| = (\mathbf{a} \cdot \mathbf{a})^{1/2}.$$

The array

$$\mathbf{A} = \begin{pmatrix} a_{11} & a_{12} & a_{13} \\ a_{21} & a_{22} & a_{23} \\ a_{31} & a_{32} & a_{33} \end{pmatrix}, \tag{2.23}$$

where a_{ij}, $i,j = 1,2,3$, are real numbers is called a (real 3×3) *matrix*. The element a_{ij} is said to be in row i and column j. The vectors $\mathbf{a}_1 = (a_{11}, a_{12}, a_{13})$, $\mathbf{a}_2 = (a_{21}, a_{22}, a_{23})$, and $\mathbf{a}_3 = (a_{31}, a_{32}, a_{33})$ are called the *row vectors* of $\mathbf{A}$. The vectors

$$\hat{\mathbf{a}}_1 = (a_{11}, a_{21}, a_{31}), \quad \hat{\mathbf{a}}_2 = (a_{12}, a_{22}, a_{32}),$$
$$\hat{\mathbf{a}}_3 = (a_{13}, a_{23}, a_{33})$$

are called the *column vectors* of $\mathbf{A}$.

If $\mathbf{A}$ and $\mathbf{B}$ are matrices we define their sum by

$$\mathbf{A} + \mathbf{B} = \begin{pmatrix} a_{11} + b_{11} & a_{12} + b_{12} & a_{13} + b_{13} \\ a_{21} + b_{21} & a_{22} + b_{22} & a_{23} + b_{23} \\ a_{31} + b_{31} & a_{32} + b_{32} & a_{33} + b_{33} \end{pmatrix}. \tag{2.24}$$

Note that in order to perform the foregoing addition it was necessary for $\mathbf{A}$ and $\mathbf{B}$ to have the identical number of rows and columns. If a is a scalar, we define

$$a\mathbf{A} = \begin{pmatrix} aa_{11} & aa_{12} & aa_{13} \\ aa_{21} & aa_{22} & aa_{23} \\ aa_{31} & aa_{32} & aa_{33} \end{pmatrix}. \tag{2.25}$$

The zero matrix is defined to be the matrix all of whose elements are zero.

That is, the only solution of the equation $\mathbf{A} = \mathbf{0}$ is $a_{ij} = 0$, $i,j = 1, 2, 3$. It therefore follows, as with vectors, that any matrix $\mathbf{A}$ has the property $\mathbf{A} + \mathbf{0} = \mathbf{A}$.

We wish next to define certain products of fundamental importance. First suppose $\mathbf{A}$ is a matrix and $\mathbf{a}$ is a vector. Then we define

$$\mathbf{A}\mathbf{a} = (\mathbf{a}_1 \cdot \mathbf{a}, \mathbf{a}_2 \cdot \mathbf{a}, \mathbf{a}_3 \cdot \mathbf{a}) \tag{2.26}$$

so that $\mathbf{A}\mathbf{a}$ is again a vector. The components of this vector are just the scalar products of the row vectors of $\mathbf{A}$ with the vector $\mathbf{a}$.

We may extend this concept to define the product of two matrices $\mathbf{A}$ and $\mathbf{B}$. The product $\mathbf{A}\mathbf{B}$ will be a new 3×3 matrix $\mathbf{C}$,

$$\mathbf{C} = \begin{pmatrix} c_{11} & c_{12} & c_{13} \\ c_{21} & c_{22} & c_{23} \\ c_{31} & c_{32} & c_{33} \end{pmatrix} \tag{2.27}$$

in which the element c_{ij} in row i and column j of $\mathbf{C}$ is simply the scalar product

$$c_{ij} = \mathbf{a}_i \cdot \hat{\mathbf{b}}_j. \tag{2.28}$$

In words, c_{ij} is the scalar product of the ith row vector of $\mathbf{A}$ with the jth column vector of $\mathbf{B}$.

At this point let us call attention to the fact that the product $\mathbf{B}\mathbf{A}$ is not necessarily the same as $\mathbf{A}\mathbf{B}$. For we have

$$\mathbf{B}\mathbf{A} = \begin{pmatrix} d_{11} & d_{12} & d_{13} \\ d_{21} & d_{22} & d_{23} \\ d_{31} & d_{32} & d_{33} \end{pmatrix}$$

where d_{ij} is the scalar product of the ith row vector of $\mathbf{B}$ with the jth column vector of $\mathbf{A}$,

$$d_{ij} = \mathbf{b}_i \cdot \hat{\mathbf{a}}_j.$$

This multiplication of matrices does not enjoy one of the basic properties of multiplication of real numbers. We express this fact by saying that matrix multiplication is not commutative. In the special case when

$$\mathbf{A}\mathbf{B} = \mathbf{B}\mathbf{A}$$

we say that $\mathbf{A}$ and $\mathbf{B}$ commute (see Problem 10 at the end of Section 2.4).

Before returning to the formulation of the problem of cascaded loops let us focus on some further facts. From the algebraic point of view the fact that a vector has three components is not of basic importance. Clearly we may define an ordered n-tuple of real numbers by writing

$$\mathbf{a} = (a_1, a_2, \ldots, a_n), \quad n \geq 1.$$

Then the manipulations we have defined above for 3-tuples can just as well be defined for n-tuples. Thus if $\mathbf{a}$ and $\mathbf{b}$ are ordered n-tuples, we define

$$aa = (aa_1, aa_2, \ldots, aa_n).$$

Similarly the zero n-tuple is defined to be $\mathbf{0} = (0,0, \ldots, 0)$, that is, the ordered n-tuple with n zeros. In the case where $n = 1$ these operations are the familiar operations with real numbers.

For n-tuples $\mathbf{a}$ and $\mathbf{b}$ the scalar product is simply

$$\mathbf{a}\cdot\mathbf{b} = a_1 b_1 + a_2 b_2 + \cdots + a_n b_n.$$

With these concepts at hand we may define the $(n \times n)$ matrix

$$\mathbf{A} = \begin{pmatrix} a_{11} & a_{12} & \cdots & a_{1n} \\ a_{21} & a_{22} & \cdots & a_{2n} \\ \cdots & \cdots & \cdots & \cdots \\ a_{n1} & a_{n2} & \cdots & a_{nn} \end{pmatrix}.$$

Obviously the matrix already defined in the 3×3 case can be equally well defined in the $n \times n$ case. For example, if $\mathbf{a}_1, \ldots, \mathbf{a}_n$ are the row vectors of $\mathbf{A}$ and if $\mathbf{a}$ is a vector, then

$$\mathbf{A}\mathbf{a} = (\mathbf{a}_1\cdot\mathbf{a}, \mathbf{a}_2\cdot\mathbf{a}, \ldots, \mathbf{a}_n\cdot\mathbf{a}).$$

Similarly, if both $\mathbf{A}$ and $\mathbf{B}$ are $n \times n$ matrices, the product $\mathbf{AB}$ is the $n \times n$ matrix whose element in row i and column j is the scalar product $\mathbf{a}_i \cdot \hat{\mathbf{b}}_j$ of the ith row vector of $\mathbf{A}$ and the jth column vector of $\mathbf{B}$.

Now let us return to the problem of cascaded loops and define the ordered triple of functions

$$\mathbf{i}(t) = (i_1(t), i_2(t), i_3(t)) \tag{2.29}$$

consisting of the three currents introduced in Figure 2.10. For a fixed value of t, $\mathbf{i}(t)$ is an ordered triple of real numbers, that is, a vector. Thus we may refer to $\mathbf{i}$ as a *vector-valued function* of t. More simply we shall call it a *vector*. We define the derivative of the vector $\mathbf{i}$ to be the vector

$$\frac{d\mathbf{i}}{dt} = \left(\frac{di_1}{dt}, \frac{di_2}{dt}, \frac{di_3}{dt}\right). \tag{2.30}$$

Now by introducing the vector

$$\mathbf{h}(t) = (\dot{e}(t), 0, 0)$$

and the matrix

$$\mathbf{S} = \begin{pmatrix} \dfrac{1}{R_1 C_1} & -\dfrac{1}{R_1 C_1} & 0 \\ -\dfrac{1}{R_2 C_2} & \dfrac{1}{R_2}\left(\dfrac{1}{C_1} + \dfrac{1}{C_2}\right) & -\dfrac{1}{R_2 C_2} \\ 0 & -\dfrac{1}{R_3 C_2} & \dfrac{1}{R_3}\left(\dfrac{1}{C_2} + \dfrac{1}{C_3}\right) \end{pmatrix},$$

70 ORDINARY DIFFERENTIAL EQUATIONS—SIMPLE SYSTEMS

the system (2.17) of three equations can then be written in the form

$$\frac{d\mathbf{i}}{dt} + \mathbf{S}\mathbf{i} = \mathbf{h}, \tag{2.31}$$

and the initial conditions (2.18) become

$$\mathbf{i}(0) = \mathbf{a}, \tag{2.32}$$

where $\mathbf{a}$ is a given vector. Once we become accustomed to operating with vectors and matrices according to the above rules we shall find (2.31) and (2.32) a far more expedient form with which to handle the problem.

2.4 SOLUTION OF THE HOMOGENEOUS PROBLEMS OF CASCADED LOOPS

Our experience in Chapter 1 suggests that we should begin by seeking solutions of the homogeneous problem

$$\frac{d\mathbf{i}}{dt} + \mathbf{S}\mathbf{i} = \mathbf{0}, \quad \mathbf{i}(0) = \mathbf{a} \tag{2.33}$$

as a first step toward solving the nonhomogeneous problem represented by (2.31) and (2.32). Moreover, by analogy with our treatment of the linear double-pendulum problem, let us try to find solutions of (2.33) having the form

$$\mathbf{i}(t) = e^{\lambda t}\mathbf{v} \tag{2.34}$$

in which λ is a constant to be determined and $\mathbf{v} = (v_1, v_2, v_3)$ is a constant vector to be determined.

From (2.34) we see at once that

$$\frac{d\mathbf{i}}{dt} = \lambda e^{\lambda t}\mathbf{v}.$$

Accordingly, substitution of (2.34) into (2.33) yields the relation

$$e^{\lambda t}(\lambda \mathbf{v} + \mathbf{S}\mathbf{v}) = \mathbf{0}.$$

Since the scalar $e^{\lambda t} \neq 0$, this equation reduces to

$$\mathbf{S}\mathbf{v} + \lambda \mathbf{v} = \mathbf{0}. \tag{2.35}$$

The values of λ, if there are any, for which the vector equation (2.35) has a solution $\mathbf{v} \neq \mathbf{0}$, are called *eigenvalues of Problem* (2.33) and the numbers $-\lambda$ are called *eigenvalues of the matrix* $\mathbf{S}$. The corresponding nonzero vectors $\mathbf{v}$ are called *eigenvectors*.

Using the definitions of the last section we may write the vector equation (2.35) as three scalar equations, namely,

SOLUTION OF THE HOMOGENEOUS PROBLEMS

$$\left(\frac{1}{R_1C_1} + \lambda\right)v_1 - \frac{1}{R_1C_1}v_2 = 0,$$

$$-\frac{1}{R_2C_2}v_1 + \left[\frac{1}{R_2}\left(\frac{1}{C_1} + \frac{1}{C_2}\right) + \lambda\right]v_2 - \frac{1}{R_2C_2}v_3 = 0, \quad (2.36)$$

$$-\frac{1}{R_3C_2}v_2 + \left[\frac{1}{R_3}\left(\frac{1}{C_2} + \frac{1}{C_3}\right) + \lambda\right]v_3 = 0.$$

Now observe that this is a homogeneous system of equations for the determination of v_1, v_2, and v_3, that is, for the determination of the components of the vector $\mathbf{v}$. Consequently, by the result of Problem 16, Section 1.4, there exists a solution vector $\mathbf{v} \neq \mathbf{0}$ for this system if and only if

$$\begin{vmatrix} \frac{1}{R_1C_1} + \lambda & -\frac{1}{R_1C_1} & 0 \\ -\frac{1}{R_2C_1} & \frac{1}{R_2}\left(\frac{1}{C_1} + \frac{1}{C_2}\right) + \lambda & -\frac{1}{R_2C_2} \\ 0 & -\frac{1}{R_3C_2} & \frac{1}{R_3}\left(\frac{1}{C_2} + \frac{1}{C_3}\right) + \lambda \end{vmatrix} = 0. \quad (2.37)$$

When the determinant (2.37) is expanded the resulting equation is a polynomial of degree three in λ. For the moment let us simply denote the three roots (*zeros*) obtained from equating this polynomial to zero by λ_1, λ_2, and λ_3, and let us also assume that λ_1, λ_2, and λ_3 are real, distinct numbers.

Returning to Equation (2.35) and setting $\lambda = \lambda_j$, $j = 1, 2, 3$, gives us the vector equations

$$\mathbf{Sv} + \lambda_j \mathbf{v} = \mathbf{0}, \quad j = 1, 2, 3. \quad (2.38)$$

As we have just remarked, these equations have nonzero solution vectors because each λ_j is a zero of Equation (2.37). Let us denote by $\mathbf{v}_j$, $j = 1, 2, 3$, solutions of these equations. These vectors are not uniquely determined, for we observe that if $\mathbf{v} \neq \mathbf{0}$ is a solution of (2.35) for a fixed value of λ and if α is any scalar, then

$$\mathbf{S}\alpha\mathbf{v} + \lambda\alpha\mathbf{v} = \alpha(\mathbf{Sv} + \lambda\mathbf{v}) = \alpha\mathbf{0} = \mathbf{0}.$$

Thus, if $\mathbf{v}_i \neq \mathbf{0}$ satisfies

$$\mathbf{Sv}_j + \lambda_j \mathbf{v}_j = \mathbf{0}.$$

so does $\alpha \mathbf{v}_j$ for any α.

With vectors $\mathbf{v}_1$, $\mathbf{v}_2$, $\mathbf{v}_3$ and corresponding λ_1, λ_2, λ_3 determined we now have three solutions for the system (2.33) as given by (2.34). They are

$$\begin{aligned} \mathbf{i}_1(t) &= e^{\lambda_1 t}\mathbf{v}_1, \\ \mathbf{i}_2(t) &= e^{\lambda_2 t}\mathbf{v}_2, \\ \mathbf{i}_3(t) &= e^{\lambda_3 t}\mathbf{v}_3. \end{aligned} \quad (2.39)$$

It follows from what we have presented above that $\alpha_1\mathbf{i}_1(t)$, $\alpha_2\mathbf{i}_2(t)$, and $\alpha_3\mathbf{i}_3(t)$ are also solutions of (2.33). At this point we note that the principle of superposition holds for the equation

$$\frac{d\mathbf{i}}{dt} + \mathbf{S}\mathbf{i} = \mathbf{0}.$$

We formalize this in:

Proposition 2.2

Suppose that $\mathbf{i}_1$, $\mathbf{i}_2$, and $\mathbf{i}_3$ are solution vectors for the system (2.33) and α_1, α_2, α_3 are real or complex numbers. Then

$$\alpha_1\mathbf{i}_1 + \alpha_2\mathbf{i}_2 + \alpha_3\mathbf{i}_3$$

is also a solution of (2.33).

It should be noted that we have not previously defined the product $\gamma\mathbf{a}$ when γ is a complex number and $\mathbf{a}$ is a vector. This product may be defined in a straightforward way as follows. Write $\gamma = \alpha + i\beta$ and set

$$(\alpha + i\beta)\mathbf{a} = \alpha\mathbf{a} + i\beta\mathbf{a}.$$

Now we can simply treat α and β as before. We shall say more about this situation at the end of this section. For now, let us prove Proposition (2.2) when α_1, α_2, and α_3 are real. By the remarks preceding the proposition we have

$$\mathbf{0} = \frac{d}{dt}\alpha_j\mathbf{i}_j + \mathbf{S}\alpha_j\mathbf{i}_j, \quad j = 1, 2, 3.$$

Hence, by addition

$$\mathbf{0} = \sum_{j=1}^{3} \frac{d}{dt}\alpha_j\mathbf{i}_j + \sum_{j=1}^{3} \mathbf{S}\alpha_j\mathbf{i}_j.$$

By (2.19), the properties of the scalar product, and (2.26), it follows that

$$\sum_{j=1}^{3} \mathbf{S}\alpha_j\mathbf{i}_j = \mathbf{S}\sum_{j=1}^{3} \alpha_j\mathbf{i}_j,$$

and by (2.30) we have

$$\sum_{j=1}^{3} \frac{d}{dt}\alpha_j\mathbf{i}_j = \frac{d}{dt}\sum_{j=1}^{3} \alpha_j\mathbf{i}_j.$$

In other words,

SOLUTION OF THE HOMOGENEOUS PROBLEMS

$$0 = \frac{d}{dt}\sum_{j=1}^{3} \alpha_j \mathbf{i}_j + \mathbf{S}\sum_{j=1}^{3}\alpha_j \mathbf{i}_j,$$

which is what we set out to prove.

We are now in a position to assert that the function

$$\mathbf{i}(t) = \alpha_1 \mathbf{i}_1(t) + \alpha_2 \mathbf{i}_2(t) + \alpha_3 \mathbf{i}_3(t), \tag{2.40}$$

where $\mathbf{i}_1$, $\mathbf{i}_2$, and $\mathbf{i}_3$ are given in (2.39) and α_1, α_2, and α_3 are arbitrary scalars, is a solution of our homogeneous differential equation. The expression on the right-hand side of (2.40) is called a *linear combination* of the vectors $\mathbf{i}_1$, $\mathbf{i}_2$, and $\mathbf{i}_3$. Thus the assertion amounts to saying that an arbitrary linear combination of solution vectors is again a solution vector.

We next choose a solution which satisfies the initial condition $\mathbf{i}(0) = \mathbf{a}$. From (2.40) we see that we must select α_1, α_2, and α_3 so as to satisfy

$$\mathbf{a} = \alpha_1 \mathbf{i}_1(0) + \alpha_2 \mathbf{i}_2(0) + \alpha_3 \mathbf{i}_3(0),$$

or

$$\mathbf{a} = \alpha_1 \mathbf{v}_1 + \alpha_2 \mathbf{v}_2 + \alpha_3 \mathbf{v}_3. \tag{2.41}$$

If we write

$$\mathbf{v}_1 = (v_{11}, v_{21}, v_{31}), \quad \mathbf{v}_2 = (v_{12}, v_{22}, v_{32}), \quad \mathbf{v}_3 = (v_{13}, v_{23}, v_{33}),$$

then the vector equation (2.41) is equivalent to the three scalar equations

$$\begin{aligned} a_1 &= v_{11}\alpha_1 + v_{12}\alpha_2 + v_{13}\alpha_3, \\ a_2 &= v_{21}\alpha_1 + v_{22}\alpha_2 + v_{23}\alpha_3, \\ a_3 &= v_{31}\alpha_1 + v_{32}\alpha_2 + v_{33}\alpha_3. \end{aligned} \tag{2.41a}$$

This system can be solved (say by Cramer's rule) to determine a unique triple α_1, α_2, α_3 corresponding to the given initial vector $\mathbf{a} = (a_1, a_2, a_3)$ if and only if the determinant

$$D = \begin{vmatrix} v_{11} & v_{12} & v_{13} \\ v_{21} & v_{22} & v_{23} \\ v_{31} & v_{32} & v_{33} \end{vmatrix} \neq 0. \tag{2.42}$$

The vectors $\mathbf{v}_1$, $\mathbf{v}_2$, and $\mathbf{v}_3$ are the column vectors of the matrix

$$\mathbf{V} = \begin{pmatrix} v_{11} & v_{12} & v_{13} \\ v_{21} & v_{22} & v_{23} \\ v_{31} & v_{32} & v_{33} \end{pmatrix}.$$

If we define the vector $\boldsymbol{\alpha} = (\alpha_1, \alpha_2, \alpha_3)$, Equation (2.41) can also be written in the form

$$\mathbf{a} = \mathbf{V}\boldsymbol{\alpha}.$$

Note that in order to perform the foregoing multiplication the number of columns in **V** must be equal to the number of rows in **α**, as can be seen in (2.41a).

If Equation (2.41) is to have a unique solution, that is, if for a given **a** there is only one vector **α** which will satisfy (2.41), it must be true that the only solution of the problem

$$0 = \alpha_1 \mathbf{v}_1 + \alpha_2 \mathbf{v}_2 + \alpha_3 \mathbf{v}_3$$

is $\alpha_1 = \alpha_2 = \alpha_3 = 0$. For suppose that **α** and **β** $= (\beta_1, \beta_2, \beta_3)$ are both solutions of (2.41) corresponding to the same vector **a**. Then we have

$$\mathbf{a} = \beta_1 \mathbf{v}_1 + \beta_2 \mathbf{v}_2 + \beta_3 \mathbf{v}_3$$

and subtracting this from (2.41) yields

$$0 = (\alpha_1 - \beta_1)\mathbf{v}_1 + (\alpha_2 - \beta_2)\mathbf{v}_2 + (\alpha_3 - \beta_3)\mathbf{v}_3.$$

If we set $\gamma_i = \alpha_i - \beta_i$, $i = 1,2,3$, then

$$0 = \gamma_1 \mathbf{v}_1 + \gamma_2 \mathbf{v}_2 + \gamma_3 \mathbf{v}_3. \tag{2.43}$$

But uniqueness is equivalent to the assertion that **α** ≡ **β**, or that **γ** is the zero vector. We see at once that (2.43) is equivalent to a homogeneous system of equations for $\gamma_1, \gamma_2, \gamma_3$. Thus $\gamma_1 = \gamma_2 = \gamma_3 = 0$ is the only solution of (2.43) if and only if the determinant D satisfies the condition (2.42). This cycle of ideas shows us that we can solve the initial-value problem (2.33) if and only if the vectors $\mathbf{v}_1, \mathbf{v}_2, \mathbf{v}_3$ satisfy either the condition (2.42) or the condition that the only solution of the vector equation

$$0 = \alpha_1 \mathbf{v}_1 + \alpha_2 \mathbf{v}_2 + \alpha_3 \mathbf{v}_3 \tag{2.44}$$

is $\alpha_1 = \alpha_2 = \alpha_3 = 0$. Thus (2.42) and (2.44) are equivalent.

This condition is so important that we formalize it in

Definition 2.1

The vectors $\mathbf{v}_1, \mathbf{v}_2$, and $\mathbf{v}_3$ are said to be linearly independent if the only solution of (2.44) is $\alpha_1 = \alpha_2 = \alpha_3 = 0$.

Thus the notion that the vectors $\mathbf{v}_i$, $i = 1,2,3$, must be linearly independent in order to obtain a solution of the initial-value problem (2.33) is the analogue of the notion of linear independence of functions which we used in Chapter 1. The determinant condition embodied in the inequality (2.42) furnishes a practical test for the linear independence of $\mathbf{v}_1, \mathbf{v}_2$, and $\mathbf{v}_3$.

Recall that we assumed above that λ_1, λ_2, and λ_3 are real numbers. But the zeros of a polynomial need not be real. When a complex value of λ occurs as a zero of a polynomial with real coefficients, the complex conjugate of λ also occurs as a zero. Thus if $\lambda = \mu + i\nu$ occurs, so does $\bar{\lambda} = \mu - i\nu$ occur. In this event two of the equations

SOLUTION OF THE HOMOGENEOUS PROBLEMS

$$\mathbf{S}\mathbf{v}_k + \lambda_k \mathbf{v}_k = \mathbf{0}$$

become equations with complex coefficients, namely the equations corresponding to the complex eigenvalues λ and $\bar{\lambda}$. Now we are confronted with the possibility that the corresponding solution vector $\mathbf{v}_k$ will have complex components rather than real components. Thus we are led to the study of complex vectors, that is, to vectors having the form

$$\mathbf{v} = (v_1' + iv_1'', v_2' + iv_2'', v_3' + iv_3'')$$

where v_k', v_k'', $k = 1,2,3$, are all real numbers. In view of the definitions of Section 2.3 it makes good sense to write $\mathbf{v}$ as

$$\mathbf{v} = \mathbf{v}' + i\mathbf{v}''.$$

This merely requires us to agree to treat i in the same way as we treat a real scalar when we multiply a vector by a scalar.

Now suppose that $\lambda = \mu + i\nu$ is an eigenvalue for a problem such as we have been considering and $\mathbf{v} = \mathbf{v}' + i\mathbf{v}''$ a corresponding eigenvector so that we have

$$\mathbf{S}(\mathbf{v}' + i\mathbf{v}'') + (\mu + i\nu)(\mathbf{v}' + i\mathbf{v}'') = \mathbf{0},$$

where $\mathbf{S}$ is a real 3×3 matrix. Treating i as a scalar this equation becomes

$$\mathbf{S}\mathbf{v}' + \mu\mathbf{v}' - \nu\mathbf{v}'' + i\{\mathbf{S}\mathbf{v}'' + \nu\mathbf{v}' + \mu\mathbf{v}''\} = \mathbf{0}.$$

Equating real and imaginary parts of this equation to zero yields the equations

$$\begin{aligned} \mathbf{S}\mathbf{v}' + \mu\mathbf{v}' - \nu\mathbf{v}'' &= \mathbf{0}, \\ \mathbf{S}\mathbf{v}'' + \nu\mathbf{v}' + \mu\mathbf{v}'' &= \mathbf{0}. \end{aligned} \qquad (2.45)$$

Next consider the equation

$$\mathbf{S}(\mathbf{v}' - i\mathbf{v}'') + (\mu - i\nu)(\mathbf{v}' - i\mathbf{v}'') = \mathbf{0}.$$

This becomes

$$\mathbf{S}\mathbf{v}' + \mu\mathbf{v}' - \nu\mathbf{v}'' - i\{\mathbf{S}\mathbf{v}'' + \nu\mathbf{v}' + \mu\mathbf{v}''\} = \mathbf{0},$$

and equating the real and imaginary parts to zero we once again obtain Equations (2.45). In other words, if $\mathbf{v} = \mathbf{v}' + i\mathbf{v}''$ is a complex eigenvector belonging to the complex eigenvalue $\lambda = \mu + i\nu$, then $\bar{\mathbf{v}} = \mathbf{v}' - i\mathbf{v}''$ is a complex eigenvector belonging to $\bar{\lambda} = \mu - i\nu$.

Some assumptions about the algebra of complex vectors are buried in the argument we have just used. Hence, let us pause a moment to formalize the situation. If $a_1, \ldots, a_n$ are complex numbers the ordered n-tuple

$$\mathbf{a} = (a_1, a_2, \ldots, a_n), \quad n \geq 1,$$

is called a *complex vector*. Complex vectors are added in the same way as are real vectors, namely,

$$\mathbf{a} + \mathbf{b} = (a_1 + b_1, a_2 + b_2, \ldots, a_n + b_n).$$

Thus $\mathbf{a} + \mathbf{b}$ is a new complex vector. Similarly if a is a complex number and $\mathbf{a}$ a complex vector, we define

$$a\mathbf{a} = (aa_1, aa_2, \ldots, aa_n)$$

so that $a\mathbf{a}$ is a new complex vector. The zero complex vector is

$$\mathbf{0} = (0, 0, \ldots, 0)$$

and we have $\mathbf{a} + \mathbf{0} = \mathbf{a}$ for any complex vector $\mathbf{a}$. The vector $\mathbf{a} = \mathbf{0}$ implies $a_1 = a_2 = \cdots = a_n = 0$, since $\mathbf{a} = \mathbf{0}$ is equivalent to the n equations $a_j = 0, j = 1, 2, \ldots, n$.

Now every complex vector $\mathbf{a}$ can obviously be written in the form

$$\mathbf{a} = \boldsymbol{\alpha} + i\boldsymbol{\beta}$$

where $\boldsymbol{\alpha} = (\alpha_1, \alpha_2, \ldots, \alpha_n)$ and $\boldsymbol{\beta} = (\beta_1, \beta_2, \ldots, \beta_n)$. Further, we may write $a_j = \alpha_j + i\beta_j$, where a_j is the jth component of $\mathbf{a}$. Then the operations defined above can be expressed in complex notation as follows:

$$\mathbf{a} + \mathbf{b} = \boldsymbol{\alpha}_1 + i\boldsymbol{\beta}_1 + \boldsymbol{\alpha}_2 + i\boldsymbol{\beta}_2 = \boldsymbol{\alpha}_1 + \boldsymbol{\alpha}_2 + i(\boldsymbol{\beta}_1 + \boldsymbol{\beta}_2)$$

and

$$\alpha \mathbf{a} = (\alpha + i\beta)(\boldsymbol{\alpha} + i\boldsymbol{\beta}) = \alpha\boldsymbol{\alpha} - \beta\boldsymbol{\beta} + i(\alpha\boldsymbol{\beta} + \beta\boldsymbol{\alpha}),$$

where α is a complex scalar.

When we come to complex matrices, that is, matrices whose elements are complex numbers, the extension is again straightforward and we may safely leave it to the student to formulate the proper definitions for himself, with one noted exception. This exception occurs when we compute products, for then complications begin to arise because the scalar product of two complex vectors is not given by the formula (2.22). In order to avoid difficulties which at present are of no concern to us we define the product of the $n \times n$ complex matrix $\mathbf{A}$ and the complex vector $\mathbf{a}$ in the following way. Let the row vectors of $\mathbf{A}$ be $\mathbf{a}_j = \boldsymbol{\alpha}_j + i\boldsymbol{\beta}_j, j = 1, \ldots, n$, and let $\mathbf{a} = \boldsymbol{\alpha} + i\boldsymbol{\beta}$. Then $\mathbf{A}\mathbf{a}$ is the vector whose jth component is simply

$$(\boldsymbol{\alpha}_j + i\boldsymbol{\beta}_j) \cdot (\boldsymbol{\alpha} + i\boldsymbol{\beta}) = \boldsymbol{\alpha}_j \cdot \boldsymbol{\alpha} - \boldsymbol{\beta}_j \cdot \boldsymbol{\beta} + i(\boldsymbol{\alpha}_j \cdot \boldsymbol{\beta} + \boldsymbol{\beta}_j \cdot \boldsymbol{\alpha}).$$

In the calculations we made where the matrix $\mathbf{A}$ was real, the foregoing simplifies to $\boldsymbol{\alpha}_j \cdot \boldsymbol{\alpha} + i\boldsymbol{\alpha}_j \cdot \boldsymbol{\beta}$.

Clearly we can define the product $\mathbf{AB}$ in a similar manner in terms of the row vectors of $\mathbf{A}$ and the column vectors of $\mathbf{B}$.

Returning now to the solution of the basic equation (2.33) we find that

SOLUTION OF THE HOMOGENEOUS PROBLEMS

$$\mathbf{i}_1(t) = e^{(\mu+i\nu)t}(\mathbf{v}' + i\mathbf{v}''),$$

and

$$\mathbf{i}_2(t) = e^{(\mu-i\nu)t}(\mathbf{v}' - i\mathbf{v}'')$$

are (complex) solution vectors. We may rewrite these in the form

$$\begin{aligned}\mathbf{i}_1(t) &= e^{\mu t}(\cos \nu t + i \sin \nu t)(\mathbf{v}' + i\mathbf{v}'') \\ &= e^{\mu t}[(\cos \nu t)\mathbf{v}' - (\sin \nu t)\mathbf{v}''] + ie^{\mu t}[(\sin \nu t)\mathbf{v}' + (\cos \nu t)\mathbf{v}'']\end{aligned}$$

and

$$\mathbf{i}_2(t) = e^{\mu t}[(\cos \nu t)\mathbf{v}' - (\sin \nu t)\mathbf{v}''] - ie^{\mu t}[(\sin \nu t)\mathbf{v}' + (\cos \nu t)\mathbf{v}''].$$

Forming

$$\frac{1}{2}(\mathbf{i}_1(t) + \mathbf{i}_2(t)) = e^{\mu t}[(\cos \nu t)\mathbf{v}' - (\sin \nu t)\mathbf{v}''],$$

and (2.46)

$$\frac{1}{2i}(\mathbf{i}_1(t) - \mathbf{i}_2(t)) = e^{\mu t}[(\sin \nu t)\mathbf{v}' + (\cos \nu t)\mathbf{v}''],$$

we obtain two real vectors which are solutions to the differential equations by virtue of the fact that they are linear combinations of solutions. This situation is entirely similar to that which we encountered in Chapter 1 when complex eigenvalues appeared.

The interesting case in which an eigenvalue λ occurs twice as a zero of the polynomial (2.37) will not be treated here.

PROBLEMS

1. Find the sum and difference of the vectors

$$(1,3,-\tfrac{1}{2}), \quad (2,1,-4).$$

Find the scalar product of these vectors.

2. Find the vector $\mathbf{v}$ defined by

$$\mathbf{v} = \mathbf{A}\mathbf{u},$$

where

$$\mathbf{A} = \begin{pmatrix} 1 & -1 & 0 \\ 0 & -1 & 3 \\ \tfrac{1}{4} & 2 & 2 \end{pmatrix}, \quad \mathbf{u} = (1,1,-2).$$

*3. The matrix $\mathbf{I}$ is defined by

$$\mathbf{I} = \begin{pmatrix} 1 & 0 & 0 \\ 0 & 1 & 0 \\ 0 & 0 & 1 \end{pmatrix}.$$

Show that if **v** is any vector, **Iv** = **v**. **I** is called the (3 × 3) *identity matrix*.

4. The matrix
$$\mathbf{D} = \begin{pmatrix} d_1 & 0 & 0 \\ 0 & d_2 & 0 \\ 0 & 0 & d_3 \end{pmatrix}.$$

Prove that if **v** is any vector $\mathbf{Dv} = (d_1 v_1, d_2 v_2, d_3 v_3)$. **D** is called a *diagonal matrix*.

5. Show that if **u** is any vector $\mathbf{u} \cdot \mathbf{u} \geq 0$ with equality holding only if **u** = **0**.

6. Show that the vectors (1,1,1), (1,1,0), and (1,0,0) are linearly independent. How many sets of three linearly independent sets of vectors can you find in the set (1,0,0), (0,1,0), (0,0,1), (1,1,1)?

7. Consider the differential equation
$$\ddot{u} + a\dot{u} + bu = 0, \tag{2.47}$$

a, b constants. Let $v = \dot{u}$ so that the equation becomes
$$\dot{v} + av + bu = 0.$$

If $\mathbf{w} = (u,v)$, then (2.47) is equivalent to the system
$$\dot{\mathbf{w}} + \mathbf{Aw} = 0, \tag{2.48}$$
where
$$\mathbf{A} = \begin{pmatrix} 0 & -1 \\ b & a \end{pmatrix}.$$

Investigate the solutions of (2.48) having the form
$$\mathbf{w}(t) = e^{\lambda t}\mathbf{z},$$

where $\mathbf{z} = (x,y)$, x,y constants. Compare your results with the results of Chapter 1.

8. Find the eigenvalues of the matrices

(a) $\begin{pmatrix} 2 & 1 \\ -1 & 3 \end{pmatrix},$

(b) $\begin{pmatrix} 0 & -3 \\ 3 & 0 \end{pmatrix},$

(c) $\begin{pmatrix} 1 & \alpha \\ \alpha & 2 \end{pmatrix},$ α real,

(d) $\begin{pmatrix} 0 & 0 & 1 \\ 0 & 0 & -3 \\ 1 & 1 & 3 \end{pmatrix}.$

9. Find real vectors associated with the eigenvalues of the matrices in Problem 8. That is, write **v** in the form $\mathbf{v}_1 + i\mathbf{v}_2$.

10. Compute **AB** and **BA** where

$$\mathbf{A} = \begin{pmatrix} 2 & -1 & 0 & 0 \\ -1 & 2 & -1 & 0 \\ 0 & -1 & 2 & -1 \\ 0 & 0 & -1 & 2 \end{pmatrix},$$

$$\mathbf{B} = \begin{pmatrix} 1 & 1 & 1 & 0 \\ 1 & 1 & 0 & 0 \\ 1 & 0 & 0 & 0 \\ 0 & 0 & 0 & 1 \end{pmatrix}.$$

*11. Show that if **A** is any real $n \times n$ matrix then $\mathbf{A}^2$ is well-defined. In other words show that **A** commutes with itself. Is $\mathbf{A}^p$, where p is a positive integer, also well-defined? How would you define $\mathbf{A}^0$?

*12. Prove Proposition 2.2 for the case where α_1, α_2, and α_3 are complex.

13. Let **A** be an $n \times n$ matrix and **D** an $n \times n$ diagonal matrix

$$\mathbf{D} = \begin{pmatrix} d_1 & 0 & \cdots & 0 \\ 0 & d_2 & \cdots & 0 \\ \vdots & & & \vdots \\ 0 & 0 & \cdots & d_n \end{pmatrix}.$$

Show that **AD** is the matrix with column vectors $d_j \hat{\mathbf{a}}_j$, where $\hat{\mathbf{a}}_j$ is the jth column vector of **A**, and **DA** is the matrix with row vectors $d_j \mathbf{a}_j$, where $\mathbf{a}_j$ is the jth row vector of **A**.

14. Compute the vector **Aa** where

$$\mathbf{A} = \begin{pmatrix} 1+i & -1 & -i \\ 2 & i & 3-i \\ 0 & 1 & 6 \end{pmatrix}, \quad \mathbf{a} = (1-i, 2i, -2+i).$$

15. Solve the system

$$\dot{\mathbf{u}} + \mathbf{A}\mathbf{u} = 0$$

where

$$\mathbf{A} = \begin{pmatrix} 0 & -3 \\ 3 & 0 \end{pmatrix}$$

subject to the initial condition $\mathbf{u}(0) = (1, -1)$.

16. Solve the system of Problem 15 when

$$\mathbf{A} = \begin{pmatrix} 0 & 0 & 1 \\ 0 & 0 & -3 \\ 1 & 1 & 3 \end{pmatrix}$$

and the initial condition is $\mathbf{u}(0) = (1, 0, 0)$.

17. Let
$$\mathbf{A} = \begin{pmatrix} 0 & 1 \\ 0 & 0 \end{pmatrix}.$$

Show that $\mathbf{A}^2 = \begin{pmatrix} 0 & 0 \\ 0 & 0 \end{pmatrix}$. What does this imply about the possibility of concluding from the matrix equations $\mathbf{AB} = \mathbf{0}$, $\mathbf{B} \neq \mathbf{0}$, that $\mathbf{A} = \mathbf{0}$?

Problems 18 through 21 require that you consider the double-pendulum problem in vector-matrix form.

18. Define the vector $\varphi = \varphi(t) = (\varphi_1(t), \varphi_2(t))$. Then the linear double-pendulum problem can be put in the form
$$\ddot{\varphi} + \mathbf{S}\varphi = \mathbf{0}, \tag{2.49}$$
where $\mathbf{S}$ is the matrix
$$\mathbf{S} = \begin{pmatrix} \nu_1 + M(\nu_1 + \nu_2) & -M\nu_1 \\ -\mu\nu_2 & \nu_2 \end{pmatrix},$$
as was shown in Problem 3, Section 2.2. Let $\mathbf{v} = (v_1, v_2)$ be an arbitrary constant vector. Show that the vector
$$\varphi(t) = \sin(\sqrt{\omega}\, t + \delta)\mathbf{v} \tag{2.50}$$
is a solution of (2.49) for all t if and only if ω and $\mathbf{v}$ are solutions of
$$\mathbf{Sv} - \omega\mathbf{v} = \mathbf{0}. \tag{2.51}$$

19. Prove that the eigenvalues ω_1, ω_2 of the matrix $\mathbf{S}$ of Problem 18 [the numbers ω_1, ω_2 for which there exists nonzero vectors $\mathbf{v}$ satisfying (2.51)] are both positive so that $\sqrt{\omega}$ is a real number.

20. Show that eigenvectors $\mathbf{v}_k$, $k = 1, 2$, corresponding to the eigenvalues ω_k, $k = 1, 2$, of the matrix $\mathbf{S}$ of Problem 18 can be taken in the form
$$\mathbf{v}_k = \left(1, \frac{\mu\nu_2}{\nu_2 - \omega_k}\right),$$
and prove that these vectors are linearly independent.

21. Solve the double-pendulum problem in the form
$$\varphi(t) = \alpha_1 \sin(\sqrt{\omega_1}\, t + \delta_1)\mathbf{v}_1 + \alpha_2 \sin(\sqrt{\omega_2}\, t + \delta_2)\mathbf{v}_2.$$
Compare this solution with that obtained in Problem 6, Section 2.2.

Problems 22 through 25 require that you consider the transverse motions of a taut string of beads.

22. Assume that a thread of length l is stretched between two stationary points in a horizontal plane producing a constant tension σ. Three beads of masses m_k, $k = 1, 2, 3$, are fixed to the thread and they subdivide the thread into 4 segments of length l_k, $k = 0, 1, 2, 3$, when

the thread and beads are in the undisturbed (equilibrium) position. Denote by y_k, $k = 1, 2, 3$, the displacement of the beads in the horizontal plane perpendicular to the line of equilibrium. Find the kinetic and potential-energy functions for this system.

23. Show that the potential-energy function for the small transverse vibrations of the system of beads is given by

$$V = \frac{\sigma}{2} \sum_{k=0}^{3} \frac{1}{l_k} (y_{k+1} - y_k)^2, \quad y_0 = 0 = y_4,$$

and that the equations for the motion can be put into the form

$$\ddot{\mathbf{y}} + \sigma \mathbf{S} \mathbf{y} = \mathbf{0}, \tag{2.52}$$

where $\mathbf{y} = (y_1, y_2, y_3)$, and

$$\mathbf{S} = \begin{pmatrix} \frac{1}{m_1}\left(\frac{1}{l_0} + \frac{1}{l_1}\right) & -\frac{1}{m_1 l_1} & 0 \\ -\frac{1}{m_2 l_1} & \frac{1}{m_2}\left(\frac{1}{l_1} + \frac{1}{l_2}\right) & -\frac{1}{m_2 l_2} \\ 0 & -\frac{1}{m_3 l_2} & \frac{1}{m_3}\left(\frac{1}{l_2} + \frac{1}{l_3}\right) \end{pmatrix}. \tag{2.53}$$

24. Let $\mathbf{v}$ be an arbitrary constant vector $\mathbf{v} = (v_1, v_2, v_3)$. Find the eigenvalue problem which must be solved in order that the vector

$$\mathbf{y}(t) = \sin(\sqrt{\omega}\, t + \delta)\mathbf{v}$$

is a solution to the three-bead problem.

25. Compare the matrix in (2.53) with the matrix given above (2.31) by making the identifications $m_k = R_k$, $k = 1, 2, 3$, and $l_k = C_k$, $k = 1, 2, 3$. Can the problem of cascaded loops be modified so as to have a matrix which can be completely identified with the matrix in (2.53)?

26. An initial angular velocity (rotation) is imparted to a circular shaft to which is fixed three rigid disks. If the shaft, which is housed in smooth bearings, were absolutely rigid the angular velocities of all of the disks would be the same as that of the shaft as the system rotates. But, since the shaft is not absolutely rigid the disks will rotate relative to each other and there will be superimposed on the rotation of the shaft as a whole relative oscillations (torques) of the disks. Denote by φ_k and I_k, $k = 1, 2, 3$, the angles of rotation and the polar moments of inertia of the disks, and by c_k, $k = 1, 2$, the torsional rigidity of the portions of the shaft between successive disks. Find the kinetic energy and the potential energy for small oscillations of the system considering only the mass of the disks.

27. Show that the equations of motion for the shaft-disk system of Problem 26 are

$$I_1\ddot{\varphi}_1 + c_1(\varphi_1 - \varphi_2) = 0,$$
$$I_2\ddot{\varphi}_2 - c_1(\varphi_1 - \varphi_2) + c_2(\varphi_2 - \varphi_3) = 0,$$
$$I_3\ddot{\varphi}_3 - c_2(\varphi_2 - \varphi_3) = 0.$$

28. Let the initial conditions for the shaft-disk problem be given in the form $\varphi = (\varphi_0, \varphi_0, \varphi_0) = \varphi_0(1,1,1)$

$$\dot{\varphi} = (\omega, \omega, \omega) = \omega(1,1,1).$$

Show that there exists a solution vector

$$\varphi = (\omega t + \varphi_0)(1,1,1)$$

satisfying these initial conditions. Explain the physical significance of this solution.

2.5 EIGENVALUES AND EIGENVECTORS OF THE PROBLEM OF CASCADED LOOPS

In Section 2.4 we conjectured, on the basis of previous experience, that a major step forward in determining the general solution of the problem of cascaded loops was first to determine the solution to the homogeneous problem (2.33). In so doing we developed the characteristic equation in the determinant form (2.37). Obviously, the determination of the eigenvalues λ_k, $k = 1, 2, 3$, of (2.37) will bring us yet another step closer to the general solution.

Writing out Equation (2.37), we obtain

$$\lambda^3 + \left[\frac{1}{R_1C_1} + \frac{1}{R_2C_1} + \frac{1}{R_2C_2} + \frac{1}{R_3C_2} + \frac{1}{R_3C_3}\right]\lambda^2$$
$$+ \left[\frac{1}{R_1R_2C_1C_2} + \frac{1}{R_1R_3C_1C_2} + \frac{1}{R_1R_3C_1C_3} + \frac{1}{R_2R_3C_1C_2}\right.$$
$$\left. + \frac{1}{R_2R_3C_1C_3} + \frac{1}{R_2R_3C_2C_3}\right]\lambda \quad (2.54)$$
$$+ \frac{1}{R_1R_2R_3C_1C_2C_3} = 0.$$

This cubic equation does not simplify much, except for very special choices of the six circuit constants R_k and C_k, $k = 1, 2, 3$. It may, therefore, require a considerable amount of numerical labor, or computer time, to obtain a solution with any significant accuracy. In view of the fact that all of its coefficients are positive, Equation (2.54) will always be a complete cubic. (A complete cubic is one in which all of the coefficients are different from zero). Thus we see that obtaining the λ_k is a very difficult problem in the general case. Disregarding for the moment the difficulty inherent in finding the λ_k, let us prove that they must be real and negative. To do this we use a method which consists of operating with the terms of the determinant in (2.37). Write the determinant's successive principal minors in the following scheme:

and

$$1,\ \frac{1}{R_1C_1} + \lambda,\ \begin{vmatrix} \dfrac{1}{R_1C_1} + \lambda & -\dfrac{1}{R_1C_1} \\ -\dfrac{1}{R_2C_1} & \dfrac{1}{R_2}\left(\dfrac{1}{C_1} + \dfrac{1}{C_2}\right) + \lambda \end{vmatrix},$$

$$\begin{vmatrix} \dfrac{1}{R_1C_1} + \lambda & -\dfrac{1}{R_1C_1} & 0 \\ -\dfrac{1}{R_2C_1} & \dfrac{1}{R_2}\left(\dfrac{1}{C_1} + \dfrac{1}{C_2}\right) + \lambda & -\dfrac{1}{R_2C_2} \\ 0 & -\dfrac{1}{R_3C_1} & \dfrac{1}{R_3}\left(\dfrac{1}{C_2} + \dfrac{1}{C_3}\right) + \lambda \end{vmatrix},$$

and denote the terms of this sequence as $f_0(\lambda)$, $f_1(\lambda)$, $f_2(\lambda)$, and $f_3(\lambda)$, respectively. Thus $f_3(\lambda)$ is the polynomial (2.54). Expanding $f_2(\lambda)$ in terms of $f_0(\lambda)$ and $f_1(\lambda)$ we obtain

$$f_2(\lambda) = \left[\frac{1}{R_2}\left(\frac{1}{C_1} + \frac{1}{C_2}\right) + \lambda\right] f_1(\lambda) - \frac{1}{R_1R_2C_1^2} f_0(\lambda), \qquad (2.55)$$

and similarly

$$f_3(\lambda) = \left[\frac{1}{R_3}\left(\frac{1}{C_2} + \frac{1}{C_3}\right) + \lambda\right] f_2(\lambda) - \frac{1}{R_2R_3C_2^2} f_1(\lambda). \qquad (2.56)$$

Suppose that λ_0 is a zero of f_2, that is, $f_2(\lambda_0) = 0$. Let us consider the product $f_3(\lambda_0)f_1(\lambda_0)$ which, using (2.56), is

$$f_3(\lambda_0)f_1(\lambda_0) = \frac{1}{R_2R_3C_2^2} f_1^2(\lambda_0).$$

Hence, either

$$f_3(\lambda_0)f_1(\lambda_0) < 0 \quad \text{or} \quad f_1(\lambda_0) = 0. \qquad (2.57)$$

However, $f_1(\lambda_0) = 0$ is not possible since if $f_2(\lambda_0) = f_1(\lambda_0) = 0$, (2.55) implies $f_0(\lambda_0) = 0$ in contradiction to the fact that $f_0(\lambda_0) \equiv 1$. Thus the first of (2.57) is valid if λ_0 is a zero of f_2.

We may argue in exactly the same way if λ_0 is a zero of f_1, $f_1(\lambda_0) = 0$. From (2.55) we obtain

$$f_2(\lambda_0)f_0(\lambda_0) = -\frac{1}{R_1R_2C_1^2} f_0^2(\lambda_0) = -\frac{1}{R_1R_2C_1^2} < 0. \qquad (2.58)$$

Now the only zero of f_1 occurs when $\lambda = -1/R_1C_1$ and, by (2.58) we must have

$$f_2\left(-\frac{1}{R_1C_1}\right)f_0\left(-\frac{1}{R_1C_1}\right) = f_2\left(-\frac{1}{R_1C_1}\right) < 0.$$

On the other hand,

$$f_2(\lambda) = \lambda^2 + \left(\frac{1}{R_1C_1} + \frac{1}{R_2C_1} + \frac{1}{R_2C_2}\right)\lambda + \frac{1}{R_1R_2C_1C_2}.$$

Consequently $f_2(0) = 1/R_1R_2C_1C_2 > 0$ and it is clear that $f_2(\lambda) > 0$ for λ sufficiently large and negative. Thus f_2 is positive for large negative λ and for $\lambda = 0$, and negative for $\lambda = -1/R_1C_1$. It follows that the graph of $f_2(\lambda) = 0$ crosses the λ axis once to the left of $-1/R_1C_1$ and once between $-1/R_1C_1$ and zero. If we denote the zeros of f_2 by λ_{21} and λ_{22} we have proved that

$$\lambda_{21} < -\frac{1}{R_1C_1} < \lambda_{22} < 0. \tag{2.59}$$

To complete our argument we will establish the inequalities

$$\lambda_1 < \lambda_{21} < \lambda_2 < \lambda_{22} < \lambda_3 < 0 \tag{2.60}$$

where λ_1, λ_2, and λ_3 are zeros of f_3. In the first place we see directly from (2.54) that f_3 is negative for sufficiently large negative values of λ and positive for $\lambda = 0$. Consequently the graph of the cubic equation $f_3(\lambda) = 0$ must cross the negative λ-axis either once or three times. But from (2.57) we have the inequalities

$$f_3(\lambda_{21})f_1(\lambda_{21}) < 0, \quad \text{and} \quad f_3(\lambda_{22})f_1(\lambda_{22}) < 0,$$

and (2.59) together with the definition of f_1 shows that

$$f_1(\lambda_{21}) < 0, \quad \text{and} \quad f_1(\lambda_{22}) > 0.$$

Consequently

$$f_3(\lambda_{21}) > 0, \quad \text{and} \quad f_3(\lambda_{22}) < 0.$$

We have thus established that f_3 is negative for sufficiently large negative λ, positive for $\lambda = \lambda_{21}$, negative for $\lambda = \lambda_{22}$ and positive for $\lambda = 0$. But this clearly implies the validity of inequalities (2.60).

In summary we have:

Proposition 2.3

The eigenvalues for the homogeneous system corresponding to (2.17) are real, distinct, and negative.

We now have before us the task of computing the eigenvectors corresponding to $\lambda_1 < \lambda_2 < \lambda_3 < 0$. Equations (2.36) are

$$\left(\frac{1}{R_1C_1} + \lambda_k\right)v_1 - \frac{1}{R_1C_1}v_2 = 0,$$

$$-\frac{1}{R_2C_1}v_1 + \left[\frac{1}{R_2}\left(\frac{1}{C_1} + \frac{1}{C_2}\right) + \lambda_k\right]v_2 - \frac{1}{R_2C_2}v_3 = 0,$$

$$-\frac{1}{R_3C_2}v_2 + \left[\frac{1}{R_3}\left(\frac{1}{C_2} + \frac{1}{C_3}\right) + \lambda_k\right]v_3 = 0,$$

$k = 1, 2, 3$. The fact that for each of the values λ_k, $k = 1, 2, 3$, the determinant of this system is zero, means that we can seek to solve two of these equations, say for v_2 and v_3, in terms of v_1. Now $1/R_1C_1 \neq 0$, and $1/R_1C_1 + \lambda_k = f_1(\lambda_k)$ so that we may write

$$v_2 = R_1C_1f_1(\lambda_k)v_1.$$

Similarly we have

$$v_3 = R_2C_2\left\{\left[\frac{1}{R_2}\left(\frac{1}{C_1} + \frac{1}{C_2}\right) + \lambda_k\right]R_1C_1f_1(\lambda_k) - \frac{1}{R_3C_2}\right\}v_1$$

or

$$v_3 = R_1R_2C_1C_2f_2(\lambda_k)v_1.$$

Thus eigenvectors $\mathbf{v}_1$, $\mathbf{v}_2$, $\mathbf{v}_3$ of our problem are given by

$$\begin{aligned}\mathbf{v}_1 &= v_{11}(1, R_1C_1f_1(\lambda_1), R_1R_2C_1C_2f_2(\lambda_1)),\\ \mathbf{v}_2 &= v_{12}(1, R_1C_1f_1(\lambda_2), R_1R_2C_1C_2f_2(\lambda_2)),\\ \mathbf{v}_3 &= v_{13}(1, R_1C_1f_1(\lambda_3), R_1R_2C_1C_2f_2(\lambda_3)),\end{aligned} \quad (2.61)$$

where v_{11}, v_{12}, and v_{13} are arbitrary real numbers. Now we note from (2.60) that the zeroes of $f_2(\lambda)$, namely λ_{21} and λ_{22} satisfy

$$\lambda_1 < \lambda_{21} < \lambda_2 < \lambda_{22} < \lambda_3,$$

hence $f_2(\lambda_k) \neq 0$ for $k = 1, 2, 3$, and since $f_2(0) = 0$, we also have $f_2(\lambda_1) > 0$, $f_2(\lambda_2) < 0$, $f_2(\lambda_3) > 0$. Moreover by virtue of (2.59) and (2.60) we have

$$f_1(\lambda_1) < 0, \quad f_1(\lambda_3) > 0.$$

Let us agree to choose $v_{1k} = 1$, $k = 1, 2, 3$. These numbers are positive so that we may assert that the vectors $\mathbf{v}_1$ and $\mathbf{v}_3$ have the sign patterns

$$\begin{aligned}\mathbf{v}_1 &= (+, -, +),\\ \mathbf{v}_3 &= (+, +, +),\end{aligned} \quad (2.62a)$$

while $\mathbf{v}_2$ has the sign pattern

$$\mathbf{v}_2 = (+, ?, -), \quad (2.62b)$$

because the term $R_1C_1f_1(\lambda_2)$ can be positive, negative, or zero.

Next let us show that these vectors are linearly independent. To do this we will prove that if $\mathbf{A}$ is a 3×3 matrix with $\lambda_1, \lambda_2, \lambda_3$ distinct numbers and $\mathbf{v}_1 \neq \mathbf{0}$, $\mathbf{v}_2 \neq \mathbf{0}$, $\mathbf{v}_3 \neq \mathbf{0}$ are vectors such that

$$\mathbf{A}\mathbf{v}_i = -\lambda_i \mathbf{v}_i, \quad i = 1, 2, 3,$$

then $\mathbf{v}_1$, $\mathbf{v}_2$, and $\mathbf{v}_3$ are linearly independent. Thus we consider the linear combination

$$\alpha_1 \mathbf{v}_1 + \alpha_2 \mathbf{v}_2 + \alpha_2 \mathbf{v}_3 = \mathbf{0} \qquad (2.63)$$

and prove that $\alpha_1 = \alpha_2 = \alpha_3 = 0$.

Applying $\mathbf{A}$ to (2.63) yields

$$\alpha_1 \mathbf{A}\mathbf{v}_1 + \alpha_2 \mathbf{A}\mathbf{v}_2 + \alpha_3 \mathbf{A}\mathbf{v}_3 = -\lambda_1 \alpha_1 \mathbf{v}_1 - \lambda_2 \alpha_2 \mathbf{v}_2 - \lambda_3 \alpha_3 \mathbf{v}_3.$$

Hence

$$\lambda_1 \alpha_1 \mathbf{v}_1 + \lambda_2 \alpha_2 \mathbf{v}_2 + \lambda_3 \alpha_3 \mathbf{v}_3 = \mathbf{0}. \qquad (2.64)$$

On the other hand, multiplying (2.63) by λ_1, and subtracting the result from (2.64) yields

$$(\lambda_2 - \lambda_1)\alpha_2 \mathbf{v}_2 + (\lambda_3 - \lambda_1)\alpha_3 \mathbf{v}_3 = \mathbf{0}. \qquad (2.65)$$

We now apply $\mathbf{A}$ to (2.65) to obtain the equation

$$(\lambda_2 - \lambda_1)\lambda_2 \alpha_2 \mathbf{v}_2 + (\lambda_3 - \lambda_1)\lambda_3 \alpha_3 \mathbf{v}_3 = \mathbf{0}. \qquad (2.66)$$

Finally we multiply (2.65) by λ_2 and subtract the result from (2.66) to obtain

$$(\lambda_3 - \lambda_1)(\lambda_3 - \lambda_2)\alpha_3 \mathbf{v}_3 = \mathbf{0}.$$

Since $\lambda_3 \neq \lambda_1$, $\lambda_3 \neq \lambda_2$, and $\mathbf{v}_3 \neq \mathbf{0}$, this equation implies that $\alpha_3 = 0$. Since $\alpha_3 = 0$, $\lambda_2 \neq \lambda_1$, and $\mathbf{v}_2 \neq \mathbf{0}$, Equation (2.65) implies that $\alpha_2 = 0$. Finally the fact that $\alpha_2 = \alpha_3 = 0$ and $\mathbf{v}_1 \neq \mathbf{0}$ enables us to deduce from (2.63) that $\alpha_1 = 0$. This proves that $\mathbf{v}_1$, $\mathbf{v}_2$, and $\mathbf{v}_3$ are linearly independent.

We have now established all of the necessary results to conclude that the solution to the problem (2.33) can be written in the form

$$\mathbf{i}(t) = \alpha_1 e^{\lambda_1 t}\mathbf{v}_1 + \alpha_2 e^{\lambda_2 t}\mathbf{v}_2 + \alpha_3 e^{\lambda_3 t}\mathbf{v}_3, \qquad (2.67)$$

where α_1, α_2, and α_3 are uniquely determined by the initial vector $\mathbf{a}$ by the equation

$$\mathbf{a} = \alpha_1 \mathbf{v}_1 + \alpha_2 \mathbf{v}_2 + \alpha_3 \mathbf{v}_3,$$

and we have

$$\lambda_1 < \lambda_2 < \lambda_3 < 0.$$

Proposition 2.4

Every solution of the homogeneous problem of cascaded loops satisfies

$$\lim_{t \to \infty} \mathbf{i}(t) = \mathbf{0}.$$

PROOF. The proof follows immediately from the formula (2.67) in view of the fact that the eigenvalues are all negative.

2.6 THE NONHOMOGENEOUS PROBLEM OF CASCADED LOOPS

In Sections 2.3, 2.4, and 2.5 we have set up the problem of the three cascaded loops shown in Figure 2.10, introduced a variety of new algebraic concepts, and with a sufficient amount of hard work, obtained important information about the associated homogeneous problem. We still have not solved the problem originally formulated in Equations (2.17) and (2.18). Thus we will now show how the solution of the homogeneous problem can be used to solve the original nonhomogeneous problem.

Let us consider the problem as reformulated with the aid of vectors and matrices, namely

$$\frac{d\mathbf{i}}{dt} + \mathbf{S}\mathbf{i} = \mathbf{h}, \tag{2.68}$$

and the initial condition

$$\mathbf{i}(0) = \mathbf{a}. \tag{2.69}$$

Suppose that the vector $\mathbf{i}'(t)$ satisfies the homogeneous equation

$$\frac{d\mathbf{i}'}{dt} + \mathbf{S}\mathbf{i}' = \mathbf{0}$$

and

$$\mathbf{i}'(0) = \mathbf{a},$$

and that $\mathbf{i}''(t)$ is a vector satisfying

$$\frac{d\mathbf{i}''}{dt} + \mathbf{S}\mathbf{i}'' = \mathbf{h},$$

and

$$\mathbf{i}''(0) = \mathbf{0}.$$

Then the vector

$$\mathbf{i}(t) = \mathbf{i}'(t) + \mathbf{i}''(t) \tag{2.70}$$

satisfies the nonhomogeneous differential equation and $\mathbf{i}(0) = \mathbf{a} + \mathbf{0} = \mathbf{a}$. In other words, $\mathbf{i}$ is the solution we seek for the problem (2.68), (2.69).

Our task, therefore, is to solve the system (2.68) with the special initial condition $\mathbf{a} = \mathbf{0}$. We shall do this by the method of variation of constants employed successfully in Chapter 1 to solve nonhomogeneous problems. For this purpose we can return to the solution as given in (2.40) except that we now put

$$\mathbf{i}''(t) = \alpha_1(t)\mathbf{i}_1(t) + \alpha_2(t)\mathbf{i}_2(t) + \alpha_3(t)\mathbf{i}_3(t). \tag{2.71}$$

That is, we allow the values of α_1, α_2, and α_3 to vary as functions of t in such a way as to force $\mathbf{i}''(t)$ to satisfy (2.68). Differentiating, we have

$$\frac{d\mathbf{i}''}{dt} = \alpha_1(t)\frac{d\mathbf{i}_1}{dt} + \alpha_2(t)\frac{d\mathbf{i}_2}{dt} + \alpha_3(t)\frac{d\mathbf{i}_3}{dt} + \dot{\alpha}_1\mathbf{i}_1(t) + \dot{\alpha}_2\mathbf{i}_2(t) + \dot{\alpha}_3\mathbf{i}_3(t).$$

Placing this and (2.71) into (2.68) we have

$$\alpha_1\frac{d\mathbf{i}_1}{dt} + \alpha_2\frac{d\mathbf{i}_2}{dt} + \alpha_3\frac{d\mathbf{i}_3}{dt} + \dot{\alpha}_1\mathbf{i}_1 + \dot{\alpha}_2\mathbf{i}_2 + \dot{\alpha}_3\mathbf{i}_3 + \alpha_1 S\mathbf{i}_1 + \alpha_2 S\mathbf{i}_2 + \alpha_3 S\mathbf{i}_3 = \mathbf{h}.$$

But $\mathbf{i}_k(t)$, $k = 1, 2, 3$, is a solution of the homogeneous equation so that we are left with

$$\dot{\alpha}_1(t)\mathbf{i}_1(t) + \dot{\alpha}_2(t)\mathbf{i}_2(t) + \dot{\alpha}_3(t)\mathbf{i}_3(t) = \mathbf{h}(t). \tag{2.72}$$

In addition, we also require that

$$\mathbf{i}(0) = \mathbf{0} = \alpha_1(0)\mathbf{i}_1(0) + \alpha_2(0)\mathbf{i}_2(0) + \alpha_3(0)\mathbf{i}_3(0)$$

or

$$\mathbf{0} = \alpha_1(0)\mathbf{v}_1 + \alpha_2(0)\mathbf{v}_2 + \alpha_3(0)\mathbf{v}_3. \tag{2.73}$$

Since the vectors $\mathbf{v}_1$, $\mathbf{v}_2$, and $\mathbf{v}_3$ are linearly independent, it follows at once from (2.73) that

$$\alpha_1(0) = \alpha_2(0) = \alpha_3(0) = 0. \tag{2.74}$$

Now writing $\mathbf{i}_k(t) = (i_{1k}(t), i_{2k}(t), i_{3k}(t))$, $k = 1, 2, 3$, we see that (2.72) is equivalent to the three scalar equations

$$\begin{aligned} i_{11}(t)\dot{\alpha}_1 + i_{12}(t)\dot{\alpha}_2 + i_{13}(t)\dot{\alpha}_3 &= h_1(t), \\ i_{21}(t)\dot{\alpha}_1 + i_{22}(t)\dot{\alpha}_2 + i_{23}(t)\dot{\alpha}_3 &= h_2(t), \\ i_{31}(t)\dot{\alpha}_1 + i_{32}(t)\dot{\alpha}_2 + i_{33}(t)\dot{\alpha}_3 &= h_3(t). \end{aligned} \tag{2.75}$$

For each fixed value of t this is a system of three linear equations for the determination of $\dot{\alpha}_1, \dot{\alpha}_2, \dot{\alpha}_3$. In matrix form we have

$$\mathcal{I}(t)\dot{\boldsymbol{\alpha}}(t) = \mathbf{h}(t) \tag{2.76}$$

where $\dot{\boldsymbol{\alpha}}(t) = (\dot{\alpha}_1(t), \dot{\alpha}_2(t), \dot{\alpha}_3(t))$, and $\mathcal{I}(t)$ is the matrix of coefficients with element $i_{jk}(t)$ in the jth row and kth column. This problem can be solved by Cramer's rule provided the determinant of $\mathcal{I}(t)$ is different from zero for each value of t. Denoting this determinant by $D(t)$ and recognizing that we are in the case where the eigenvalues of the homogeneous system are real and distinct, we find that

$$\begin{aligned} D(t) &= \begin{vmatrix} e^{\lambda_1 t}v_{11} & e^{\lambda_2 t}v_{12} & e^{\lambda_3 t}v_{13} \\ e^{\lambda_1 t}v_{21} & e^{\lambda_2 t}v_{22} & e^{\lambda_3 t}v_{23} \\ e^{\lambda_1 t}v_{31} & e^{\lambda_2 t}v_{32} & e^{\lambda_3 t}v_{33} \end{vmatrix} \\ &= e^{\lambda_1 t}e^{\lambda_2 t}e^{\lambda_3 t}\begin{vmatrix} v_{11} & v_{12} & v_{13} \\ v_{21} & v_{22} & v_{23} \\ v_{31} & v_{32} & v_{33} \end{vmatrix}. \end{aligned}$$

Since the vectors $\mathbf{v}_1$, $\mathbf{v}_2$, and $\mathbf{v}_3$ are linearly independent, the last written determinant is different from zero. But $e^{\lambda_k t} \neq 0$, $k = 1, 2, 3$, hence $D(t) \neq 0$ for all t. This means that we can solve (2.75) or, equivalently, (2.76) for the vector $\dot{\boldsymbol{\alpha}}(t)$ for all t.

It will prove to be instructive for us to carry out the computation outlined above and relate the result to the concepts introduced in Section 2.3. Assume now that t is fixed and suppose we can find a matrix $\mathbf{M}(t)$ such that $\mathbf{M}(t)\boldsymbol{\mathcal{I}}(t) = \mathbf{I}$, the 3×3 identity matrix. Multiplying equation (2.76) on the left by $\mathbf{M}(t)$ yields

$$\mathbf{I}\dot{\boldsymbol{\alpha}}(t) = \mathbf{M}(t)\boldsymbol{\mathcal{I}}(t)\dot{\boldsymbol{\alpha}}(t) = \mathbf{M}(t)\mathbf{h}(t).$$

By Problem 3 of Section 2.4 we have

$$\mathbf{I}\dot{\boldsymbol{\alpha}}(t) = \dot{\boldsymbol{\alpha}}(t).$$

Consequently the above equation reduces to

$$\dot{\boldsymbol{\alpha}}(t) = \mathbf{M}(t)\mathbf{h}(t)$$

and we have succeeded in solving for $\dot{\boldsymbol{\alpha}}(t)$ at the given value of t. Thus, the problem at hand can be put in the following form. Given a matrix $\mathbf{A}$, find $\mathbf{B}$ such that $\mathbf{BA} = \mathbf{I}$.

We wish to develop a formula to guide us in computing the solution to the latter problem, but first it will prove to be useful to present a few tools we will need and their associated terminology.

Without loss of generality we consider a 3×3 case. The determinant

$$D = \begin{vmatrix} a_{11} & a_{12} & a_{13} \\ a_{21} & a_{22} & a_{23} \\ a_{31} & a_{32} & a_{33} \end{vmatrix}$$

is called the *determinant of the matrix* $\mathbf{A}$, written $|\mathbf{A}|$. If a_{ij} is an element of $\mathbf{A}$, then the matrix obtained from $\mathbf{A}$ by striking out row i and column j of $\mathbf{A}$ is a 2×2 matrix and its determinant is called the minor of $\mathbf{A}$ complementary to a_{ij}. Thus the minor of $\mathbf{A}$ complementary to a_{23} is $\begin{vmatrix} a_{11} & a_{12} \\ a_{31} & a_{32} \end{vmatrix}$ and the minor complementary to a_{11} is $\begin{vmatrix} a_{22} & a_{23} \\ a_{32} & a_{33} \end{vmatrix}$. We also need the concept of the signed minor of a_{ij}, often called the *algebraic complement* of a_{ij}. The signed minor of a_{ij} is $(-1)^{i+j}$ times the complementary minor to a_{ij}. Let $\hat{\mathbf{a}}_i$ be the ith column vector of $\mathbf{A}$ and let $\mathbf{A}_i$ be the vector whose components are the signed complementary minors of $\mathbf{A}$ to the components of $\hat{\mathbf{a}}_i$. Thus corresponding to $\hat{\mathbf{a}}_1$ we have the vector

$$\mathbf{A}_1 = \left(\begin{vmatrix} a_{22} & a_{23} \\ a_{32} & a_{33} \end{vmatrix}, -\begin{vmatrix} a_{12} & a_{13} \\ a_{32} & a_{33} \end{vmatrix}, \begin{vmatrix} a_{12} & a_{13} \\ a_{22} & a_{23} \end{vmatrix} \right).$$

Returning now to the problem of finding $\mathbf{B}$ we see that it is essentially solved by using:

Lemma 2.1

For the matrix $\mathbf{A}$ we have

$$\delta_{ij}|\mathbf{A}| = \mathbf{A}_i \cdot \hat{\mathbf{a}}_j, \tag{2.77}$$

where

$$\delta_{ij} = \begin{matrix} 1, & \text{if } i = j, \\ 0, & \text{if } i \neq j, \end{matrix} \quad i,j = 1, 2, 3.$$

(The symbol δ_{ij} is called the *Kronecker delta*.)

The proof of Lemma 2.1 is left to the student as a problem.

We shall show now that the formula (2.77) permits us to find a matrix $\mathbf{B}$ such that $\mathbf{BA} = \mathbf{I}$. In fact, let $\mathbf{b}_i$ be the ith row vector of $\mathbf{B}$, then we must have

$$\mathbf{b}_i \cdot \hat{\mathbf{a}}_j = \delta_{ij}, \quad i,j = 1, 2, 3.$$

But by (2.77) we have

$$\frac{1}{|\mathbf{A}|} \mathbf{A}_i \cdot \hat{\mathbf{a}}_j = \delta_{ij}, \quad i,j = 1, 2, 3$$

provided $|\mathbf{A}| \neq 0$. It follows that if $\mathbf{B}$ is the matrix with row vectors $(1/|\mathbf{A}|)\, \mathbf{A}_i$, then $\mathbf{BA} = \mathbf{I}$.

Definition 2.2

The matrix $\mathbf{A}$ is said to be nonsingular if $|\mathbf{A}| \neq 0$.

Thus if $\mathbf{A}$ is a nonsingular matrix we have solved our problem. The matrix $\mathbf{B}$ is called a *left inverse* of $\mathbf{A}$ and is denoted by $\mathbf{A}^{-1}$. Let us show also that $\mathbf{B}$ or, equivalently, $\mathbf{A}^{-1}$ also satisfies

$$\mathbf{AA}^{-1} = \mathbf{I}. \tag{2.78}$$

For this purpose we must use the information we obtained from problem 3 of Section 2.4, that is, if $\mathbf{C}$ is any matrix

$$\mathbf{CI} = \mathbf{IC} = \mathbf{C}.$$

It immediately follows that

$$\mathbf{A}^{-1}\mathbf{A}\mathbf{A}^{-1} = \mathbf{I}\mathbf{A}^{-1} = \mathbf{A}^{-1}. \tag{2.79}$$

Now suppose that we can show that

$$|\mathbf{A}^{-1}| = \frac{1}{|\mathbf{A}|}. \tag{2.80}$$

Then $\mathbf{A}^{-1}$ is nonsingular and has an inverse itself, call it $\mathbf{C}$. Multiplying (2.79) on the left by $\mathbf{C}$ yields

$$\mathbf{IAA}^{-1} = \mathbf{I}$$

or

$$\mathbf{AA}^{-1} = \mathbf{I}.$$

as we set out to prove. We leave the task of showing that (2.80) is true to the student in the form of a problem.

Summarizing at this point, we have shown that

$$\mathbf{A}^{-1}\mathbf{A} = \mathbf{A}\mathbf{A}^{-1} = \mathbf{I}.$$

If it is true that the inverse of a matrix is unique, then this equation implies that $(\mathbf{A}^{-1})^{-1} = \mathbf{A}$. Let us therefore establish:

Lemma 2.2

Assume that $\mathbf{A}$ is nonsingular and that

$$\mathbf{AB} = \mathbf{AC} \quad \text{or} \quad \mathbf{BA} = \mathbf{CA},$$

then $\mathbf{B} = \mathbf{C}$.

PROOF. Since $\mathbf{A}$ is nonsingular, $\mathbf{A}^{-1}$ exists. By multiplying both sides of the first equation above on the left by $\mathbf{A}^{-1}$ we obtain at once $\mathbf{B} = \mathbf{C}$. Similarly multiplying both sides of the second equation on the right by $\mathbf{A}^{-1}$ yields the same result.

Lemma 2.2 also enables us to conclude that if $\mathbf{A}$ has an inverse, it must be unique. For suppose $\mathbf{B}$ and $\mathbf{C}$ are such that

$$\mathbf{BA} = \mathbf{I} \quad \text{and} \quad \mathbf{CA} = \mathbf{I}.$$

By subtraction $\mathbf{BA} = \mathbf{CA}$ and, since $\mathbf{A}$ has an inverse, it follows from the lemma that $\mathbf{B} = \mathbf{C}$.

With all this background at hand, let us once again return to our original problem, that of cascaded loops, and compute its general solution. Left multiplication of (2.76) by $\mathcal{G}^{-1}$ yields

$$\dot{\boldsymbol{\alpha}}(t) = \mathcal{G}^{-1}(t)\mathbf{h}(t). \tag{2.81}$$

On the basis of the previous calculations we know that $\mathcal{G}(t)$ is nonsingular for each value of t.

Now if $\mathbf{a}(t) = (a_1(t), a_2(t), a_3(t))$ is a vector, the definite integral of $\mathbf{a}(t)$, written $\int_{t_0}^{t_1} \mathbf{a}(t)\, dt$ is defined to be

$$\int_{t_0}^{t_1} \mathbf{a}(t)\, dt = \left(\int_{t_0}^{t_1} a_1(t)\, dt,\ \int_{t_0}^{t_1} a_2(t)\, dt,\ \int_{t_0}^{t_1} a_3(t)\, dt \right).$$

Thus $\int_{t_0}^{t_1} \mathbf{a}(t) \, dt$ is again a vector. Integrating (2.81) from $t_0 = 0$ to $t_1 = t$, and making use of (2.74), we obtain

$$\boldsymbol{\alpha}(t) = \int_0^t \boldsymbol{\mathcal{S}}^{-1}(\tau)\mathbf{h}(\tau) \, d\tau. \tag{2.82}$$

The solution we seek is obtained by substituting this vector $\boldsymbol{\alpha}$ back into (2.71).

Let us next note that (2.71) can be written in the form

$$\mathbf{i}''(t) = \boldsymbol{\mathcal{S}}(t)\boldsymbol{\alpha}(t).$$

Consequently from (2.82) we derive the representation

$$\mathbf{i}''(t) = \boldsymbol{\mathcal{S}}(t) \int_0^t \boldsymbol{\mathcal{S}}^{-1}(\tau)\mathbf{h}(\tau) \, d\tau.$$

Finally the solution of the homogeneous problem with the given initial conditions is given by

$$\mathbf{i}'(t) = \boldsymbol{\mathcal{S}}(t)\mathbf{b},$$

where the vector $\mathbf{b}$ is uniquely determined by the initial vector $\mathbf{a}$. It follows that the vector

$$\mathbf{i}(t) = \boldsymbol{\mathcal{S}}(t) \left(\mathbf{b} + \int_0^t \boldsymbol{\mathcal{S}}^{-1}(\tau)\mathbf{h}(\tau) \, d\tau \right) \tag{2.83}$$

is the general solution to the nonhomogeneous problem.

Let us pause to reflect on what were the essential parts in the development we have just been through. The basic problem is to use (2.72) and (2.73) to obtain the vector $\boldsymbol{\alpha}(t)$. What we found was that everything will work if for each value of t the determinant of $\boldsymbol{\mathcal{S}}(t)$ is different from zero. This is the same as saying that the solution vectors $\mathbf{i}_1(t)$, $\mathbf{i}_2(t)$, and $\mathbf{i}_3(t)$ must be linearly independent for all values of t. Note, in this connection, that the determinant of $\boldsymbol{\mathcal{S}}(t)$ plays essentially the same role for the system as the Wronski determinant plays for the second-order equation of Chapter 1. As long as we can find three linearly independent solution vectors—linearly independent for all values of t—for the homogeneous system, the method of variation of constants will work and the nonhomogeneous problem will be solved by formula (2.83).

PROBLEMS

1. Determine the general solutions for the following systems. Discuss the behavior of the solutions as $t \to +\infty$.

    (a) $\dfrac{di_1}{dt} = i_1 + 2i_2,\ \dfrac{di_2}{dt} = 12i_1 - i_2.$

    (b) $\dfrac{dy}{dt} = -2x + 5y,\ \dfrac{dx}{dt} = x + 2y.$

(c) $\dfrac{dx}{dt} = x - 2y + t^2 + 2t, \dfrac{dy}{dt} = 5x - y - 4t^2 + 2t.$

(d) $\dfrac{dx}{dt} = x - y, \dfrac{dy}{dt} = y - z, \dfrac{dz}{dt} = z - x.$

(e) $\dfrac{di_1}{dt} = i_1 + 2i_2 + e^t, \dfrac{di_2}{dt} = -i_1 + 4i_2 - 2.$

(f) $\dfrac{di_1}{dt} = 2i_1 - i_2 + 3i_3 + t, \dfrac{di_2}{dt} = -i_1 + i_2 - i_3 - 1, \dfrac{di_3}{dt} = i_2 - i_3.$

2. Prove Cramer's rule for the linear system

$$a_{11}x_1 + a_{12}x_2 + a_{13}x_3 = b_1,$$
$$a_{21}x_1 + a_{22}x_2 + a_{23}x_3 = b_2,$$
$$a_{31}x_1 + a_{32}x_2 + a_{33}x_3 = b_3.$$

3. Determine the general solution of the equation

$$\dfrac{d^2q}{dt^2} + 2\dfrac{dq}{dt} - 3q = 3t^2 - 4t - 2$$

by solving the equivalent system

$$\dfrac{dq}{dt} = i, \dfrac{di}{dt} = 3q - 2i + 3t^2 - 4t - 2.$$

Discuss the behavior of the solution as $t \to +\infty$.

4. Show that the analysis of Section 2.4 can be applied to the eigenvectors and eigenvalues for the problem of the transverse vibrations of a string of three beads. (Refer to Problems 22 through 25 of Section 2.4.)

5. Consider the problem of the string of three beads. Assume that $m_1 = m_2 = m_3 = m$ and $l_0 = l_1 = l_2 = l_3 = \frac{1}{4}$. Set

$$2 \cos \theta = 2 - \dfrac{ml}{4\sigma} \omega.$$

Show that the solution vector

$$\mathbf{y}(t) = \sin(\sqrt{\omega}\, t + S)\mathbf{v}$$

must satisfy

$$v_{k-1} - 2v_k \cos \theta + v_{k+1} = 0, \quad k = 1, 2, 3, \quad v_0 = v_4 = 0. \quad (2.84)$$

6. Show that if the components v_k, $k = 1, 2, 3$, of $\mathbf{v}$ satisfy

$$v_k = C_1 \cos k\theta + C_2 \sin k\theta, \quad (2.85)$$

then Equations (2.84) are satisfied if C_1 is properly chosen, and if

$$C_2 \sin 4\theta = 0.$$

7. Deduce from the result of Problem 4 that $\theta = k\pi/4$, $k = \pm 1, \pm 2$, and obtain the eigenvalues of the bead problem explicitly using these values of θ. That is, obtain ω_1, ω_2, and ω_3.

8. Show that the eigenvectors for the three-bead problem presented in Problem 5 are given by

$$\mathbf{v}_k = C_k\left(\sin\frac{\pi k}{4},\ \sin\frac{2\pi k}{4},\ \sin\frac{3\pi k}{4}\right),\quad k = 1, 2, 3.$$

*9. Establish Lemma 2.1 in the 3 × 3 case.

10. If

$$A = \begin{pmatrix} 2 & -1 & 0 \\ -1 & 2 & -1 \\ 0 & -1 & 2 \end{pmatrix},$$

compute the inverse of A.

11. Give an example of a (3 × 3) matrix for which the inverse does not exist.

*12. Prove that (2.80) is true for a nonsingular 3 × 3 matrix.

13. Verify directly that zero is an eigenvalue for the shaft-disk problem (Problems 26 through 28) of Section 2.4.

14. Consider the system of equations

$$\ddot{\mathbf{u}} + S\mathbf{u} = \mathbf{0} \tag{2.86}$$

where $\mathbf{u} = (u_1, u_2, u_3)$ and

$$S = \begin{pmatrix} s_{11} & s_{12} & s_{13} \\ s_{21} & s_{22} & s_{23} \\ s_{31} & s_{32} & s_{33} \end{pmatrix}.$$

Assume that there exist constants α, β, γ, not all zero, such that $\alpha S_1 + \beta S_2 + \gamma S_3 = \mathbf{0}$, where S_k is the kth row vector of the matrix S. Show that in this case the system (2.86) has a solution of the form

$$\mathbf{u}(t) = \mathbf{a}t + \mathbf{b} \tag{2.87}$$

where $\mathbf{a}$ and $\mathbf{b}$ are constant vectors. How are these vectors related to the initial values of the system?

15. Suppose the system (2.86) is replaced by the system of first-order equations

$$\dot{\mathbf{u}} + S\mathbf{u} = \mathbf{0}.$$

If the condition of Problem 14 on the matrix S is retained, how is the solution (2.87) modified? Relate the solution to the initial data.

16. Let a chemical reaction take place between the three species A_k, $k = 1, 2, 3$. Denote the concentration of the species A_k by $a_k = a_k(t)$, then the corresponding reaction rate is da_k/dt. If $\mathbf{a} = (a_1, a_2, a_3)$ is the concentration vector for the reaction and the reaction is monomolecular, it is described by the system

REMARKS ABOUT LINEAR FIRST-ORDER SYSTEMS

$$\dot{\mathbf{a}} + \mathbf{C}\mathbf{a} = \mathbf{0}, \qquad (2.88)$$

where

$$\mathbf{C} = \begin{pmatrix} C_{21} + C_{31} & -C_{12} & -C_{13} \\ -C_{21} & C_{12} + C_{32} & -C_{23} \\ -C_{31} & -C_{32} & C_{13} + C_{23} \end{pmatrix}.$$

The numbers C_{jk} in this matrix are called the *reaction rates*. That is, the rate at which the jth species is converted into the kth species, $A_j \to A_k$, is associated with C_{jk}. Show that the system (2.88) has a solution having the form

$$\mathbf{a}(t) = \mathbf{b},$$

where $\mathbf{b}$ is a constant vector.

17. Let $\mathbf{C}$ be the matrix of Problem 16 and assume that $C_{12}C_{23}C_{31}/C_{13}C_{32}C_{21} = 1$. Show that a diagonal matrix $\mathbf{P}$ can be found such that the matrix

$$\mathbf{D} = \mathbf{P}\mathbf{C}\mathbf{P}^{-1} = \begin{pmatrix} d_{11} & d_{12} & d_{13} \\ d_{12} & d_{22} & d_{23} \\ d_{13} & d_{23} & d_{33} \end{pmatrix}.$$

That is, elements symmetrically placed in $\mathbf{D}$ are equal. Here $\mathbf{P}^{-1}$ is the inverse of $\mathbf{P}$.

18. Prove that for any vector $\mathbf{u} = (u_1, u_2, u_3)$ the scalar product

$$\mathbf{u} \cdot \mathbf{D}\mathbf{u} \leq 0,$$

where $\mathbf{D}$ is the matrix of Problem 17.

19. If λ is an eigenvalue of the matrix $\mathbf{D}$ of Problem 17, show that

$$\lambda \leq 0.$$

[*Hint*: Use the result of Problem 18 and show that if $\mathbf{u}$ is not the zero vector and $\mathbf{u}$ satisfies $\mathbf{D}\mathbf{u} - \lambda\mathbf{u} = 0$, then

$$\lambda = \frac{\mathbf{u} \cdot \mathbf{D}\mathbf{u}}{\mathbf{u} \cdot \mathbf{u}}.$$

The desired result follows from the fact that $\mathbf{u} \cdot \mathbf{u} > 0$. Verify this.]

2.7 REMARKS ABOUT LINEAR FIRST-ORDER SYSTEMS

Through our work with the problem of cascaded loops in the past three sections we have obtained an initial exposure to the theory of first-order systems of linear equations. By this time it should be clear that it is by no means a trivial problem to solve such systems. Our purpose in this section is to show that for the study of applied problems it is essential to have a comprehensive theory of linear equations.

Let
$$\mathbf{u}(t) = (u_1(t), \ldots, u_n(t))$$
be a vector-valued function of t and let $\mathbf{A}$ be an $n \times n$ matrix of real numbers

$$\mathbf{A} = \begin{pmatrix} a_{11} & \cdots & a_{1n} \\ \cdots & \cdots & \cdots \\ a_{n1} & \cdots & a_{nn} \end{pmatrix}, \quad a_{ij} \text{ real}, \quad i, j = 1, 2, \ldots, n.$$

Then the general first-order homogeneous linear system with constant coefficients is

$$\dot{\mathbf{u}} = \mathbf{A}\mathbf{u}. \tag{2.89}$$

We say (2.89) has constant coefficients because the elements of the matrix $\mathbf{A}$ are constants and not functions of t. The corresponding nonhomogeneous system is

$$\dot{\mathbf{u}} = \mathbf{A}\mathbf{u} + \mathbf{f}(t) \tag{2.90}$$

where
$$\mathbf{f}(t) = (f_1(t), \ldots, f_n(t))$$

is a given vector-valued function of t.

Using (2.89) we can formulate the following homogeneous initial-value problem.

PROBLEM H

Given the real vector
$$\mathbf{u}_0 = (u_{01}, \ldots, u_{0n})$$
and the matrix $\mathbf{A}$, find a solution $\mathbf{u}$ satisfying Equation (2.89) for all $t > 0$ and the initial condition

$$\mathbf{u}(0) = \mathbf{u}_0, \tag{2.91}$$

at $t = 0$.

The corresponding nonhomogeneous problem is

PROBLEM NH

Given the real vector $\mathbf{u}_0$, the matrix $\mathbf{A}$, and the vector valued function $\mathbf{f}$ defined for all $t > 0$, find a solution vector $\mathbf{u}$ satisfying (2.90) for $t > 0$ and the initial condition (2.91).

At this point it might prove instructive (and perhaps reassuring) to show that Equation (2.89) is actually linear. Suppose $\mathbf{u}$ and $\mathbf{v}$ both satisfy (2.89) so that

REMARKS ABOUT LINEAR FIRST-ORDER SYSTEMS   97

$$\dot{\mathbf{u}} = \mathbf{A}\mathbf{u} \quad \text{and} \quad \dot{\mathbf{v}} = \mathbf{A}\mathbf{v}. \tag{2.92}$$

If α and β are arbitrary real or complex numbers, define

$$\mathbf{w} = \alpha\mathbf{u} + \beta\mathbf{v}.$$

Then

$$\dot{\mathbf{w}} = \alpha\dot{\mathbf{u}} + \beta\dot{\mathbf{v}} = \alpha\mathbf{A}\mathbf{u} + \beta\mathbf{A}\mathbf{v} = \mathbf{A}\alpha\mathbf{u} + \mathbf{A}\beta\mathbf{v} = \mathbf{A}(\alpha\mathbf{u} + \beta\mathbf{v})$$

verifying that (2.89) is indeed linear.

Next we shall consider some examples.

EXAMPLE 1

The general second-order equation with constant coefficients studied in Chapter 1 has the form

$$\ddot{u} + a\dot{u} + bu = f(t) \tag{2.93}$$

where f is a given function defined for $t > 0$ and a,b are given constants. We set

$$\begin{aligned} u_1 &= u_1(t) = u(t), \\ u_2 &= u_2(t) = \dot{u}(t) = \dot{u}_1(t). \end{aligned} \tag{2.94}$$

Then the differential equation (2.93) becomes

$$\dot{u}_2 + au_2 + bu_1 = f. \tag{2.95}$$

The second of Equations (2.94) and Equation (2.95) form the system

$$\dot{\mathbf{u}} = \mathbf{A}\mathbf{u} + \mathbf{f}$$

where

$$\mathbf{u} = (u_1, u_2), \quad \mathbf{A} = \begin{pmatrix} 0 & 1 \\ -b & -a \end{pmatrix}, \quad \mathbf{f} = (0, f(t)).$$

Thus we have replaced the single second-order equation (2.93) by an equivalent system of two first-order equations. It follows then that the theory of linear first-order systems includes the theory of Equation (2.93) as a special case.

EXAMPLE 2

The linear double-pendulum problem is a special case of the system of second-order equations

$$\begin{aligned} m_{11}\ddot{\theta}_1 + m_{12}\ddot{\theta}_2 + S_{11}\theta_1 + S_{12}\theta_2 &= 0, \\ m_{21}\ddot{\theta}_1 + m_{22}\ddot{\theta}_2 + S_{21}\theta_1 + S_{22}\theta_2 &= 0. \end{aligned} \tag{2.96}$$

Again we define $\mathbf{u}$ as follows:

$$u_1 = \theta_1, \quad u_2 = \theta_2, \quad u_3 = \dot{\theta}_1 = \dot{u}_1, \quad u_4 = \dot{\theta}_2 = \dot{u}_2, \tag{2.97}$$

so that (2.96) becomes

$$m_{11}\dot{u}_3 + m_{12}\dot{u}_4 + S_{11}u_1 + S_{12}u_2 = 0,$$
$$m_{21}\dot{u}_3 + m_{22}\dot{u}_4 + S_{21}u_1 + S_{22}u_2 = 0.$$

Now define the matrices

$$\mathbf{M} = \begin{pmatrix} m_{11} & m_{12} \\ m_{21} & m_{22} \end{pmatrix}, \quad \mathbf{S} = \begin{pmatrix} s_{11} & s_{12} \\ s_{21} & s_{22} \end{pmatrix}$$

and the vectors $\mathbf{v} = (u_3, u_4)$, $\mathbf{w} = (u_1, u_2)$. Then the last system can be written

$$\mathbf{M}\dot{\mathbf{v}} + \mathbf{S}\mathbf{w} = \mathbf{0}$$

or

$$\mathbf{M}\dot{\mathbf{v}} = -\mathbf{S}\mathbf{w}.$$

Assume that $\mathbf{M}^{-1}$ exists so that we have

$$\dot{\mathbf{v}} = -\mathbf{M}^{-1}\mathbf{S}\mathbf{w}.$$

This equation together with the last two of Equations (2.97) yields a system of the form (2.89) with

$$\mathbf{A} = \begin{pmatrix} 0 & 0 & 1 & 0 \\ 0 & 0 & 0 & 1 \\ p_{11} & p_{12} & 0 & 0 \\ p_{21} & p_{22} & 0 & 0 \end{pmatrix},$$

where the 2×2 matrix

$$\mathbf{P} = \begin{pmatrix} p_{11} & p_{12} \\ p_{21} & p_{22} \end{pmatrix} = -\mathbf{M}^{-1}\mathbf{S}.$$

Here we note that the system of two second-order equations (2.96) was replaced by an equivalent system of four first-order equations. This example shows that the theory of linear first-order equations includes the theory of the linear double pendulum as a special case.

EXAMPLE 3

We next consider the circuit problem of Section 1.6 with a variable inductance. The differential equation is

$$\ddot{q} + \frac{R}{L_0(1 + \alpha t)}\dot{q} + \frac{1}{CL_0(1 + \alpha t)}q = 0, \quad t \geq 0. \tag{2.98}$$

As in Example 1 we set

$$u_1 = q, \quad u_2 = \dot{q} = \dot{u}_1 \tag{2.99}$$

so that (2.98) becomes

$$\dot{u}_2 + \frac{R}{L_0(1 + \alpha t)}u_2 + \frac{1}{CL_0(1 + \alpha t)}u_1 = 0. \tag{2.100}$$

Define the matrix-valued function $\mathbf{A}(t)$ by

$$\mathbf{A}(t) = \begin{pmatrix} 0 & 1 \\ -\dfrac{1}{CL_0(1+\alpha t)} & -\dfrac{R}{L_0(1+\alpha t)} \end{pmatrix}.$$

Equation (2.100) and the second of Equations (2.99) therefore define the system

$$\dot{\mathbf{u}} = \mathbf{A}(t)\mathbf{u}. \tag{2.101}$$

This example leads us to a linear first-order homogeneous system with variable coefficients because the matrix $\mathbf{A}$ has elements which are functions of t. The corresponding nonhomogeneous system has the form

$$\dot{\mathbf{u}} = \mathbf{A}(t)\mathbf{u} + \mathbf{f}(t). \tag{2.102}$$

Problems entirely similar to Problems H and NH can be formulated for these linear systems with variable coefficients.

We are now in a position where we can generalize and note that the theory of the general linear first-order system (2.102) will cover all of the linear problems we have met thus far as well as many additional problems. The theory of such systems depends, in an essential way, upon the development of a considerable amount of linear algebra and matrix theory. To this end, we have already seen in the last three sections and in the present section that concepts from these fields are extremely useful in dealing with linear systems.

Even though the general theory of linear first-order systems in principle at least includes most of our previous work as a special case, many specific results are more easily obtained by working directly with a problem as it is given. Equally important, the variables of the original problem are most often more readily interpreted physically than those of the transformed system. For these reasons, in practice we often do not reduce second-order equations or systems of second-order equations to first-order systems. Thus the student need not fear that we intend to lead him into the abstract for the sake of a meaningless consistency. As a matter of fact, in the next section we shall give a further illustration of a linear system of second-order equations for which no advantage will be gained by reducing it to a first-order system.

PROBLEMS

1. Using the forms (2.89) and (2.90) write out the equivalent system of first-order linear equations for:
   (a) the nonhomogeneous linear damped oscillator (2.93)
   (b) the linear double-pendulum system (2.96)
   (c) the circuit problem (2.98), and
   (d) the nonhomogeneous linear damped system of Problem 5 below.

2. Show that the problem of finding a solution of (2.89) for all $t > t_0$ satisfying $\mathbf{u}(t_0) = \mathbf{u}_0$ at $t = t_0$ can always be reduced to Problem H.
3. Show that if $\mathbf{u}$ and $\mathbf{v}$ are both solutions of (2.90), then $\mathbf{u} - \mathbf{v}$ is a solution of (2.89).
4. Compute the matrix $\mathbf{M}^{-1}$ for the system (2.96).
5. The second-order damped system

$$m_{11}\ddot{x}_1 + m_{12}\ddot{x}_2 + d_{11}\dot{x}_1 + d_{12}\dot{x}_2 + S_{11}x_1 + S_{12}x_2 = f_1(t)$$
$$m_{21}\ddot{x}_1 + m_{22}\ddot{x}_2 + d_{21}\dot{x}_1 + d_{22}\dot{x}_2 + s_{21}x_1 + S_{22}x_2 = f_2(t)$$

arises frequently in problems in mechanics. Show that if $\mathbf{M}^{-1}$ exists, where

$$\mathbf{M} = \begin{pmatrix} m_{11} & m_{12} \\ m_{21} & m_{22} \end{pmatrix},$$

this system can be put into the form (2.90).
6. Compute the matrix $\mathbf{M}^{-1}$ for the system of Problem 5.

2.8 EIGENVALUES AND EIGENVECTORS OF A CLASS OF DAMPED DYNAMIC SYSTEMS OF EQUATIONS

Up to this point the systems of second-order differential equations we have studied have not contained any first-derivative terms. The presence of such terms usually makes the behavior of the solutions more complicated. In this section we shall illustrate such behavior by investigating a special case of the whirling motion of a rotating shaft.

Next to the wheel, the rotating shaft is perhaps the simplest moving part imaginable and hence has received a great deal of attention in the engineering literature. In the past it would have been regarded as too difficult a problem to be used as an illustrative example in a textbook, but we shall show by a very simple and direct mathematical device how to reduce this fairly complex problem to a series of routine calculations.

Consider a system of coordinates (x,y) rotating with the shaft. The shaft is symmetric, massless, and carries a heavy central disk. It is acted upon by both internal (material) and external damping. The equations of motion for such a system are

$$\ddot{x} + 2(c + \epsilon)\dot{x} - 2\omega\dot{y} + (\bar{\omega}^2 - \omega^2)x - 2\epsilon\omega y = 0,$$
$$\ddot{y} + 2\omega\dot{x} + 2(c + \epsilon)\dot{y} + 2\epsilon\omega x + (\bar{\omega}^2 - \omega^2)y = 0.$$
(2.103)

In this system c is the coefficient of internal damping, ϵ is the coefficient of external (viscous) damping, ω is the impressed angular velocity of the shaft and $\bar{\omega}$ is the critical angular velocity of the shaft. It is easily shown that $\bar{\omega}$ is exactly equal to the transverse natural frequency of vibration of the same system. The constants c, ω, ϵ, and $\bar{\omega}$ are all positive.

We reformulate the system (2.103) with the help of the vector
$$\mathbf{z} = (x, y) \tag{2.104}$$
and the matrices
$$\mathbf{D} = \begin{pmatrix} c+\epsilon & -\omega \\ \omega & c+\epsilon \end{pmatrix}, \quad \mathbf{S} = \begin{pmatrix} \bar{\omega}^2 - \omega^2 & -2\epsilon\omega \\ 2\epsilon\omega & \bar{\omega}^2 - \omega^2 \end{pmatrix}. \tag{2.105}$$

The system now assumes the form
$$\ddot{\mathbf{z}} + 2\mathbf{D}\dot{\mathbf{z}} + \mathbf{S}\mathbf{z} = \mathbf{0}. \tag{2.106}$$

In theoretical mechanics the matrix $\mathbf{S}$ is called the *stiffness matrix* and $\mathbf{D}$ is called the *damping matrix* for the system.

Now we will develop a fairly elegant method for determining, in a very simple manner, the eigenvalues and eigenvectors for the system (2.106). The eigenvalues of the matrix $\mathbf{D}$ obviously satisfy
$$(c + \epsilon - \mu)^2 + \omega^2 = 0$$
or
$$\begin{aligned} c + \epsilon - \mu_1 + i\omega &= 0, \\ c - \epsilon - \mu_2 - i\omega &= 0. \end{aligned} \tag{2.107}$$

Similarly, the eigenvalues of $\mathbf{S}$ satisfy
$$(\bar{\omega}^2 - \omega^2 - \nu)^2 + 4\epsilon^2\omega^2 = 0$$
or
$$\begin{aligned} \bar{\omega}^2 - \omega^2 - \nu_1 + 2i\epsilon\omega &= 0, \\ \bar{\omega}^2 - \omega^2 - \nu_2 - 2i\epsilon\omega &= 0. \end{aligned} \tag{2.108}$$

Let $\mathbf{u} = (u_1, u_2)$ be an eigenvector of $\mathbf{D}$ corresponding to μ_1. Therefore we have
$$-i\omega u_1 - \omega u_2 = 0,$$
from which $u_2 = -i u_1$ so that we may write
$$\mathbf{u} = u(1, -i), \tag{2.109}$$
where u is a constant which is arbitrary. For convenience we set $u = 1$, and use the vector
$$\mathbf{u} = (1, -i).$$

Now a remarkable fact we wish to exploit is that $\mathbf{u}$ is also an eigenvector of $\mathbf{S}$ corresponding to ν_1, for we obtain
$$-2i\epsilon\omega + 2i\epsilon\omega = 0$$
when we substitute $\mathbf{u}$ into the equation
$$S\mathbf{u} - \nu_1\mathbf{u} = \mathbf{0}.$$

Next let $\mathbf{v} = (v_1, v_2)$ be an eigenvector of $\mathbf{D}$ corresponding to μ_2. We have
$$i\omega v_1 - \omega v_2 = 0$$
from which $v_2 = iv_1$ so that we may take
$$\mathbf{v} = (1, i). \tag{2.110}$$

$\mathbf{v}$ is also an eigenvector of $\mathbf{S}$ corresponding to ν_2.

In summary, then, the vectors $\mathbf{u}$ and $\mathbf{v}$ satisfy the following equations
$$\begin{aligned} \mathbf{Du} &= \mu_1 \mathbf{u}, & \mathbf{Su} &= \nu_1 \mathbf{u}, \\ \mathbf{Dv} &= \mu_2 \mathbf{v}, & \mathbf{Sv} &= \nu_2 \mathbf{v}. \end{aligned} \tag{2.111}$$

We now exploit these equations by seeking to solve (2.106) in the form
$$\mathbf{z} = e^{\lambda t} \mathbf{u}. \tag{2.112}$$

Since $\mathbf{u}$ is a constant vector, placing (2.112) into (2.106) and using (2.111) we obtain
$$e^{\lambda t} \{\lambda^2 + 2\mu_1 \lambda + \nu_1\} \mathbf{u} = \mathbf{0}.$$

Since $e^{\lambda t} \neq 0$ and $\mathbf{u}$ is not the zero vector this equation can only be satisfied if the bracketed term
$$\lambda^2 + 2\mu_1 \lambda + \nu_1 = 0. \tag{2.113}$$

The quadratic equation (2.113) obviously has the solutions
$$\begin{aligned} \lambda_1 &= -\mu_1 + \sqrt{\mu_1^2 - \nu_1}, \\ \lambda_2 &= -\mu_1 - \sqrt{\mu_1^2 - \nu_1}, \end{aligned} \tag{2.114}$$

which, in turn, yield two solution vectors for our system when substituted back into (2.112). In precisely the same manner we seek solutions to (2.106) of the form
$$\mathbf{z} = e^{\lambda t} \mathbf{v}. \tag{2.115}$$

This leads to the eigenvalues
$$\begin{aligned} \lambda_3 &= -\mu_2 + \sqrt{\mu_2^2 - \nu_2}, \\ \lambda_4 &= -\mu_2 - \sqrt{\mu_2^2 - \nu_2}. \end{aligned} \tag{2.116}$$

Since the numbers μ_1, μ_2, ν_1, and ν_2 are themselves complex numbers, the formulas (2.114) and (2.116) do not express λ_j, $j = 1, 2, 3, 4$, directly in any of the standard forms for a complex number. It is to our advantage to put λ in the form $\hat{\lambda} + i\lambda^*$ with $\hat{\lambda}$ and λ^* real. Therefore, we must find the square roots of the complex numbers $\mu_1^2 - \nu_1$ and $\mu_2^2 - \nu_2$.

Using the first of (2.107) and (2.108) formulate
$$\begin{aligned} \sigma = \mu_1^2 - \nu_1 &= (c + \epsilon + i\omega)^2 - \bar{\omega}^2 + \omega^2 - 2i\epsilon\omega \\ &= (c + \epsilon)^2 - \bar{\omega}^2 + 2ic\omega. \end{aligned}$$

Therefore the imaginary part of σ is positive and, consequently, in polar form we have
$$\sigma = re^{i\theta}$$
with
$$r = \{[(c+\epsilon)^2 - \bar{\omega}^2]^2 + 4c^2\omega^2\}^{1/2} \tag{2.117}$$
and
$$0 < \theta < \pi. \tag{2.118}$$

Now the square roots of σ are given by the formulas
$$\sqrt{r}\left[\cos\frac{\theta}{2} + i\sin\frac{\theta}{2}\right],$$
and
$$-\sqrt{r}\left[\cos\frac{\theta}{2} + i\sin\frac{\theta}{2}\right],$$
where $0 < \theta/2 < \pi/2$. In other words, $\theta/2$ is a first-quadrant angle and therefore $\sin\theta/2$ and $\cos\theta/2$ are positive. But $\cos\theta = [(c+\epsilon)^2 - \bar{\omega}^2]/r$ and recalling the well-known trigonometric identities
$$\cos\frac{\theta}{2} = \pm\sqrt{\frac{1+\cos\theta}{2}} \quad \text{and} \quad \sin\frac{\theta}{2} = \pm\sqrt{\frac{1-\cos\theta}{2}}$$
it follows at once that
$$\cos\frac{\theta}{2} = \frac{1}{\sqrt{2}}\frac{1}{\sqrt{r}}\sqrt{r + (c+\epsilon)^2 - \bar{\omega}^2},$$
and
$$\sin\frac{\theta}{2} = \frac{1}{\sqrt{2}}\frac{1}{\sqrt{r}}\sqrt{r - (c+\epsilon)^2 + \bar{\omega}^2},$$
where the positive square roots are required. Let us set
$$\hat{c} = \frac{1}{\sqrt{2}}\sqrt{r + (c+\epsilon)^2 - \bar{\omega}^2},$$
and
$$\hat{\omega} = \frac{1}{\sqrt{2}}\sqrt{r - (c+\epsilon)^2 + \bar{\omega}^2}. \tag{2.119}$$

Then placing (2.107) and (2.119) into (2.114) we get
$$\begin{aligned}\lambda_1 &= -(c+\epsilon-\hat{c}) - i(\omega-\hat{\omega}),\\ \lambda_2 &= -(c+\epsilon+\hat{c}) - i(\omega+\hat{\omega}).\end{aligned} \tag{2.120}$$

In the next section we will show that $\hat{c}$ and $\hat{\omega}$ are positive numbers. This means that the formulas (2.120) do yield the representation of λ_1 and λ_2 in standard complex form.

Turning next to $\mu_2{}^2 - \nu_2$, and proceeding as before we get

$$\mu_2{}^2 - \nu_2 = \bar{\sigma} = (c + \epsilon)^2 - \bar{\omega}^2 - 2ic\omega.$$

The imaginary part of $\bar{\sigma}$ is therefore negative and $\pi < \theta < 2\pi$. Consequently $\pi/2 < \theta/2 < \pi$ and in using the formulas above we must take the negative square root to compute $\cos \theta/2$ and the positive square root to compute $\sin \theta/2$. Therefore

$$\cos \frac{\theta}{2} = -\frac{1}{\sqrt{2}} \frac{1}{\sqrt{r}} \sqrt{r + (c + \epsilon)^2 - \bar{\omega}^2},$$

and

$$\sin \frac{\theta}{2} = \frac{1}{\sqrt{2}} \frac{1}{\sqrt{r}} \sqrt{r - (c + \epsilon)^2 + \bar{\omega}^2}.$$

Accordingly, again following the foregoing procedure, we have

$$\begin{aligned} \lambda_3 &= -(c + \epsilon + \hat{c}) + i(\omega + \hat{\omega}), \\ \lambda_4 &= -(c + \epsilon - \hat{c}) + i(\omega - \hat{\omega}). \end{aligned} \tag{2.121}$$

The rest is now relatively straightforward. Returning to (2.112) and (2.115) we obtain the four solution vectors

$$\begin{aligned} \mathbf{z}_1(t) &= e^{-(c+\epsilon-\hat{c})t}(\cos(\omega - \hat{\omega})t - i\sin(\omega - \hat{\omega})t)\mathbf{u}, \\ \mathbf{z}_2(t) &= e^{-(c+\epsilon+\hat{c})t}(\cos(\omega + \hat{\omega})t - i\sin(\omega + \hat{\omega})t)\mathbf{u}, \\ \mathbf{z}_3(t) &= e^{-(c+\epsilon+\hat{c})t}(\cos(\omega + \hat{\omega})t + i\sin(\omega + \hat{\omega})t)\mathbf{v}, \\ \mathbf{z}_4(t) &= e^{-(c+\epsilon-\hat{c})t}(\cos(\omega - \hat{\omega})t + i\sin(\omega - \hat{\omega})t)\mathbf{v}. \end{aligned}$$

At this point we have complex solution vectors. Writing out $\mathbf{z}_1$, for example, yields

$$e^{-(c+\epsilon-\hat{c})t}(\cos(\omega - \hat{\omega})t, -\sin(\omega - \hat{\omega})t)$$
$$-ie^{-(c+\epsilon-\hat{c})t}(\sin(\omega - \hat{\omega})t, \cos(\omega - \hat{\omega})t).$$

Similarly $\mathbf{z}_4$ turns out to be

$$e^{-(c+\epsilon-\hat{c})t}(\cos(\omega - \hat{\omega})t, -\sin(\omega - \hat{\omega})t)$$
$$+ie^{-(c+\epsilon-\hat{c})t}(\sin(\omega - \hat{\omega})t, \cos(\omega - \hat{\omega})t).$$

Since a linear combination of solution vectors is again a solution vector, $\frac{1}{2}(\mathbf{z}_1 + \mathbf{z}_4)$ and $(1/2i)(\mathbf{z}_4 - \mathbf{z}_1)$ are also solutions. In other words, the real and imaginary parts of $\mathbf{z}_4$ are real solutions. In the same way one finds that $\mathbf{z}_2$ and $\mathbf{z}_3$ are conjugate complex vectors so that the real and imaginary parts of $\mathbf{z}_3$ are solution vectors. We thus arrive at last at the four real solution vectors

$$\begin{aligned} \zeta_1(t) &= e^{-(c+\epsilon-\hat{c})t}(\cos(\omega - \hat{\omega})t, -\sin(\omega - \hat{\omega})t), \\ \zeta_2(t) &= e^{-(c+\epsilon-\hat{c})t}(\sin(\omega - \hat{\omega})t, \cos(\omega - \hat{\omega})t), \\ \zeta_3(t) &= e^{-(c+\epsilon+\hat{c})t}(\cos(\omega + \hat{\omega})t, -\sin(\omega + \hat{\omega})t), \\ \zeta_4(t) &= e^{-(c+\epsilon+\hat{c})t}(\cos(\omega + \hat{\omega})t, \cos(\omega + \hat{\omega})t). \end{aligned} \tag{2.122}$$

We shall conclude this section by showing that the initial value problem for the system (2.106) can be solved with the help of these vectors. Let C_j, $j = 1, 2, 3, 4$, be arbitrary constants and form the linear combination

$$\mathbf{z}(t) = C_1 \zeta_1(t) + C_2 \zeta_2(t) + C_3 \zeta_3(t) + C_4 \zeta_4(t). \tag{2.123}$$

Prescribe the initial conditions to be

$$\begin{aligned} \mathbf{z}(0) &= (x_0, y_0), \\ \dot{\mathbf{z}}(0) &= (x_1, y_1). \end{aligned} \tag{2.124}$$

From the first of (2.124) we obtain

$$\begin{aligned}(x_0, y_0) &= C_1 \zeta_1(0) + C_2 \zeta_2(0) + C_3 \zeta_3(0) + C_4 \zeta_4(0) \\ &= C_1(1,0) + C_2(0,1) + C_3(1,0) + C_4(0,1)\end{aligned}$$

which yields the two equations

$$x_0 = C_1 + C_3,$$

and

$$y_0 = C_2 + C_4.$$

In computing the derivatives it is useful to observe that

$$\begin{aligned}\dot{\zeta}_1(t) &= -(c + \epsilon - \hat{c})\zeta_1(t) - (\omega - \hat{\omega})\zeta_2(t), \\ \dot{\zeta}_2(t) &= -(c + \epsilon - \hat{c})\zeta_2(t) + (\omega - \hat{\omega})\zeta_1(t), \\ \dot{\zeta}_3(t) &= -(c + \epsilon + \hat{c})\zeta_3(t) - (\omega + \hat{\omega})\zeta_4(t), \\ \dot{\zeta}_4(t) &= -(c + \epsilon + \hat{c})\zeta_4(t) + (\omega + \hat{\omega})\zeta_3(t).\end{aligned} \tag{2.125}$$

Then from the second of Equations (2.124) we get

$$\begin{aligned}(x_1, y_1) = &[-(c + \epsilon - \hat{c})(1,0) - (\omega - \hat{\omega})(0,1)]C_1 \\ &+ [-(c + \epsilon - \hat{c})(0,1) + (\omega - \hat{\omega})(1,0)]C_2 \\ &+ [-(c + \epsilon + \hat{c})(1,0) - (\omega + \hat{\omega})(0,1)]C_3 \\ &+ [-(c + \epsilon + \hat{c})(0,1) + (\omega + \hat{\omega})(1,0)]C_4.\end{aligned}$$

Consequently we obtain the equations

$$x_1 = -(c + \epsilon - \hat{c})C_1 + (\omega - \hat{\omega})C_2 - (c + \epsilon + \hat{c})C_3 + (\omega + \hat{\omega})C_4,$$

and

$$y_1 = -(\omega - \hat{\omega})C_1 - (c + \epsilon - \hat{c})C_2 - (\omega + \hat{\omega})C_3 - (c + \epsilon + \hat{c})C_4.$$

Thus, we have four linear equations to solve for the constants C_j, $j = 1, 2, 3, 4$. This system will always have a unique solution for nonzero initial data. That is, either $\mathbf{z}(0) \neq 0$ or $\dot{\mathbf{z}}(0) \neq 0$, or both, if and only if

$$D = \begin{vmatrix} 1 & 0 & 1 & 0 \\ 0 & 1 & 0 & 1 \\ -(c + \epsilon - \hat{c}) & \omega - \hat{\omega} & -(c + \epsilon + \hat{c}) & \omega + \hat{\omega} \\ -(\omega - \hat{\omega}) & -(c + \epsilon - \hat{c}) & -(\omega + \hat{\omega}) & -(c + \epsilon + \hat{c}) \end{vmatrix} \neq 0.$$

A simple calculation shows that

$$D = 4(\hat{c}^2 + \hat{\omega}^2) = 4r \neq 0. \tag{2.126}$$

2.9 WHIRLING MOTION OF A ROTATING SHAFT

Now that we have derived an explicit solution for the problem of the whirling motion of a rotating shaft we wish to examine some of its properties. For this purpose we must learn more about the numbers $\hat{c}$ and $\hat{\omega}$, particularly $\hat{c}$.

From Section 2.8 we have

$$\hat{c} = \frac{1}{\sqrt{2}}\sqrt{r + (c + \epsilon)^2 - \bar{\omega}^2} \quad \text{and} \quad \hat{\omega} = \frac{1}{\sqrt{2}}\sqrt{r - (c + \epsilon)^2 + \bar{\omega}^2}, \quad (2.127)$$

where

$$r = [((c + \epsilon)^2 - \bar{\omega}^2)^2 + 4c^2\omega^2]^{1/2}. \quad (2.128)$$

From (2.128) it is immediately obvious that

$$r > |(c + \epsilon)^2 - \bar{\omega}^2|$$

for all $\omega > 0$, which means that the numbers $\hat{c}$ and $\hat{\omega}$ are well-defined as positive real numbers if $\omega > 0$. Expanding Equation (2.128) yields

$$r = [(c + \epsilon)^4 + \bar{\omega}^4 - 2(c + \epsilon)^2\bar{\omega}^2 + 4c^2\omega^2]^{1/2}$$
$$= [((c + \epsilon)^2 + \bar{\omega}^2)^2 - 4(c + \epsilon)^2\bar{\omega}^2 + 4c^2\omega^2]^{1/2}.$$

It follows immediately that:

(a) if $\omega = \left(1 + \frac{\epsilon}{c}\right)\bar{\omega}$, then $r = (c + \epsilon)^2 + \bar{\omega}^2$;

(b) if $\omega > \left(1 + \frac{\epsilon}{c}\right)\bar{\omega}$, then $r > (c + \epsilon)^2 + \bar{\omega}^2$;

(c) if $\omega < \left(1 + \frac{\epsilon}{c}\right)\bar{\omega}$, then $r < (c + \epsilon)^2 + \bar{\omega}^2$.

With this result at hand, consider the solution $\zeta_1(t)$ given in (2.122). The exponential function in this solution is $\exp[-(c + \epsilon - \hat{c})t]$. We see at once that if condition (a) holds, then $c + \epsilon - \hat{c} = 0$, if (b) holds then $c + \epsilon - \hat{c} < 0$, and if (c) holds then $c + \epsilon - \hat{c} > 0$. This means that for the critical angular velocity,

$$\omega_{cr} = \left(1 + \frac{\epsilon}{c}\right)\bar{\omega}, \quad (2.129)$$

the exponential term in the solution $\zeta_1(t)$ is identically equal to 1 and ζ_1 is therefore a periodic solution of the system. If $\omega < \omega_{cr}$ the exponential in ζ_1 is negative and this solution damps out exponentially. Finally, if $\omega > \omega_{cr}$ the exponential is positive and ζ_1 becomes unbounded as $t \to +\infty$.

Obviously the solution $\zeta_2(t)$ has exactly the same properties as $\zeta_1(t)$.

The solutions $\zeta_3(t)$ and $\zeta_4(t)$ are quite different. They represent transients no matter what the impressed angular velocity is.

We can see directly from (2.129) that as ϵ increases the critical angular velocity, after which instability sets in, becomes larger. This is in accord with our intuition about the nature of linear viscous damping. The type of external damping which our model permits can be realized by rotating the shaft in a fluid which is at rest. (If the fluid is permitted to move, or more specifically, if fluid motion is significant, the model would have to be altered to reflect this condition.) The precise role played by the internal damping is more difficult to study. Of course, it follows at once from (2.129) that the critical velocity varies inversely with c. Indeed, if one simply sets $c = 0$ in (2.129) we obtain $\omega_{cr} = +\infty$. That is, in the limit as $c \to 0$ $\omega_{cr} \to +\infty$. This result makes sense because when $c = 0$ (2.128) yields

$$r = |\epsilon^2 - \bar{\omega}^2|,$$

and (2.127) becomes

$$\hat{c} = \frac{1}{\sqrt{2}} \sqrt{|\epsilon^2 - \bar{\omega}^2| + \epsilon^2 - \bar{\omega}^2} \quad \text{and} \quad \hat{\omega} = \frac{1}{\sqrt{2}} \sqrt{|\epsilon^2 - \bar{\omega}^2| - \epsilon^2 + \bar{\omega}^2}.$$

It follows that $\hat{c} < \epsilon$ and hence

$$\epsilon - \hat{c} > 0.$$

Consequently if $c = 0$ every solution of the whirling shaft problem damps out exponentially as $t \to +\infty$. We shall leave further results about the system (2.103) for the problems.

Let us now turn to a simple nonlinear analog of the whirling shaft. Suppressing the effect of external damping by setting $\epsilon = 0$ the nonlinear model will take the form

$$\begin{aligned} \ddot{x} + 2c\dot{x} - 2\omega\dot{y} + (\bar{\omega}^2 - \omega^2)x + \gamma(x^2 + y^2)x &= 0, \\ \ddot{y} + 2\omega\dot{x} + 2c\dot{y} + (\bar{\omega}^2 - \omega^2)y + \gamma(x^2 + y^2)y &= 0. \end{aligned} \quad (2.130)$$

Here the terms $\gamma(x^2 + y^2)x$ and $\gamma(x^2 + y^2)y$ take into account certain nonlinear elastic effects in the shaft. The constant γ is taken to be positive.

As one would expect, the detailed results obtained for the linear case do not apply to this system. Therefore we resort to the energy method which was used with considerable success in Chapter 1. It should be pointed out that it would be exceedingly difficult to apply any other method to the system (2.130). For this purpose multiply the first of Equations (2.130) by $\dot{x}$ and the second by $\dot{y}$ and note that

$$\dot{x}\ddot{x} = \frac{1}{2}\frac{d}{dt}(\dot{x}^2), \quad \dot{y}\ddot{y} = \frac{1}{2}\frac{d}{dt}(\dot{y}^2),$$

$$x\dot{x} = \frac{1}{2}\frac{d}{dt}(x^2), \quad y\dot{y} = \frac{1}{2}\frac{d}{dt}(y^2).$$

$$\dot{x}x^3 = \frac{1}{4}\frac{d}{dt}(x^4), \quad \dot{x}xy^2 = \frac{y^2}{2}\frac{d}{dt}(x^2),$$

$$\dot{y}y^3 = \frac{1}{4}\frac{d}{dt}(y^4), \quad \dot{y}yx^2 = \frac{x^2}{2}\frac{d}{dt}(y^2).$$

Placing the foregoing identities into (2.130) we obtain the equations

$$\frac{1}{2}\frac{d}{dt}\left[\dot{x}^2 + (\bar{\omega}^2 - \omega^2)x^2 + \frac{\gamma}{2}x^4\right] + 2c\dot{x}^2 - 2\omega\dot{x}y + \frac{\gamma}{2}y^2\frac{d}{dt}(x^2) = 0,$$

and

$$\frac{1}{2}\frac{d}{dt}\left[\dot{y}^2 + (\bar{\omega}^2 - \omega^2)y^2 + \frac{\gamma}{2}y^4\right] + 2\omega\dot{x}y + 2c\dot{y}^2 + \frac{\gamma}{2}x^2\frac{d}{dt}(y^2) = 0.$$

Adding these equations and noting that

$$y^2\frac{d}{dt}(x^2) + x^2\frac{d}{dt}(y^2) = \frac{d}{dt}(x^2 y^2)$$

we have

$$\frac{1}{2}\frac{d}{dt}\left[\dot{x}^2 + \dot{y}^2 + (\bar{\omega}^2 - \omega^2)(x^2 + y^2) + \frac{\gamma}{2}(x^2 + y^2)^2\right] \\ + 2c(\dot{x}^2 + \dot{y}^2) = 0. \quad (2.131)$$

Now let us set

$$\mathcal{E}(x,y;t) = \dot{x}^2(t) + \dot{y}^2(t) + (\bar{\omega}^2 - \omega^2)(x^2(t) + y^2(t))$$
$$+ \frac{\gamma}{2}(x^2(t) + y^2(t))^2, \quad (2.132)$$

and integrate (2.131) from time 0 to some time $t > 0$. This yields

$$\mathcal{E}(x,y;t) + 2c\int_0^t (\dot{x}^2(\sigma) + \dot{y}^2(\sigma))\, d\sigma = \mathcal{E}(x,y;0). \quad (2.133)$$

The function $\mathcal{E}(x,y;t)$ is proportional to the total energy in the system at time t and (2.133) is the basic energy identity which states that the energy of the system at time t equals the energy of the system at time $t = 0$ less the energy dissipated by the forces of internal friction, if indeed it does dissipate energy. Thus if $c = 0$, the system is conservative. Here our task is to determine what conditions, if any, exist for which the internal friction force dissipates energy.

Using (2.133) we are going to show first that every solution of (2.133) is bounded and has a bounded derivative no matter what the angular velocity ω is. This is in contrast to the linear model of Section 2.8, which we have seen has exponential growth for sufficiently large ω.

A glance at (2.132) shows that if $\bar{\omega} \geq \omega$ each term in the function $\mathcal{E}(x,y;t)$ is nonnegative, but that if $\bar{\omega} < \omega$ the second term will always be nonpositive. In particular, $\mathcal{E}(x,y;0)$ involves only the initial data as given, say, by Equation (2.124). Accordingly we have

$$\mathcal{E}(x,y;0) = x_1{}^2 + y_1{}^2 + (\bar{\omega}^2 - \omega^2)(x_0{}^2 + y_0{}^2) + \frac{\gamma}{2}(x_0{}^2 + y_0{}^2)^2. \quad (2.134)$$

Thus, for example, if we are given the data $x_1 = y_1 = 0$,

$$x_0{}^2 + y_0{}^2 = \frac{\omega^2 - \bar{\omega}^2}{\gamma}$$

when $\omega > \bar{\omega}$, we will have

$$\mathcal{E}(x,y;0) = \frac{\omega^2 - \bar{\omega}^2}{\gamma} \frac{\bar{\omega}^2 - \omega^2}{2} < 0.$$

In other words, if $\bar{\omega} < \omega$ there are a variety of choices of the initial data for which the initial energy in the system is negative.

These remarks make it clear that we must consider separately the cases where $\bar{\omega} \geq \omega$ and $\bar{\omega} < \omega$. We start with the case $\bar{\omega} \geq \omega$.

Case 1. ($\bar{\omega} \geq \omega$)

In this case the constant $\mathcal{E}(x,y;0)$ is positive unless all of the initial data is zero. Moreover it is clear that

$$\int_0^t (\dot{x}^2(\sigma) + \dot{y}^2(\sigma))\, d\sigma \geq 0,$$

and, from (2.132), we have the inequalities

$$\mathcal{E}(x,y;t) \geq \dot{x}^2(t) + \dot{y}^2(t).$$

and

$$\mathcal{E}(x,y;t) \geq \frac{\gamma}{2}(x^2(t) + y^2(t))^2.$$

Using these results in the basic identity (2.133), we obtain

$$\dot{x}^2(t) + \dot{y}^2(t) \leq \mathcal{E}(x,y;0), \quad (2.135)$$

and

$$x^2(t) + y^2(t) \leq \sqrt{\frac{2}{\gamma}\mathcal{E}(x,y;0)}. \quad (2.136)$$

These inequalities, (2.135) and (2.136), imply that the solution and its derivative with respect to time are both bounded for all $t > 0$.

Case 2. ($\bar{\omega} < \omega$)

In order to treat this case let us suppose at first that $x^2(t) + y^2(t) \neq 0$. Then we can rewrite (2.132) in the form

$$\mathcal{E}(x,y;t) = \dot{x}^2(t) + \dot{y}^2(t) + [x^2(t) + y^2(t)]^2 \left(\frac{\bar{\omega}^2 - \omega^2}{x^2(t) + y^2(t)} + \frac{\gamma}{2} \right).$$

Thus if
$$x^2(t) + y^2(t) > \frac{4}{\gamma}(\omega^2 - \bar{\omega}^2),$$
we have
$$\frac{1}{x^2 + y^2} < \frac{\gamma}{4(\omega^2 - \gamma^2)}$$
and
$$\frac{\bar{\omega}^2 - \omega^2}{x^2 + y^2} > -\frac{\gamma}{4}.$$
Accordingly, either
$$x^2(t) + y^2(t) \leq \frac{4}{\gamma}(\omega^2 - \bar{\omega}^2), \tag{2.137}$$
which furnishes a positive upper bound for the solution or
$$x^2 + y^2 > \frac{4}{\gamma}(\omega^2 - \bar{\omega}^2)$$
and
$$\mathcal{E}(x;y;t) > \frac{\gamma}{4}[x^2(t) + y^2(t)]^2. \tag{2.138}$$

But if (2.138) holds, from (2.133) we must have the inequality
$$\frac{\gamma}{4}[x^2(t) + y^2(t)]^2 \leq \mathcal{E}(x,y;0),$$
which is only valid if $\mathcal{E}(x,y;0) \geq 0$. Thus we obtain
$$x^2(t) + y^2(t) \leq \sqrt{\frac{4}{\gamma}\mathcal{E}(x,y;0)}. \tag{2.139}$$

The two estimates (2.137) and (2.139) show that when $\bar{\omega} < \omega$ the solution is again bounded.

It remains for us to prove that every solution of (2.130) also has a bounded derivative when $\omega > \bar{\omega}$. To do this we note that
$$\mathcal{E}(x,y;t) \geq \dot{x}^2(t) + \dot{y}^2(t) + (\bar{\omega}^2 - \omega^2)(x^2(t) + y^2(t))$$
and conclude from (2.133) that
$$\dot{x}^2(t) + \dot{y}^2(t) \leq \mathcal{E}(x,y;0) + (\omega^2 - \bar{\omega}^2)(x^2(t) + y^2(t)).$$
Hence
$$\dot{x}^2(t) + \dot{y}^2(t) \leq |\mathcal{E}(x,y;0)| + (\omega^2 - \bar{\omega}^2)(x^2(t) + y^2(t)). \tag{2.140}$$

The boundedness of the derivative therefore follows at once from the boundedness of the solution itself.

Let us summarize these results formally.

Theorem 2.1

Every solution of the nonlinear model (2.130) of the symmetric whirling shaft remains bounded and has a bounded derivative for all $t > 0$ for each positive value of ω.

It is possible to deduce many additional interesting facts about the behavior of solutions of the system (2.130) by a further study of the energy identity (2.133). We shall, however, leave those within reach of the student to the problems below.

We have presented a relatively simple illustration of the important fact that solutions of nonlinear equations can exhibit very different behavior from the solutions of linear equations. The next question is, which is the more reliable mathematical model? Such questions are best answered in the laboratory. In this case, as experiments prove, the nonlinear model clearly is the better mathematical model.

PROBLEMS

1. In the linear whirling-shaft problem prove that if there is no external damping, $\epsilon = 0$, then

$$\lim_{c \to \infty} c - \hat{c} = 0.$$

   $\left[Hint: \text{Write } c - \hat{c} = \dfrac{c - \hat{c}}{c + \hat{c}}(c + \hat{c}). \right]$

2. Solve Problem 1 when $\epsilon > 0$ by evaluating

$$\lim_{c \to \infty} c + \epsilon - \hat{c}.$$

3. Compute the constants c_1, c_2, c_3, and c_4 in Equation (2.123). Are there any solutions of the linear whirling-shaft problem which remain bounded as $t \to \infty$ when $\omega > (1 + \epsilon/c)\bar{\omega}$?

4. Show that the matrices $\mathbf{D}$ and $\mathbf{S}$ given in (2.105) commute. That is, show that they satisfy

$$\mathbf{DS} - \mathbf{SD} = \mathbf{0}.$$

   Give examples of 2×2 matrices for which

$$\mathbf{AB} - \mathbf{BA} \neq \mathbf{0}.$$

5. Using the formulas (2.122) and (2.125) verify directly by substitution that the four vectors $\zeta_j(t)$ are solutions of the system (2.106).

6. In the linear whirling-shaft problem introduce the vectors

112   ORDINARY DIFFERENTIAL EQUATIONS—SIMPLE SYSTEMS

$$\xi_1(t) = (\cos(\omega - \hat{\omega})t, -\sin(\omega - \hat{\omega})t),$$
$$\mathbf{n}_1(t) = (\sin(\omega - \hat{\omega})t, \cos(\omega - \hat{\omega})t),$$
$$\xi_2(t) = (\cos(\omega + \hat{\omega})t, -\sin(\omega + \hat{\omega})t),$$
$$\mathbf{n}_2(t) = (\sin(\omega + \hat{\omega})t, \cos(\omega + \hat{\omega})t),$$
$$\hat{\zeta} = (\hat{c}, -\hat{\omega}).$$

Prove that the function

$$\mathbf{z}(t) = \alpha_1(t)\zeta_1(t) + \alpha_2(t)\zeta_2(t) + \alpha_3(t)\zeta_3(t) + \alpha_4(t)\zeta_4(t), \qquad (2.141)$$

solves the nonhomogeneous system obtained from (2.106) when the vector $\mathbf{F} = (F_1, F_2)$ replaces $\mathbf{0}$, where

$$\alpha_1(t) = \frac{1}{4}\int_0^t e^{(c+\epsilon-\hat{c})s}\{(\hat{\zeta}\cdot\xi_1(s))F_1(s) - (\hat{\zeta}\cdot\mathbf{n}_1(s))F_2(s)\}\,ds,$$

$$\alpha_2(t) = \frac{1}{4}\int_0^t e^{(c+\epsilon+\hat{c})s}\{(\hat{\zeta}\cdot\mathbf{n}_1(s))F_1(s) + (\hat{\zeta}\cdot\xi_1(s))F_2(s)\}\,ds,$$

$$\alpha_3(t) = \frac{1}{4}\int_0^t e^{(c+\epsilon+\hat{c})s}\{-(\hat{\zeta}\cdot\xi_2(s))F_1(s) + (\hat{\zeta}\cdot\mathbf{n}_2(s))F_2(s)\}\,ds,$$

$$\alpha_4(t) = \frac{1}{4}\int_0^t e^{(c+\epsilon+\hat{c})s}\{-(\hat{\zeta}\cdot\mathbf{n}_2(s))F_1(s) - (\hat{\zeta}\cdot\xi_2(s))F_2(s)\}\,ds.$$

What initial conditions does the vector $\mathbf{z}(t)$ satisfy?

7. An important nonhomogeneous case of the whirling-shaft problem occurs when the shaft is not perfectly balanced, such as when the center of mass of any cross section does not coincide with the geometric center (say for a circular shaft). In this case the vector $\mathbf{F}$ has the form

$$\mathbf{F} = (e\omega^2, e\omega^2),$$

$e \neq 0$. Another important case occurs when the shaft rotates in a horizontal plane. In this case we get the gravitational effects

$$\mathbf{F} = (+g\cos\omega t, -g\sin\omega t).$$

Write out the corresponding solutions to these nonhomogeneous problems.

8. Equations (2.130) for the nonlinear shaft can be looked upon as nonhomogeneous linear equations by writing them in the form

$$\ddot{x} + 2c\dot{x} - 2\omega\dot{y} + (\bar{\omega}^2 - \omega^2)x = -\gamma(x^2 + y^2)x,$$
$$\ddot{y} + 2\omega\dot{x} + 2c\dot{y} + (\bar{\omega}^2 - \omega^2)y = -\gamma(x^2 + y^2)y.$$

Thus the vector $\mathbf{F}$ has the form

$$\mathbf{F} = (-\gamma(x^2+y^2)x, -\gamma(x^2+y^2)y) = -\gamma(x^2+y^2)(x,y).$$

Write out solution (2.141) for this vector (remember now that $\epsilon = 0$). What sort of equations do you obtain for the functions $x(t)$ and $y(t)$?

9. Let A be a 2×2 matrix with eigenvalues μ_1 and μ_2 and B a 2×2 matrix with eigenvalues ν_1 and ν_2. Suppose u satisfies $Au = \mu_1 u$, $Bu = \nu_1 u$ and v satisfies $Av = \mu_2 v$, $Bv = \nu_2 v$. Prove that

$$(AB - BA)u = 0 = (AB - BA)v.$$

This shows that A and B commute relative to the vectors u and v. Assume further that every vector w can be written as a linear combination of u and v, namely

$$w = \alpha u + \beta v, \quad \alpha, \beta \text{ scalars}.$$

Prove that $(AB - BA)w = 0$ and deduce from this that the matrix equation $AB - BA = 0$ is therefore valid.

10. Let A and B be 2×2 matrices with distinct eigenvalues. A has eigenvalues $\mu_1 \neq \mu_2$ and B has eigenvalues $\nu_1 \neq \nu_2$. Prove that if $AB - BA = 0$, then there exist vectors u and v satisfying

$$Au = \mu_1 u, \quad Bu = \nu_1 u,$$
$$Av = \mu_2 v, \quad Bv = \nu_2 V,$$

or

$$Au = \mu_1 u, \quad Bu = \nu_2 u,$$
$$Av = \mu_2 v, \quad Bv = \nu_1 v.$$

What happens if we insist $\mu_1 \neq \mu_2$ but permit $\nu_1 = \nu_2$? (Note that Problems 9 and 10 show that it is the commutativity of S and D which enabled us to solve the linear whirling-shaft problem so easily.)

11. Suppose the linear whirling-shaft problem (2.106) is changed by setting

$$S = \begin{pmatrix} \omega_1^2 - \omega^2 & -2\epsilon\omega \\ 2\epsilon\omega & \omega_2^2 - \omega^2 \end{pmatrix}, \quad (2.142)$$

while D remains unchanged. Show that the method we employed to solve the problem is no longer applicable. Why? What essential difficulty is encountered if we seek a solution $z(t) = e^t u$, where u is a fixed constant vector?

12. The matrix S of Problem 11 is that obtained if we suppose the shaft to be nonsymmetric, say with an elliptic cross section. For such a shaft we may also expect the internal damping to be nonsymmetric so that with S given by (2.142) we should replace D by

$$D = \begin{pmatrix} c_1 + \epsilon & -\omega \\ \omega & c_2 + \epsilon \end{pmatrix}. \quad (2.143)$$

Show that S and D commute if and only if

$$\omega_1^2 - \omega_2^2 = 2\epsilon(c_1 - c_2) \quad (2.144)$$

and that when this condition is satisfied this more complex problem can be again solved by the method of Section 2.8.

13. Let the nonlinear shaft equations be replaced by

$$\ddot{x} + 2c\dot{x} - 2\omega\dot{y} + (\bar{\omega}^2 - \omega^2)x + \gamma(x^2 + y^2)^P x = 0 \quad (\gamma > 0)$$
$$\ddot{y} + 2\omega\dot{x} + 2c\dot{y} + (\bar{\omega}^2 - \omega^2)y + \gamma(x^2 + y^2)^P y = 0.$$
(2.145)

Apply the energy method of Section 2.9 to the study of this system.

14. Prove that for the nonlinear shaft problem (2.130) a solution corresponding to $\bar{\omega} < \omega$ with negative initial energy $\mathcal{E}(x,y,0) < 0$, satisfies

$$x^2(t) + y^2(t) \geq \frac{-\mathcal{E}(x,y,0)}{\omega^2 - \bar{\omega}^2}.$$

Are such solutions bounded away from zero? Is the same result true for (2.145)? Compute

$$\lim_{\omega \to +\infty} \frac{-\mathcal{E}(x,y,0)}{\omega^2 - \bar{\omega}^2}.$$

What peculiar role is played by solutions satisfying the initial condition $\mathbf{z}(0) = \mathbf{0}$?

Chapter 3

PARTIAL DIFFERENTIAL EQUATIONS

In this chapter we begin the study of a broad class of problems whose formulations lead to partial differential equations. Such equations, and more important their analysis, are substantially different in character from those we studied in the previous two chapters. Nevertheless, we will be able to draw upon the knowledge gained with ordinary differential equations to establish starting points in the analysis of partial differential equations. In Chapter 5 we will relate the two concepts by developing techniques with which we can reduce a partial differential equation to a related system of ordinary differential equations. The motivation for the study of partial differential equations in this chapter is provided through problems concerning heat conduction and diffusion.

3.1 HEAT CONDUCTION IN A ROD

Consider a rod (or bar) of constant cross section and of length L, thermally insulated along its surface and sufficiently thin so that at any instant of time it is reasonable to assume that the temperature is a constant at all points through its cross section. The bar is made of a material whose elastic properties are the same at all points in the rod (homogeneous). From the foregoing assumptions we see that our problem can be represented by one Cartesian coordinate, say x. Let x be zero at one end of the rod and increase to L at the other end.

It has been experimentally established that the temperature varies linearly along the rod when the ends of the rod are maintained at the constant temperatures u_1 at $x = 0$ and u_2 at $x = L$. Thus

$$u(x) = u_1 + \frac{u_2 - u_1}{L} x. \tag{3.1}$$

Another quantity which can be measured experimentally is the quantity of heat Q flowing through a cross-sectional area A in unit time. It is

$$Q = -k \frac{u_2 - u_1}{L} A = -k \frac{\partial u}{\partial x} A, \tag{3.2}$$

where the physical constant k is associated with the material of the rod and has the name *coefficient of thermal conductivity*. In addition to the assumptions given above (3.2) has the further restriction of being valid only for comparatively small local variations in temperature.

The concept of heat propagation in the rod is described by the function u, where $u(x,t)$ represents the temperature at a cross section located x units from the origin at a time t. The value of the heat flow is considered positive when the heat flows in the direction of increasing x. It now follows that our first objective is to establish the equation which the function $u(x,t)$ must satisfy.

A natural generalization of (3.2) to obtain the amount of heat flow through a cross section of the bar at a coordinate x in a time interval $t_1 \leq t \leq t_2$ is

$$Q = -A \int_{t_1}^{t_2} k \frac{\partial u}{\partial x} (x,\tau) \, d\tau. \tag{3.3}$$

Now consider a finite element of a rod $x_1 \leq x \leq x_2$. In order to increase the temperature at a cross section x from the value $u(x,t_1)$ to the value $u(x,t_2)$ the heat required is

$$Q = A \int_{x_1}^{x_2} c\rho [u(\xi,t_2) - u(\xi,t_1)] \, d\xi, \tag{3.4}$$

where the physical constants c and ρ are the specific heat and density of the rod respectively. It should be noted for clarity that in (3.4) the positions x_i need not in any way be associated with the times t_i.

Another source of heat which we wish to take into account is the creation (or absorption) of heat at some cross section of the rod such as could occur by chemical reaction or the passage of current. We characterize such generation of heat by a function f whose values $f(x,t)$ represent the density of the heat source at the point x at a time t. Accordingly the quantity of heat generated in the length of the bar $x_1 \leq x \leq x_2$ in the time interval $t_1 \leq t \leq t_2$ is

$$Q = A \int_{x_1}^{x_2} \int_{t_1}^{t_2} f(\xi,\tau) \, d\xi \, d\tau. \tag{3.5}$$

We are now in a position to calculate the balance of heat in an arbitrary length (x_1,x_2) of the rod during an arbitrary time interval (t_1,t_2). Noting that the heat flowing through the ends of the segment plus the heat generated within the segment must equal the amount of heat necessary to change the temperature from $u(x,t_1)$ to $u(x,t_2)$, we write

$$\int_{t_1}^{t_2} \left[k \frac{\partial u}{\partial x}(x,\tau)\bigg|_{x=x_2} - k \frac{\partial u}{\partial x}(x,\tau)\bigg|_{x=x_1} \right] d\tau + \int_{x_1}^{x_2} \int_{t_1}^{t_2} f(\xi,\tau) \, d\xi \, d\tau$$
$$= \int_{x_1}^{x_2} c\rho[u(\xi,t_2) - u(\xi,t_1)] \, d\xi. \quad (3.6)$$

The integral equation (3.6) is the equation of heat conduction and is valid in every interval (x_1,x_2) where $0 \leq x_1$, $x_2 \leq L$ and in every time interval (t_1,t_2). It turns out that (3.6) is not the most suitable form for working with the heat equation, and accordingly we shall transform it into a differential equation. We do this by making the assumption that the temperature function u has the nice mathematical property of continuous derivatives, $u_{xx} \equiv \partial^2 u/\partial x^2$ and $u_t \equiv \partial u/\partial t$, at all points of the rod and for all time $t \geq 0$.

With the foregoing assumption made we can now apply the mean-value theorem of the differential calculus which is:

Theorem 3.1

If g is a function of the two variables x and t and is defined and differentiable in a closed rectangle $0 \leq x \leq L$, $0 \leq t \leq T$, and if (x_1,t_1) and (x_2,t_2) are any two points in the rectangle then

$$g(x_2,t_2) - g(x_1,t_1) = \frac{\partial g}{\partial x}(\tilde{x},\tilde{t})(x_2 - x_1) + \frac{\partial g}{\partial t}(\tilde{x},\tilde{t})(t_2 - t_1) \quad (3.7)$$

where the point $(\tilde{x},\tilde{t})$ lies in the interior of the straight-line segment joining the points (x_1,t_1) and (x_2,t_2).

Let us first apply the theorem to the difference

$$\frac{\partial u}{\partial x}(x_2,\tau) - \frac{\partial u}{\partial x}(x_1,\tau) = \frac{\partial^2 u}{\partial x^2}(x_3,\tau)(x_2 - x_1).$$

Here we have held τ fixed and applied the mean-value theorem to the function $\partial u/\partial x$ considered as a function of x and with $x_1 < x_3 < x_2$. Next we consider u to be a function of t with ξ held fixed and applying the mean value theorem we obtain

$$u(\xi,t_2) - u(\xi,t_1) = \frac{\partial u}{\partial t}(\xi,t_3)(t_2 - t_1).$$

where $t_1 < t_3 < t_2$. Substituting these results into (3.6) yields

$$\int_{t_1}^{t_2} k \frac{\partial^2 u}{\partial x^2}(x_3,\tau)\, d\tau\, (x_2 - x_1) + \int_{x_1}^{x_2}\int_{t_1}^{t_2} f(\xi,\tau)\, d\xi\, d\tau$$
$$= \int_{x_1}^{x_2} c\rho \frac{\partial u}{\partial t}(\xi,t_3)\, d\xi\, (t_2 - t_1). \quad (3.8)$$

In order to further simplify (3.8) let us next invoke the mean-value theorem of integral calculus, namely:

Theorem 3.2

Let the function g be defined and continuous for $x_1 \leq x \leq x_2$, then

$$\frac{1}{x_2 - x_1}\int_{x_1}^{x_2} g(x)\, dx = g(x_0) \quad (3.9)$$

for at least one x_0 such that $x_1 < x_0 < x_2$.

Applying this theorem to the first term on the left of (3.8) and the term on the right of (3.8) we obtain the existence of points x_4, t_4 with $x_1 < x_4 < x_2$ and $t_1 < t_4 < t_2$ such that

$$k\frac{\partial^2 u}{\partial x^2}(x_3,t_4)(x_2-x_1)(t_2-t_1) + \int_{x_1}^{x_2}\int_{t_1}^{t_2} f(\xi,\tau)\, d\xi\, d\tau$$
$$= c\rho \frac{\partial u}{\partial t}(x_4,t_3)(x_2-x_1)(t_2-t_1).$$

Finally, the double integral analog of (3.9) is invoked in order to obtain a point (x_5,t_5) on the open line segment joining (x_1,t_1) to (x_2,t_2) such that

$$\int_{x_1}^{x_2}\int_{t_1}^{t_2} f(\xi,\tau)\, d\xi\, d\tau = f(x_5,t_5)(x_2-x_1)(t_2-t_1).$$

Thus the heat equation reduces to

$$ku_{xx}(x_3,t_4) + f(x_5,t_5) = c\rho u_t(x_4,t_3), \quad (3.10)$$

where the nonzero quantity $(x_2 - x_1)(t_2 - t_1)$ has been divided out.

In summary, Equation (3.10) holds for arbitrary choices of $x_1 < x_2$, $t_1 < t_2$ and, moreover, the functions u_{xx}, u_t, and f are all assumed to be continuous. Consequently we may pass to the limit as $x_1, x_2 \to x$ and $t_1, t_2 \to t$, for then all the $x_i \to x$ and all the $t_i \to t$ and (3.10) further simplifies to

$$ku_{xx} + f = c\rho u_t \quad (3.11)$$

which holds for all x such that $0 < x < L$ and for all $t > 0$.

For a heat-conduction problem to be well defined there must exist information, in addition to the differential equation (3.11), as to how the ends of the rod are exchanging heat with the surrounding environment and also what the distribution of temperature in the rod is at the initial time $t = 0$. Thus we have the initial condition

$$u(x,0) = \varphi(x), \quad 0 \leq x \leq L, \tag{3.12}$$

where $\varphi(x)$ is required to be continuous, at least for $0 \leq x \leq L$, and boundary conditions having various possible forms such as

$$u(0,t) = s(t), \quad u(L,t) = r(t) \tag{3.13}$$

or

$$u_x(0,t) = s(t), \quad u_x(L,t) = r(t),$$

or

$$u_x(0,t) + \alpha u(0,t) = s(t), \quad u_x(L,t) + \beta u(L,t) = r(t),$$

or combinations thereof, such as

$$u(0,t) = s(t), \quad u_x(L,t) = r(t).$$

The differential equation (3.11) is said to be homogeneous if $f = 0$ for $0 \leq x \leq L$, $0 \leq t$, and the boundary conditions are said to be homogeneous if $s \equiv 0 \equiv r$. The several and varied problems of solving the homogeneous equation (3.11) together with homogeneous boundary conditions of any of the possible forms suggested by (3.13) and with the general nonzero initial conditions (3.12) are referred to as *homogeneous problems*. Thus we can now formulate a variety of problems in which we seek the temperature function u as an answer. To give the student a feel for formulating such problems consider the following:

PROBLEM H_{11}

Find the function u such that

$$ku_{xx} = c\rho u_t, \quad 0 < x < L, \quad t > 0 \tag{3.14}$$

subject to the initial condition $u(x,0) = \varphi(x)$, $0 \leq x \leq L$ and the boundary conditions

$$u(0,t) = 0 = u(L,t), \quad t > 0. \tag{3.15}$$

By a solution to Problem H_{11} we shall mean the following. A solution is a function $u(x,t)$ defined and continuous in both x and t for $0 \leq t < +\infty$ and $0 \leq x \leq L$ except possibly at the points $(0,0)$ and $(0,L)$, and which (i) has a continuous second derivative with respect to x and a continuous first derivative with respect to t for $0 < t < +\infty$ and $0 < x < L$, and (ii) satisfies (3.14) for $0 < t < +\infty$ and $0 < x < L$, the boundary conditions (3.15), and the initial condition. This means that the functions $u_{xx}(x,t)$ and $u_t(x,t)$ must be continuous for $0 < t < +\infty$ and $0 < x < L$. The reason for permitting u to be discontinuous at the points $(0,0)$ and $(0,L)$ is that we do not require $\varphi(0) = 0$ and $\varphi(L) = 0$. If $\varphi(0) \neq 0$, then $u(0,0)$ may take on some such value as $\varphi(0)/2$ rather than $\varphi(0)$ itself. Hence in the event $\varphi(0) \neq 0$ and $\varphi(L) = 0$, for example, we usually require only that u take on the values $\varphi(x)$ for $0 < x \leq L$.

It is possible to solve Equation (3.14) with discontinuous initial data $\varphi(x)$. But a discussion of such problems is beyond the scope and intent of this section.

PROBLEM H_{12}

Determine u such that (3.14) and (3.12) are satisfied subject to the boundary conditions

$$u(0,t) = 0 = u_x(L,t), \quad t > 0. \tag{3.16}$$

PROBLEM H_{32}

Determine u such that (3.14) and (3.12) are satisfied subject to the boundary conditions

$$u_x(0,t) + \alpha u(0,t) = 0 = u_x(L,t), \quad \alpha \neq 0, \quad t > 0.$$

Thus we may define the several homogeneous problems H_{ij} through a permutation of the boundary conditions $i,j = 1, 2, 3$. A general procedure for obtaining solutions to these problems is the subject of the next section. We shall leave it for the student to formulate for himself exactly what is meant by a solution to Problem H_{ij} if $i,j \neq 1, 1$.

An obvious variation of the fundamental equation of heat conduction is obtained when the rod exchanges heat with its surrounding medium—that is, when we no longer suppose the rod to be thermally insulated. If the heat exchange in question obeys Newton's law of cooling the amount of heat lost by the rod per unit of length and time is

$$f = h(u - \theta),$$

where $\theta(x,t)$ is the temperature of the surrounding medium and h is the coefficient of heat exchange. We may think of this exchange as being effected by a distribution of surface sources along the rod. This is equivalent to the effect of volume sources of heat, since in our approximation the distribution of temperature through the cross section is not considered. Consequently the density function f is given by

$$f = f_1 - h(u - \theta) \tag{3.17}$$

in which $f_1(x,t)$ is the density of other sources of heat.

Therefore, if the rod is homogeneous, the equation of heat conduction with lateral heat exchange assumes the form

$$u_t = a^2 u_{xx} - bu + g \tag{3.18}$$

where

$$a^2 = \frac{k}{c\rho}, \quad b = \frac{h}{c\rho}, \quad g = b\theta + \frac{f_1}{c\rho}.$$

The boundary and initial conditions appropriate to (3.11) are also appropriate to Equation (3.18). Equation (3.18) is called *homogeneous* if $g \equiv 0$, which case yields

$$u_t = a^2 u_{xx} - bu. \tag{3.19}$$

One usually assumes that the function g is defined and continuous for $0 < t < +\infty, 0 < x < L$.

In addition to the problems we have outlined up to this point there are certain important limiting cases that are frequently encountered. For example, in the case of a very long bar the effect of the boundary conditions is very slight in the central portion of the rod, at least in a small interval of time. The length of the rod is therefore of comparatively little interest and hence, for the mathematical convenience which it fosters, one usually considers the bar to be of infinite length. Accordingly, we may formulate the *Cauchy problem of heat conduction*.

PROBLEM

Find the solution of (3.14) in the region $-\infty < x < \infty, t \geqslant 0$ satisfying the condition

$$u(x,0) = \varphi(x), \quad -\infty < x < \infty,$$

where φ is a given function. It should be noted that quite frequently one wishes to solve the Cauchy problem for discontinuous initial data.

Another important limiting case is that in which the section of the bar of interest is located near one end and far from the other. We may then formulate various boundary value problems for this semi-infinite rod. For example, find the solution of (3.14) in $0 < x < \infty$, $t_0 > t$ satisfying the conditions

$$u(x,0) = \varphi(x), \quad 0 \leq x < \infty,$$
$$u(0,t) = s(t), \quad t > 0,$$

with φ and s given functions. An equally interesting formulation is the problem in which the condition $u(0,t) = s(t)$, $t > 0$, is replaced by $u_x(0,t) = s(t), t > 0$.

It follows that these limiting problems can be formulated relative to Equations (3.17) or (3.18) or indeed for other variations of the heat equation which the student will encounter in the exercises and later in the book.

Finally, we mention a boundary condition which is not linear in u and u_x, namely

$$k\, \partial u/\partial x\, (0,t) = \alpha[u^4(0,t) - \theta^4(0,t)]. \qquad (3.20)$$

This condition corresponds to radiation from the end $x = 0$ in a medium of temperature $\theta(t)$ according to the Stefan-Boltzmann law, where, of course, θ is a given function of t.

3.2 SOME PROPERTIES OF THE HEAT EQUATION

Let us begin by considering the homogeneous partial differential equation

$$v_t = a^2 v_{xx} + b v_x + cv \qquad (3.21)$$

having constant coefficients a, b, and c. This equation appears to be more general than any of the homogeneous equations of Section 3.1. The generality is deceiving, however, as the following calculations will prove. Consider the substitution

$$v = e^{\mu x + \lambda t} u. \qquad (3.22)$$

Operating on (3.22) we have

$$v_t = e^{\mu x + \lambda t}(\lambda u + u_t),$$

$$v_x = e^{\mu x + \lambda t}(\mu u + u_x),$$

and

$$v_{xx} = e^{\mu x + \lambda t}(\mu^2 u + 2\mu u_x + u_{xx}).$$

Substitution of the above into (3.21) yields

$$\lambda u + u_t = a^2(\mu^2 u + 2\mu u_x + u_{xx}) + b(\mu u + u_x) + cu$$

after noting that we may divide out $e^{\mu x + \lambda t} \neq 0$. Rearrange this equation to read

$$u_t = a^2 u_{xx} + (2a^2\mu + b)u_x + (a^2\mu^2 + b\mu + c - \lambda)u.$$

Now setting

$$\mu = -\frac{b}{2a^2} \quad \text{and} \quad \lambda = c - \frac{b^2}{4a^2}$$

we obtain

$$v = u \exp\left\{-\frac{b}{2a^2}x + \left(c - \frac{b^2}{4a^2}\right)t\right\} \qquad (3.23)$$

with u satisfying

$$u_t = a^2 u_{xx}. \qquad (3.24)$$

Equation (3.24) is recognized as being the same as (3.14). In general, by the device used above we may reduce any second-order homogeneous equation with constant coefficients to the consideration of (3.24). Accordingly we shall concentrate our attention upon this equation.

Drawing upon our experience with ordinary differential equations leads us to expect that the study of the several boundary-value problems in the theory of heat conduction will be most effectively carried forward by starting with the examination of the corresponding homogeneous problem. With this in mind let us consider Problem H_{11} of Section 3.1. The method we shall expound is due to Fourier and is equally applicable for nearly all of the heat conduction problems mentioned thus far.

Equation (3.24) is an equation with constant coefficients and we have seen before that such equations have exponential solutions of the form

$$u = \omega e^{\lambda t}. \tag{3.25}$$

For the ordinary differential equations ω turned out to be a constant independent of t. Relying on this as a start let us seek solutions of (3.24) having the form (3.25) in which ω is a function of x alone and λ is a constant independent of both x and t. This gives

$$u_t = \lambda \omega e^{\lambda t},$$
$$u_{xx} = \omega'' e^{\lambda t}$$

where we use a prime to denote differentiation with respect to x. Thus (3.24) yields

$$\lambda \omega e^{\lambda t} = a^2 \omega'' e^{\lambda t}$$

or

$$e^{\lambda t}[a^2 \omega'' - \lambda \omega] = 0.$$

Since $e^{\lambda t} \neq 0$, we are left with

$$\omega'' - \frac{\lambda}{a^2} \omega = 0. \tag{3.26}$$

Equation (3.26) is an ordinary differential equation similar to the linear simple pendulum equation of Section 1.8. If we can find solutions of (3.26) which also satisfy the boundary conditions $\omega(0) = 0 = \omega(L)$, then the corresponding $u(x,t) = \omega(x)e^{\lambda t}$ will be such that $u(0,t) = \omega(0)e^{\lambda t} = 0$ and $u(L,t) = \omega(L)e^{\lambda t} = 0$. In this way we will find functions u satisfying the differential equation and the boundary conditions, and the only remaining task will be to satisfy the initial condition (3.12).

Following the procedure established in Chapter 1, the characteristic equation for (3.26) is (letting $\omega = e^{\mu x}$)

$$\mu^2 - \frac{\lambda}{a^2} = 0.$$

Examining this equation we see that there are three conditions to consider, namely $\lambda = 0$, $\lambda > 0$, and $\lambda < 0$.

When $\lambda = 0$ the solution of (3.26) is

$$\omega(x) = \alpha x + \beta.$$

In order to satisfy the boundary conditions (which are $\omega(0) = 0 = \omega(L)$) we must have

$$\alpha = \beta = 0.$$

Thus when $\lambda = 0$ the only solution is $\omega(x) \equiv 0$ on $[0,L]$.

When $\lambda > 0$ the solution of (3.26) is

$$\omega(x) = \alpha e^{\frac{\sqrt{\lambda}}{a}x} + \beta e^{-\frac{\sqrt{\lambda}}{a}x},$$

and in order to satisfy the boundary conditions, $\omega(0) = 0 = \omega(L)$, we must have

$$\alpha + \beta = 0 \tag{3.27}$$

and

$$\alpha e^{\frac{\sqrt{\lambda}}{a}L} + \beta e^{-\frac{\sqrt{\lambda}}{a}L} = 0.$$

This homogeneous system for α and β will have nontrivial solutions if and only if

$$\begin{vmatrix} 1 & 1 \\ e^{\frac{\sqrt{\lambda}}{a}L} & e^{-\frac{\sqrt{\lambda}}{a}L} \end{vmatrix} = e^{-\frac{\sqrt{\lambda}}{a}L} - e^{\frac{\sqrt{\lambda}}{a}L} = 0.$$

But, since $L > 0$ we have $e^{-\frac{\sqrt{\lambda}}{a}L} - e^{\frac{\sqrt{\lambda}}{a}L} < 0$, and once more the problem is reduced to the trivial solution.

Finally if $\lambda < 0$ (3.26) has solutions of the form

$$\omega(x) = \alpha \cos \frac{\sqrt{-\lambda}}{a} x + \beta \sin \frac{\sqrt{-\lambda}}{a} x. \tag{3.28}$$

Setting $-\lambda = \nu$ and once again applying the boundary conditions we obtain

$$\alpha = 0$$

$$\beta \sin \frac{\sqrt{\nu}}{a} L = 0.$$

It follows from the second of the foregoing equations that we can have nontrivial solutions if

$$\frac{\sqrt{\nu}}{a} L = n\pi, \quad n = 1, 2, \ldots.$$

In this way

$$\lambda = -\left(\frac{n\pi a}{L}\right)^2, \quad n = 1, 2, \ldots$$

yields nontrivial solutions of (3.26). Placing the foregoing value of λ into (3.27) yields $\alpha = -\beta$. Thus we see once again, as was shown for the mechanical system, that the arbitrary constants are related. On substituting λ back into (3.28) we obtain the functions

$$\omega_n(x) = \sin n\pi \frac{x}{L} \qquad (3.29)$$

as solutions to our problem. Here we have replaced β by 1 in view of the fact, and to again call attention to the fact that β is arbitrary.

Returning finally to (3.25) we find that there exists a family of functions

$$u_n(x,t) = \sin \frac{n\pi x}{L} e^{-(n\pi a/L)^2 t}, \quad n = 1, 2, \ldots, \qquad (3.30)$$

each of which satisfies the differential equation (3.24) and the boundary conditions

$$u_n(0,t) = u_n(L,t) = 0.$$

It still remains for us to satisfy the initial condition (3.12).

We have repeatedly satisfied initial conditions by forming linear combinations of solutions. This technique is based upon the principle of superposition which says for the present case that if $u_0, \ldots, u_k$ are solutions to the differential equation, then so is the linear combination $a_1 u_1 + \cdots + a_k u_k$ for arbitrary constants $a_1, \ldots, a_k$. Moreover, we also observe that if $u_1, \ldots, u_k$ satisfy the boundary conditions so does the linear combination $a_1 u_1 + \cdots + a_k u_k$. This possibility of constructing new solutions from old ones by forming linear combinations is characteristic of linear problems; in fact, one definition for a problem to be linear is that the principle of superposition be valid for it.

In the problem at hand every linear combination

$$\sum_{j=1}^{k} a_j u_j, \quad k = 0, 1, \ldots,$$

is a solution of the differential equation and the boundary conditions. We might expect that the same is true of the infinite series

$$u = \sum_{j=1}^{\infty} a_j u_j. \qquad (3.31)$$

The verification that this is indeed so is not a simple problem for all cases. Accepting (3.31) for the present we seek to use it in order to satisfy the initial conditions.

We have from (3.31) that

$$u(x,0) = \sum_{j=1}^{\infty} a_j u_j(x,0) = \sum_{j=1}^{\infty} a_j \sin j\frac{\pi x}{L}.$$

Since $u(x,0) = \varphi(x)$, we must choose $a_1, \ldots,$ so that

$$\varphi(x) = \sum_{j=1}^{\infty} a_j \sin j\frac{\pi x}{L}. \tag{3.32}$$

To do this multiply both sides of (3.32) by $\sin k\pi x/L$ and integrate with respect to x from 0 to L. Then

$$\int_0^L \varphi(x) \sin \frac{k\pi x}{L} dx = \int_0^L \sum_{j=1}^{\infty} a_j \sin j\frac{\pi x}{L} \sin \frac{k\pi x}{L} dx.$$

Since we are working formally, we assume that it is permissible to interchange integration and summation to obtain

$$\int_0^L \varphi(x) \sin \frac{k\pi x}{L} dx = \sum_{j=1}^{\infty} a_j \int_0^L \sin j\frac{\pi x}{L} \sin \frac{k\pi x}{L} dx. \tag{3.33}$$

The integrals on the right-hand side of (3.33) can be evaluated explicitly. It is easily shown that if $j \neq k$,

$$\int_0^L \sin j\frac{\pi x}{L} \sin \frac{k\pi x}{L} dx = 0.$$

We leave this calculation to the student in the form of an exercise. For the case of $j = k$ we obtain

$$\int_0^L \sin^2 \frac{k\pi x}{L} dx = \frac{1}{2} \int_0^L \left(1 - \cos \frac{2k\pi x}{L}\right) dx = \frac{L}{2}.$$

It follows that for each value of $k = 1, 2, \ldots,$ (3.33) yields

$$a_k = \frac{2}{L} \int_0^L \varphi(x) \sin \frac{k\pi x}{L} dx. \tag{3.34}$$

Hence, from (3.32) the initial condition becomes

$$\varphi(x) = \sum_{j=1}^{\infty} \frac{2}{L} \sin j\frac{\pi x}{L} \int_0^L \varphi(\xi) \sin j\frac{\pi \xi}{L} d\xi$$

$$= \frac{2}{L} \int_0^L \varphi(\xi) \sum_{j=1}^{\infty} \sin j\frac{\pi x}{L} \sin j\frac{\pi \xi}{L} d\xi. \tag{3.35}$$

It is worth reemphasizing that the foregoing work starting with (3.31) has been purely formal and must eventually be justified.

Finally, substituting all of the above results into (3.31) yields

$$u(x,t) = \frac{2}{L} \int_0^L \varphi(\xi) \sum_{j=1}^{\infty} \sin j\frac{\pi x}{L} \sin j\frac{\pi \xi}{L} e^{-(n\pi a/L)^2 t} \, d\xi \qquad (3.36)$$

as a formal solution to Problem H_{11}. If we are fortunate, the infinite series in the integrand of (3.36) will be convergent for all (x,t) such that $0 \leq x \leq L$, $t \geq 0$. The convergence of such infinite series is a problem in the theory of Fourier series.

Define the function

$$G(x,\xi,t) = \frac{2}{L} \sum_{j=1}^{\infty} \sin j\frac{\pi x}{L} \sin j\frac{\pi \xi}{L} e^{-(j\pi a/L)^2 t}. \qquad (3.37)$$

Significantly, this function G has nothing to do with the function φ, and consequently its properties can in no way depend upon those of φ. In fact, the function G is associated with Problem H_{11} itself and with its use we obtain, by integration, the solution u from the given function φ. G is called the *Green's function for Problem* H_{11}. Summarizing, the representation of φ furnished by formula (3.32) is called the *Fourier sine series* for φ. The coefficients a_j are called the *Fourier coefficients* of φ with respect to the functions $\sin j(\pi x/L)$, $j = 1, \ldots$. Once φ is given the a_j may be determined explicitly from (3.34). Observe that these numbers are well defined if, for example, φ is continuous on $0 \leq x \leq L$. It does not automatically follow that the Fourier sine series (3.32) is convergent whenever the Fourier coefficients are defined. In fact, as a consequence of our method of solution, we are confronted here with some of the basic questions in the theory of Fourier series. An example is the question of determining classes of functions φ for which (3.32) is convergent.

On the other hand, we have obtained an integral representation

$$u(x,t) = \int_0^L \varphi(\xi) G(x,\xi,t) \, d\xi \qquad (3.38)$$

for the solution of Problem H_{11}. It follows that if we can verify directly that this integral does solve the problem, then the formalities used to obtain the answer do not need further justification. This kind of approach can be, and often is, used with great success.

The entire technique of this section can be applied to all of the homogeneous problems of heat conduction formulated in Section 3.1.

Looking at the integral (3.38) we see that it provides a possible means for solving Problem H_{11} when φ is discontinuous. Actually it requires some work to do this because of the fact that the properties of $G(x,\xi,0)$ are not easy to obtain. We shall not do this work here, but content ourselves with the rough observation that the integral (3.38) does solve Problem H_{11} whenever the function φ is integrable in whatever sense is under consideration. These remarks apply equally well to problems H_{ij}.

PROBLEMS

1. Show that the function $u = e^{-n^2 t} \sin nx$ satisfies the simplified heat equation
$$u_t = u_{xx}.$$

2. Solve the boundary-value problem
$$u_t = a^2 u_{xx}, \quad 0 < x < L, \quad 0 < t,$$
$$u_x(0,t) = u_x(L,t) = 0, \quad 0 < t,$$
$$u(x,0) = f(x), \quad 0 < x < L.$$

State exactly what is meant by a solution to this problem.

3. Solve the boundary-value problem
$$u_t = a^2 u_{xx}, \quad 0 < x < L, \quad 0 < t,$$
$$u_x(0,t) = 0, \quad u_x(L,t) = p_1, \quad 0 < t,$$
$$u(x,0) = 0, \quad 0 < x < L.$$

State exactly what is meant by a solution to this problem.

4. Solve the boundary-value problem
$$u_t = a^2 u_{xx}, \quad 0 < x < L, \quad 0 < t,$$
$$u_x(0,t) - bu(0,t) = 0, \quad u_x(L,t) + bu(L,t) = 0, \quad 0 < t,$$
$$u(x,0) = u_0, \quad 0 < x < L,$$

u_0, u_1, u_2 constant, $b > 0$. In solving this boundary-value problem you will not be able to find the eigenvalues explicitly. Even so, you should convince yourself that there are infinitely many eigenvalues. In writing out the solution leave the Fourier coefficients in integral form.

5. Show that the boundary-value problem
$$u_t = a^2 u_{xx}, \quad 0 < x < L, \quad 0 < t,$$
$$u(0,t) = u_1, \quad u(L,t) = u_2, \quad 0 < t,$$
$$u(x,0) = u_0, \quad 0 < x < L,$$

u_0, u_1, u_2 constants, can be solved in the form
$$u(x,t) = v(x,t) + w(x),$$

where $w(x)$ satisfies the equations
$$w_{xx} = 0, \quad w(0) = u_1, \quad w(L) = u_2,$$

and $v(x,t)$ solves
$$v_t = a^2 v_{xx},$$
$$v(0,t) = v(L,t) = 0,$$
$$v(x,0) = u_0 - w(x).$$

State exactly what is meant by a solution to this problem. Write out the solution to this problem and compute
$$\lim_{t \to \infty} u(x,t).$$

6. Solve the boundary-value problem
$$u_t = a^2 u_{xx}, \quad 0 < x < L, \quad 0 < t,$$
$$u(0,t) = u_0, \quad u_x(L,t) = p_0, \quad 0 < t,$$
$$u(x,0) = 0.$$

7. Solve the boundary-value problem
$$u_t = a^2 u_{xx} - hu,$$
$$u_x(0,t) = u(L,t) = 0,$$
$$u(x,0) = u_0 = \text{const.}$$

8. Solve the boundary-value problem
$$u_t = a^2 u_{xx} - hu, \quad 0 < x < L, \quad 0 < t,$$
$$u_x(0,t) = -p_1, \quad u_x(L,t) = p_2, \quad 0 < t,$$
$$u(x,0) = p_0, \quad 0 < x < L,$$

by a technique similar to that of Problem 5. Compute
$$\lim_{t \to \infty} u(x,t).$$

9. Solve the boundary-value problem
$$u_t = a^2 u_{xx}, \quad 0 < x < L, \quad 0 < t,$$
$$u_x(0,t) - b[u(0,t) - u_1] = 0, \quad u_x(L,t) + b[u(L,t) - u_2] = 0, \quad 0 < t,$$
$$u_x(x,0) = f(x), \quad 0 < x < L.$$

Please see the remarks at the end of Problem 4.

10. Consider a thin ring at the surface of which a heat exchange with the surrounding medium takes place. Assume the temperature of the medium is u_0 and the temperature distribution across the wire can be considered uniform. Show that if the radius of the ring is R and θ is an angular coordinate, the temperature distribution in the ring is determined by the equations
$$u_t = au_{\theta\theta} - b(u - u_0), \quad 0 < \theta < 2\pi, \quad 0 < t,$$

where a and b are positive constants, and

$$u(0,t) = u(2\pi,t), \quad u_\theta(0,t) = u_\theta(2\pi,t), \quad 0 < t,$$
$$u(\theta,0) = f(\theta), \quad 0 < \theta < 2\pi.$$

*11. Verify that $\int_0^L \sin j(\pi x/L) \sin k(\pi x/L)\, dx = 0$ if $j \neq k$, and that this integral has the value $L/2$ if $j = k$.

12. Suppose a thin wire is heated by a constant electric current and at its surface convective heat transfer occurs obeying Newton's law. Suppose the temperature of the environment is u_0 and that the ends of the wire are fixed in clamps with given heat capacity and very large thermal conductivity. Show that the temperature satisfies the boundary-value problem

$$u_t = au_{xx} - b(u - u_0) + c, \quad 0 < x < L, \quad 0 < t,$$
$$c_1 u_t(0,t) = u_x(0,t), \quad c_2 u_t(L,t) = u_x(L,t), \quad 0 < t,$$
$$u(x,0) = f(x), \quad 0 < x < L,$$

where a, b, c, c_1, and c_2 are positive constants. State exactly what is meant by a solution to this problem.

13. A molten metal fills a vertical vessel, the walls and bottom of which are insulated. From time $t = 0$ the free surface of the metal is maintained at a temperature v_0 which is below the temperature of fusion. Suppose the initial temperature of the metal is $v_1 =$ constant, and variations of temperatures in cross sections can be ignored. Then show that the boundary-value problem for cooling and solidification of the metal has the form

$$\left. \begin{array}{l} \dfrac{\partial u_1}{\partial t} = a_1^2 \dfrac{\partial^2 u_1}{\partial x^2}, \quad 0 < x < y(t) \\[2mm] \dfrac{\partial u_2}{\partial t} = a_2^2 \dfrac{\partial^2 u_2}{\partial x^2}, \quad y(t) < x < L \end{array} \right\} \quad 0 < t < t_0,$$

$$u_1(0,t) = v_0, \quad \lambda_1 \dfrac{\partial u_1}{\partial x}\bigg|_{x=y(t)} - \lambda_2 \dfrac{\partial u_2}{\partial x}\bigg|_{x=y(t)} = c\dfrac{dy(t)}{dt} \quad 0 < t < t_0,$$

$$u_1(y(t),t) = u_2(y(t),t) = 0, \quad 0 < t < t_0,$$

$$\dfrac{\partial u_2}{\partial x}(L,t) = 0, \quad 0 < t < t_0,$$

$$u_2(x,0) = v_1, \quad 0 < x < L.$$

Here the function $y(t)$ is the depth to which solidification has penetrated by the time t, t_0 is the time at which $y(t) = L$, and $x = 0$ at the surface and L at the bottom of the vessel. State exactly what is meant by a solution to this problem.

14. Formulate the boundary-value problem for the cooling of a uniformly heated rod having the shape of a truncated cone if the ends of the rod are thermally insulated and there is a lateral heat exchange be-

tween the surface and the surrounding medium whose temperature is zero. Assume the temperature constant over a cross section.

15. Determine the solution of the heat equation for $t > 0$, $0 < x < \pi$ which is continuous for $t \geq 0$, $0 \leq x \leq \pi$ and has a continuous derivative $\partial u/\partial x$ in this region. The boundary conditions are $\partial u/\partial x = 0$ at $x = 0$ and $x = \pi$. The initial condition is $u(x,0) = f(x)$, where $f(x)$ has continuous first and second derivatives for $0 \leq x \leq \pi$. This can be interpreted as a problem in heat conduction in a slab whose faces are insulated. How does this differ from Problem 2?

16. Two slabs of metal, each r inches thick, one at temperature $T_1°$C and the other at temperature $0°$C throughout are placed in full contact with each other and their outer faces are kept at $0°$C. Determine the temperature at a point common to the two slabs t minutes after contact is made. Determine the temperature midway through each slab at that time.

3.3 REMARKS ON THE FOURIER METHOD

When we solved the homogeneous heat-conduction problem H_{11} in Section 3.2 a number of important new concepts were introduced. It therefore will prove to be expedient to pause here and study the significance of these ideas at greater length.

The basic tool with which we worked is the infinite sequence of functions

$$\left\{ \sin \frac{\pi x}{L}, \sin \frac{2\pi x}{L}, \ldots, \sin \frac{n\pi x}{L}, \ldots \right\}. \tag{3.39}$$

These functions are called the *eigenfunctions* for Problem H_{11}. For ordinary differential equations of second order we only found it necessary to work with two linearly independent functions. Problem H_{11} has infinitely many functions associated with it. This comparison leads to a few questions. Are the functions associated with H_{11} linearly independent? How shall we define linear independence for an infinite sequence of functions? We begin the work of this section by answering these questions.

Definition 3.1

Let $\{f_1, f_2, \ldots, f_n, \ldots\}$ be an infinite sequence of functions all defined on the same open interval (a,b) (the interval may be finite or infinite). We say that these functions are linearly independent if every finite subsequence is linearly independent.

Recall that a finite sequence of functions $\{f_{n_1}, f_{n_2}, \ldots, f_{n_k}\}$ is linearly independent if the only sequence of constants $a_{n_1}, a_{n_2}, \ldots, a_{n_k}$ for which

132 PARTIAL DIFFERENTIAL EQUATIONS

$$a_{n_1} f_{n_1}(x) + \cdots + a_{n_k} f_{n_k}(x) \equiv 0 \quad \text{on } (a,b) \tag{3.40}$$

is

$$a_{n_1} = a_{n_2} = \cdots = a_{n_k} = 0. \tag{3.41}$$

If there exists a sequence of constants $\{a_{n_1}, \ldots, a_{n_k}\}$ not all zero such that (3.41) holds, then the function $\{f_{n_1}, f_{n_2}, \ldots, f_{n_k}\}$ are linearly dependent.

Let us now state and prove:

Theorem 3.3

The sequence of functions $\{\sin \pi x/L, \sin 2\pi x/L, \ldots, \sin n\pi x/L, \ldots\}$ is a linearly independent set on the interval $(0,L)$.

PROOF. We must show that every equation of the form

$$a_{n_1} \sin \frac{n_1 \pi x}{L} + a_{n_2} \sin \frac{n_2 \pi x}{L} + \cdots + a_{n_k} \sin \frac{n_k \pi x}{L} = 0$$

implies

$$a_{n_1} = a_{n_2} = \cdots = a_{n_k} = 0.$$

For this purpose we make use of Problem 11 of Section 3.2. Multiply the equation by $\sin n_j \pi x/L$ where n_j is one of the integers in the sequence $\{n_1, \ldots, n_k\}$ and integrate from 0 to L. Thus

$$\int_0^L \sin \frac{n_j \pi x}{L} \left[a_{n_1} \sin \frac{n_1 \pi x}{L} + \cdots + a_{n_k} \sin \frac{n_k \pi x}{L} \right] dx = 0.$$

Since only a finite sum occurs in the integrand, we may write this equation in the form

$$a_{n_1} \int_0^L \sin \frac{n_1 \pi x}{L} \sin \frac{n_j \pi x}{L} dx + \cdots + a_{n_k} \int_0^L \sin \frac{n_j \pi x}{L} \sin \frac{n_k \pi x}{L} dx = 0. \tag{3.42}$$

By Problem 11 each term is zero except the term involving

$$\int_0^L \sin^2 \frac{n_j \pi x}{L} dx = \frac{L}{2}.$$

Therefore (3.42) reduces to

$$\frac{L}{2} a_{n_j} = 0.$$

Since $L/2 \neq 0$, we must have $a_{n_j} = 0$. Applying this argument for $n_j = n_1, \ldots, n_k$, we find that each coefficient must be zero. Thus the theorem is proved.

The fact that the integral

$$\int_0^L \sin\frac{n\pi x}{L} \sin\frac{m\pi x}{L} \, dx = 0$$

if $n \neq m$ is an important and useful piece of information. It turns out to be one of those fundamental tools which researchers, designers, and mathematicians alike use again and again. It is a special case of the following general idea.

Definition 3.2

Let f and g be real-valued functions defined on (a,b) (the interval may be finite or infinite) and such that f^2 and g^2 are integrable on (a,b). We define the scalar product of f and g by the formula

$$(f,g) = \int_a^b f(x)g(x) \, dx. \qquad (3.43)$$

This scalar product has exactly the same properties as the scalar product introduced in Section 2.3. We have

$$\begin{aligned}(f_1 + f_2, g) &= \int_a^b [f_1(x) + f_2(x)]g(x) \, dx \\ &= \int_a^b f_1(x)g(x) \, dx + \int_a^b f_2(x)g(x) \, dx \\ &= (f_1,g) + (f_2,g).\end{aligned}$$

We also have

$$(f,g) = \int_a^b f(x)g(x) \, dx = \int_a^b g(x)f(x) \, dx = (g,f).$$

If α is a real number, then

$$\begin{aligned}(\alpha f, g) &= \int_a^b \alpha f(x) g(x) \, dx = \alpha \int_a^b f(x)g(x) \, dx = \alpha(f,g) \\ &= (f, \alpha g).\end{aligned}$$

Similarly,

$$(f, g_1 + g_2) = (f,g_1) + (f,g_2).$$

Finally we have

$$(f,f) = \int_a^b f^2(x) \, dx \geq 0$$

and $(f,f) = 0$ if and only if $f = 0$ on (a,b).

The number $(f,f)^{1/2}$ is called the *norm* of f and is usually written as

$$(f,f)^{1/2} = \|f\|. \qquad (3.44)$$

We use double vertical bars to distinguish the norm of f, which is a real number, from the function $|f|$.

In terms of these concepts we can say that the method of proof of Theorem 3.3 consists in forming the scalar product of the linear combination

$$\sum_{l=1}^{k} a_{n_l} \sin \frac{n_l \pi x}{L}$$

with the function $\sin n_j \pi x / L$ and making use of the fact established in Problem 11 of Section 3.2.

Definition 3.3

Two functions f, g defined on (a,b) are said to be *orthogonal* on (a,b) if $(f,g) = 0$. The sequence of functions $\{f_1, f_2, \ldots\}$ defined on (a,b) is said to be an *orthogonal set* on (a,b) if $(f_n, f_m) = 0$ whenever $n \neq m$. We say the function f is *normalized* or normal if $\|f\| = 1$, and the sequence $\{f_1, f_2, \ldots\}$ is called an *orthonormal sequence* if $(f_n, f_m) = 0$ for $n \neq m$ and $(f_n, f_n) = 1$ for all n.

In terms of these concepts the result of Problem 11 implies that the sequence (3.39) is an orthogonal sequence and the sequence

$$\left\{ \sqrt{\frac{2}{L}} \sin \frac{\pi x}{L}, \sqrt{\frac{2}{L}} \sin \frac{2\pi x}{L}, \ldots, \sqrt{\frac{2}{L}} \sin \frac{n\pi x}{L}, \ldots \right\}, \quad (3.45)$$

is an orthonormal sequence.

From the foregoing it follows that any nonzero function can be normalized. For if $\|f\| \neq 0$, we can define $f^* = f / \|f\|$.

Let us use these ideas to prove the following result.

Theorem 3.4

Let $\{f_1, f_2, \ldots\}$ be an orthonormal sequence of functions defined on (a,b). Then the functions are linearly independent on (a,b) (the interval may be finite or infinite).

PROOF. This proof follows the reasoning of the proof to Theorem 3.3. Thus let $\sum_{l=1}^{k} a_{n_l} f_{n_l} = 0$. Our task is to show that $a_{n_l} = 0, l = 1, \ldots, k$. Let n_j be an integer in the sequence $\{n_1, n_2, \ldots, n_k\}$ and form the scalar product

$$\left(f_{n_j}, \sum_{l=1}^{k} a_{n_l} f_{n_l}\right) = \sum_{l=1}^{k} (f_{n_j}, a_{n_l} f_{n_l})$$

$$= \sum_{l=1}^{k} a_{n_l}(f_{n_j}, f_{n_l}) = 0.$$

Here we have made use of the properties of the scalar product mentioned above. Since the sequence is orthonormal, the last written sum reduces to

$$a_{n_j} = 0.$$

This proves the theorem because n_j is an arbitrary integer in the sequence.

The only difference between the proofs of Theorems 3.3 and 3.4 is that in the latter case we used an orthonormal sequence. But by virtue of the remark preceding the theorem any orthogonal sequence with nonzero norms can be made into an orthonormal sequence. Hence we are entitled to work with the orthonormal sequence if we prefer.

Returning to the specific problem H_{11} we may now refer to the functions $\sqrt{2/L}\,\sin n\pi x/L$ as the normalized eigenfunctions for Problem H_{11}. Recall that we found that the boundary-value problem could be formally solved if we could choose the sequence of numbers $\{a_1, a_2, \ldots\}$ in such a way that the initial function $\varphi(x)$ has the representation

$$\varphi(x) = \sum_{j=1}^{\infty} a_j \sin \frac{j\pi x}{L}. \tag{3.46}$$

Moreover the method of determining what value each constant a_j has consists in forming the scalar products

$$\left(\sin \frac{k\pi x}{L}, \varphi\right) = \left(\sin \frac{k\pi x}{L}, \sum_{j=1}^{\infty} a_j \sin \frac{j\pi x}{L}\right), \quad k = 1, 2, \ldots,$$

and assuming that

$$\left(\sin \frac{k\pi x}{L}, \sum_{j=1}^{\infty} a_j \sin \frac{j\pi x}{L}\right) = \sum_{j=1}^{\infty} a_j \left(\sin \frac{k\pi x}{L}, \sin \frac{j\pi x}{L}\right)$$

$$= a_k \left\|\sin \frac{k\pi x}{L}\right\|^2$$

$$= \frac{L}{2} a_k.$$

Thus, in terms of the concepts of this section

$$a_k = \frac{\left(\sin\frac{k\pi x}{L}, \varphi\right)}{\left\|\sin\frac{k\pi x}{L}\right\|^2}. \tag{3.47}$$

Let us conclude this section by leading the student in a direction in which the formal calculations we have just performed can be justified.

Let $\{f_1, f_2, \ldots, f_N\}$ be a finite orthonormal sequence defined on the interval (a,b) (finite or infinite). Let us try to approximate the function f by a linear combination of the functions $f_1, \ldots, f_N$ so as to make the norm

$$\left\|f - \sum_{j=1}^{N} b_j f_j\right\|$$

as small as possible. In other words, we seek to choose the numbers $b_1, \ldots, b_N$ so as to minimize the number

$$\left(\int_a^b \left[f(x) - \sum_{j=1}^{N} f_j f_j(x)\right]^2 dx\right)^{1/2}.$$

Now by definition

$$\left\|f - \sum_{j=1}^{N} b_j f_j\right\|^2 = \left(f - \sum_{j=1}^{N} b_j f_j, f - \sum_{k=1}^{N} b_k f_k\right)$$

$$= (f,f) - \sum_{j=1}^{N} b_j(f_j,f) - \sum_{k=1}^{N} b_k(f,f_k) + \sum_{j,k=1}^{N} b_j b_k (f_j, f_k)$$

$$= \|f\|^2 - 2\sum_{j=1}^{N} b_j(f,f_j) + \sum_{j=1}^{N} b_j^2$$

$$= \|f\|^2 - \sum_{j=1}^{N} (f,f_j)^2 + \sum_{j=1}^{N} [b_j - (f,f_j)]^2. \tag{3.48}$$

Studying the last line it becomes clear that the minimum will be attained when

$$b_j = (f,f_j) = \int_a^b f(x) f_j(x)\, dx.$$

When we substitute these values into (3.48) we obtain

$$\left\|f - \sum_{j=1}^{N} (f,f_j) f_j\right\|^2 = \|f\|^2 - \sum_{j=1}^{N} (f,f_j)^2. \tag{3.49}$$

Equation (3.49) is called *Bessel's identity*. Since the left-hand side is clearly nonnegative, we have

$$\sum_{j=1}^{N} (f,f_j)^2 \leq \|f\|^2, \qquad (3.50)$$

a result known as *Bessel's inequality*.

Definition 3.4

Given the infinite orthonormal sequence defined on the interval (a,b), we say that the sequence is *complete* if for every square integrable function f the sequence of norms

$$\left\| f - \sum_{j=1}^{N} (f,f_j)f_j \right\| \to 0$$

as $N \to \infty$.

We may now state:

Theorem 3.5

The sequence $\{\sin n\pi x/L\}^{\infty}$, is complete on the interval (a,b).

In a general sense Theorem 3.5 justifies our formal computations. It implies that

$$\left(\int_0^L \left[\varphi(x) - \sum_{j=1}^{N} \left(\int_0^L \varphi(\xi) \sin \frac{j\pi \xi}{L} d\xi \right) \sin \frac{j\pi x}{L} \right] dx \right)^{1/2} \to 0$$

as $N \to \infty$ if φ is square integrable (φ^2 is integrable) on $(0,L)$. Considerably more work is required to completely justify these manipulations in the context of the ordinary concept of convergence familiar to us from calculus.

PROBLEMS

1. Find the eigenfunctions for Problem H_{22} for the heat equation $u_t = a^2 u_{xx}$ on $(0,L)$. What are the normalized eigenfunctions? Prove that the normalized eigenfunctions are an orthonormal sequence.
2. The following is a theorem in algebra:

   A polynomial $p_N x^N + p_{N-1} x^{N-1} + \cdots + p_1 x + p_0 = 0$ has at most N zeros on any interval (a,b).

   Use this result to prove that the sequence $\{1, x, \ldots, x^N\}$ is linearly independent on any interval (a,b).
3. Use the theorem of Problem 2 to conclude that the infinite sequence $\{1, x, \ldots, x^n, \ldots\}$ is linearly independent on any interval. Is this sequence orthonormal? Normalize the sequence.

4. Find the eigenfunctions for Problem H_{12} for the heat equation on $(0,L)$. Normalize these eigenfunctions and prove they are an orthonormal set.
5. Work Problem 4 with the boundary conditions H_{13}. (You will find that some of your results must be left in integral form.)
6. Find the normalized eigenfunctions for Problem H_{21} for the differential equation $u_t = a^2 u_{xx} - hu$ on $(0,L)$. Show that they form an orthonormal sequence.
7. A generalization of Problem H_{11} for the heat equation is obtained by solving the problem

$$u_t = (p(x)u_x)_x, \quad 0 < x < L, \quad t > 0, \quad (3.51)$$

with

$$u(0,t) = u(L,t) = 0, \quad t > 0, \quad (3.52)$$

$$u(x,0) = f(x), \quad 0 < x < L.$$

Here it is assumed that p has a continuous derivative in $(0,L)$ and is positive in the sense that $p(x) \geq p_0 > 0$ in $(0,L)$. Show that if the method of Section 3.2 is used to seek solutions to this problem, one arrives at the boundary-value problem

$$\text{(BV)} \begin{cases} (p(x)W')' - \lambda W = 0, & 0 < x < L, \\ W(0) = W(L) = 0. \end{cases}$$

8. An eigenvalue of Problem (BV) is a real number λ_0 corresponding to which there is a nonzero solution W defined on $(0,L)$. Show that 0 is *not* an eigenvalue of Problem (BV), that is, that the only solution of $(p(x)W')' = 0$ satisfying $W(0) = W(L) = 0$ is $W \equiv 0$ on $[0,L]$. [*Hint*: Show that a nonzero solution of Problem (BV) must satisfy $W'(x) = a/p(x)$, where a is a constant. Since $p(x)$ is positive this means that $W(x)$ is either a strictly increasing or a strictly decreasing function on $(0,L)$.]

*9. Suppose λ_1 and λ_2 are distinct eigenvalues of Problem (BV) and $W_1(x)$, $W_2(x)$ are corresponding eigenfunctions. Prove that W_1 and W_2 are orthogonal. [*Hint*: W_i satisfies $(p(x)W_i')' - \lambda_i W_i = 0$, $W_i(0) = W_i(L) = 0$, $i = 1, 2$. Show that

$$(\lambda_2 - \lambda_1) \int_0^L W_1(x)W_2(x)\,dx$$
$$+ \int_0^L \{W_1(x)(p(x)W_2')' - W_2(x)(p(x)W_1')'\}\,dx = 0.$$

Integrate by parts and use the boundary conditions to obtain the desired result.]

10. Work Problems 7, 8, and 9 for Equation (3.51) with the boundary conditions

$$u(0,t) = u_x(L,t) = 0, \quad t > 0,$$

in place of (3.52). Can you carry out these same steps for all the boundary-value problems H_{ij}, $i,j = 1, 2, 3$?
11. Does the phrase "orthogonal sequence of functions" have any meaning with reference to a system of ordinary differential equations?

3.4 A DIFFUSION PROBLEM

Just as heat flows from regions of high temperature to regions of low temperature, so does an unevenly distributed volume of gas diffuse from points of higher concentration to points of lower concentration. The same phenomenon also occurs in chemical solutions if the concentration of the solute is not constant throughout the volume.

Let us consider a special diffusion problem. Suppose the pressure and temperature of the air in a cylinder $0 \leq x \leq l$ is the same as that of the surrounding atmosphere. Assume that the end of the cylinder $x = 0$ is opened at time $t = 0$ and that the other end remains closed for all $t \geq 0$. Let the concentration of a gas in the surrounding atmosphere equal $U_0 = $ constant. At time $t = 0$ the gas diffuses into the cylinder through the open end. Given that the initial concentration of gas in the cylinder is zero, we are to find the amount of gas in the cylinder as a function of time for all $t \geq 0$.

To solve this problem let $u(x,t)$ be the concentration of the diffusing gas in the cylinder at time t. Then the amount of gas in the cylinder is given by

$$Q_1(t) = \rho \int_0^l u(x,t) \, dx, \tag{3.53}$$

where ρ is the density. The amount of gas in the cylinder can also be determined by means of the flow of the diffusing gas through the open end, namely

$$Q_2(t) = -a^2\rho \int_0^t \frac{\partial u(0,s)}{\partial x} \, ds. \tag{3.54}$$

It now remains to be shown that $Q_1 = Q_2 \equiv Q$. This equality can be proved as follows.

Assuming that the coefficient of diffusion for the gas is constant, it can be shown that the process of diffusion in the cylinder is described by the differential equation

$$u_t = a^2 u_{xx} \tag{3.55}$$

where a is a constant related to the coefficient of diffusion and the coefficient of porosity. We also have the boundary conditions

$$u(0,t) = U_0, \quad u_x(l,t) = 0, \tag{3.56}$$

and the initial condition

$$u(x,0) = 0. \tag{3.57}$$

The student should pause and make certain that he understands exactly what is meant by a solution to this problem. Now integrate (3.55) over the rectangle $0 \leq x \leq l, 0 \leq s \leq t$. This gives

$$\int_0^l \int_0^t u_s(x,s) \, ds \, dx = a^2 \int_0^t \int_0^l u_{xx}(x,s) \, dx \, ds,$$

from which

$$\int_0^l u(x,t) \, dx - \int_0^l u(x,0) \, dx = a^2 \int_0^t u_x(l,s) \, ds - a^2 \int_0^t u_x(0,s) \, ds.$$

By virtue of the second of conditions (3.56) the first integral on the right is zero and by virtue of (3.57) the second integral on the left is zero. Thus (3.53) and (3.54) are equivalent as asserted.

Our task is now defined to be one of solving the boundary-value problem given by (3.55), (3.56), and (3.57) and thereafter obtaining Q from either (3.53) or (3.54). Moreover the basic problem of finding $u(x,t)$ is essentially a special case of Problem 6 at the end of Section 3.2 (also see Problem 5). Accordingly, we search for a solution of the form

$$u(x,t) = v(x,t) + w(x)$$

where $w(x)$ satisfies

$$w_{xx} = 0, \quad w(0) = U_0, \quad w_x(l) = 0.$$

The first condition requires that

$$w(x) = \alpha x + \beta,$$

the second that $\beta = U_0$ and the third that $\alpha = 0$. Thus the solution will have the form

$$u(x,t) = v(x,t) + U_0 \tag{3.58}$$

where V satisfies

$$(A) \quad \begin{cases} V_t = a^2 v_{xx}, \\ v(0,t) = v_x(l,t) = 0, \\ v(x,0) = -U_0. \end{cases}$$

In the notation of Section 3.1, Problem (A) is of type H_{12}.

The formal solution of Problem (A) is

$$v(x,t) = -\frac{4U_0}{\pi} \sum_{n=0}^{\infty} \frac{1}{2n+1} e^{-\frac{(2n+1)^2 \pi^2 a^2}{4l^2} t} \sin \frac{(2n+1)\pi x}{2l}.$$

Consequently, the concentration of gas in the cylinder is given by

$$u(x,t) = U_0 - \frac{4U_0}{\pi} \sum_{n=0}^{\infty} \frac{1}{2n+1} e^{-\frac{(2n+1)^2\pi^2a^2}{4l^2}t} \sin \frac{(2n+1)\pi x}{2l}. \quad (3.59)$$

Now let us use (3.53) to compute $Q(t)$. Doing so, we have

$$Q(t) = \rho \int_0^l \left[U_0 - \frac{4U_0}{\pi} \sum_{n=0}^{\infty} \frac{1}{2n+1} e^{-\frac{(2n+1)^2\pi^2a^2t}{4l^2}} \sin \frac{(2n+1)\pi x}{2l} \right] dx$$

$$= \rho l U_0 - \rho \int_0^l \frac{4U_0}{\pi} \sum_{n=0}^{\infty} \frac{1}{2n+1} e^{-\frac{(2n+1)^2\pi^2a^2t}{4l^2}} \sin \frac{(2n+1)\pi x}{2l} dx. \quad (3.60)$$

Assuming that the operations of integration and summation can be interchanged in the second term on the right, we have

$$Q(t) = \rho l U_0 \left[1 - \frac{8}{\pi^2} \sum_{n=0}^{\infty} \frac{e^{-\frac{(2n+1)^2\pi^2a^2t}{4l^2}}}{(2n+1)^2} \right]. \quad (3.61)$$

Although this expression for Q has been obtained by a series of formal manipulations, we can persuade ourselves that it is quite reasonable by the following argument. Suppose we can show that

$$\sum_{n=0}^{\infty} \frac{1}{(2n+1)^2} = \frac{\pi^2}{8}, \quad (3.62)$$

then $Q(0) = 0$. If we can also show that

$$\lim_{t \to +\infty} \sum_{n=0}^{\infty} \frac{e^{-\frac{(2n+1)^2\pi^2a^2}{4l^2}t}}{(2n+1)^2} = 0 \quad (3.63)$$

then $\lim_{t \to \infty} Q(t) = \rho l U_0$. Finally if we can show that

$$g(t) = \sum_{n=0}^{\infty} \frac{e^{-\frac{(2n+1)^2\pi^2a^2}{4l^2}t}}{(2n+1)^2}$$

is a strictly decreasing function of t, then we will know that $Q(t)$ as defined by (3.61) is a strictly increasing function of t defined for all $t \geq 0$ and increasing from zero at $t = 0$ to $\rho l U_0$ as $t \to \infty$. This agrees very well with our intuitive feeling of how the function Q must behave.

We shall begin by showing that the series

142 PARTIAL DIFFERENTIAL EQUATIONS

$$\sum_{n=0}^{\infty} \frac{e^{-\frac{(2n+1)^2\pi^2 a^2}{4l^2}t}}{(2n+1)^2} \quad (3.64)$$

converges for all values of $t \geq 0$. To see this, observe that

$$e^{-\frac{(2n+1)^2\pi^2 a^2}{4l^2}t} \leq 1, \quad \text{for all } t \geq 0.$$

Therefore the series (3.64) is term by term less than the series

$$\sum_{n=0}^{\infty} \frac{1}{(2n+1)^2} \leq 1 + \sum_{n=1}^{\infty} \frac{1}{n^2}.$$

By the comparison test it follows that (3.64) is convergent for all values of $t \geq 0$.

Next let us prove that (3.63) is valid. For this purpose we note that

$$\sum_{n=0}^{\infty} \frac{e^{-\frac{(2n+1)^2\pi^2 a^2}{4l^2}t}}{(2n+1)^2} \leq e^{-\frac{\pi^2 a^2}{4l^2}t} \sum_{n=0}^{\infty} \frac{1}{(2n+1)^2}.$$

Since the series $\sum_{n=0}^{\infty} \frac{1}{(2n+1)^2}$ converges it is bounded, say by $K > 0$. Thus

$$0 < \sum_{n=0}^{\infty} \frac{e^{-\frac{(2n+1)^2\pi^2 a^2}{4l^2}t}}{(2n+1)^2} \leq K e^{-\frac{\pi^2 a^2}{4l^2}t}.$$

The validity of (3.63) now follows at once from the fact that

$$\lim_{t \to \infty} e^{-\frac{\pi^2 a^2}{4l^2}t} = 0.$$

We next prove that $g(t)$ is strictly decreasing. We have

$$g(t) - g(t+h) = \sum_{n=0}^{\infty} \frac{e^{-\frac{(2n+1)^2\pi^2 a^2}{4l^2}t}}{(2n+1)^2} - \sum_{n=0}^{\infty} \frac{e^{-\frac{(2n+1)^2\pi^2 a^2}{4l^2}(t+h)}}{(2n+1)^2}.$$

Now

$$e^{-\frac{(2n+1)^2\pi^2 a^2}{4l^2}(t+h)} = e^{-\frac{(2n+1)^2\pi^2 a^2}{4l^2}t} e^{-\frac{(2n+1)^2\pi^2 a^2}{4l^2}h}$$

$$\leq e^{-\frac{(2n+1)^2\pi^2 a^2}{4l^2}t} e^{-\frac{\pi^2 a^2}{4l^2}h}, \quad \text{for all } n \geq 0.$$

It follows that

$$-\sum_{n=0}^{\infty} \frac{e^{-\frac{(2n+1)^2\pi^2 a^2}{4l^2}(t+h)}}{(2n+1)^2} \geq -e^{-\frac{\pi^2 a^2}{4l^2}h} g(t).$$

Thus

$$g(t) - g(t+h) \geq g(t)(1 - e^{-\frac{\pi^2 a^2}{4l^2}h}) > 0,$$

since $g(t) > 0$ and $1 - e^{-\frac{\pi^2 a^2}{4l^2}h} > 0$ for $h > 0$. This proves that $g(t)$ is strictly decreasing.

It remains for us to prove $g(0) = \pi^2/8$. To solve this problem let us, for the moment, suppose that we have established the result

$$\sum_{n=1}^{\infty} \frac{1}{n^2} = \frac{\pi^2}{6}. \tag{3.65}$$

Then we may write

$$\sum_{n=1}^{\infty} \frac{1}{n^2} = \sum_{n=1}^{\infty} \frac{1}{(2n)^2} + \sum_{n=0}^{\infty} \frac{1}{(2n+1)^2}$$

$$= \frac{1}{4} \sum_{n=1}^{\infty} \frac{1}{n^2} + \sum_{n=0}^{\infty} \frac{1}{(2n+1)^2}.$$

Hence

$$\frac{\pi^2}{6} = \frac{1}{4} \frac{\pi^2}{6} + g(0),$$

or

$$g(0) = \frac{3}{4} \frac{\pi^2}{6} = \frac{\pi^2}{8}.$$

Thus we need only establish (3.65). Incidentally, we have made essential use here of the fact that if a series converges absolutely the terms can be rearranged in any order without changing the limit. This is not true if a series is only conditionally convergent.

The easiest way to establish the validity of (3.65) is to find the Fourier series for the function $F(t) = t^2$ on, say, the interval $(-\pi,\pi)$. That is, we wish to write $F(t)$ in the form

$$F(t) = \frac{a_0}{2} + \sum_{n=1}^{\infty} (a_n \cos nt + b_n \sin nt). \tag{3.66}$$

Let us suppose that this series converges to F and in such a way that it can be integrated term by term. Then

$$\int_{-\pi}^{\pi} F(t)\, dt = \tfrac{1}{2} a_0 \int_{-\pi}^{\pi} dt + \sum_{1}^{\infty} \left\{ a_n \int_{-\pi}^{\pi} \cos nt\, dt + b_n \int_{-\pi}^{\pi} \sin nt\, dt \right\}$$

$$= \pi a_0.$$

It follows that

$$a_0 = \frac{1}{\pi} \int_{-\pi}^{\pi} F(t)\, dt. \tag{3.67}$$

If $k \geq 1$,

$$\int_{-\pi}^{\pi} F(t) \cos kt\, dt = \int_{-\pi}^{\pi} \frac{a_0}{2} \cos kt\, dt$$

$$+ \sum_{1}^{\infty} \left\{ a_n \int_{-\pi}^{\pi} \cos nt \cos kt\, dt + b_n \int_{-\pi}^{\pi} \sin nt \cos kt\, dt \right\}. \tag{3.68}$$

But we have

$$\int_{-\pi}^{\pi} \cos nt \cos kt\, dt = \begin{cases} 0, & k \neq n \\ \pi, & k = n \geq 1, \end{cases}$$

and

$$\int_{-\pi}^{\pi} \sin nt \cos kt\, dt = 0, \quad \text{for all } n, k.$$

Consequently, from (3.68) we obtain

$$a_k = \frac{1}{\pi} \int_{-\pi}^{\pi} F(t) \cos kt\, dt. \tag{3.69}$$

The values b_k can be computed by forming $\int_{-\pi}^{\pi} F(t) \sin kt\, dt$, from which

$$b_k = \frac{1}{\pi} \int_{-\pi}^{\pi} F(t) \sin kt\, dt.$$

Now if F is any integrable function the series (3.66) with a_0, a_k, b_k, $k \geq 1$, given by (3.67), (3.68), and (3.69) is called the *Fourier series* of F. This series may or may not converge and when it does converge we have at present no advance guarantee that it will converge to F.

At this point we will state a useful convergence theorem. For this we require first the following definitions.

Definition 3.5

A function F is said to be *piecewise continuous* on an interval $I = \{a \leq t \leq b\}$ if there is a sequence of points of I satisfying $t_0 = a < t_1 < t_2 < \cdots < t_p < b = t_{p+1}$ such that F is continuous in each subinterval

$t_r < t < t_{r+1}, r = 0, 1, \ldots, p$, and the limits from the right and left of F exist at each partition point t_r, $r = 1, \ldots, p$, the limit from the left exists at b, and the limit from the right exists at a.

A common notation found in most of the mathematical literature is that the limit from the left of F at t is written $F(t - 0)$ and the limit from the right is denoted by $F(t + 0)$. Thus Definition 3.5 requires that the numbers $F(t_r - 0), F(t_r + 0), r = 1, \ldots, p$, and $F(b - 0), F(a + 0)$ all exist. Of course, at any point t where F is continuous we have $F(t - 0) = F(t + 0) = F(t)$.

Definition 3.6

A function F defined on an interval $I = \{a \leq t \leq b\}$ will be called quasi-differentiable there if

(i) F is piecewise continuous on I.
(ii) At each point t satisfying $a < t < b$, the one-sided derivatives

$$F'(t + 0) = \lim_{h \to 0^+} \frac{F(t + h) - F(t + 0)}{h},$$

and

$$F'(t - 0) = \lim_{h \to 0^+} \frac{F(t - h) - F(t - 0)}{h}$$

both exist.

(iii) $$F'(a + 0) = \lim_{h \to 0^+} \frac{F(a + h) - F(a + 0)}{h},$$

and

$$F'(t - 0) = \lim_{h \to 0^+} \frac{F(b - h) - F(b - 0)}{h}$$

both exist.

Here the notation $\lim_{h \to 0^+}$ means h tends to zero through positive values. The reason for using the generalized difference quotient

$$\frac{F(t + h) - F(t + 0)}{h}$$

in place of $[F(t + h) - F(t)]/h$ is that the former is defined even at points of discontinuity for a piecewise continuous function.

Now if F is defined on the interval $I = \{a < t \leq b\}$ we can define a new function for all t called the *periodic extension* of F by getting

$$F(t + k|b - a|) = F(t) \quad \text{for all } k = 0, \pm 1, \pm 2, \ldots.$$

Theorem 3.6

Suppose F is quasidifferentiable on $\{-\pi \le t \le \pi\}$ and is extended periodically. Then the Fourier series of F converges to the value

$$F(t) = \tfrac{1}{2}[F(t+0) + F(t-0)]$$

at each point t. In particular, at π and $-\pi$ it converges to

$$F(\pi) = F(-\pi) = \tfrac{1}{2}[F(-\pi + 0) + F(\pi - 0)].$$

We can apply this theorem at once to the function $F(t) = t^2$, which is differentiable, to conclude that the Fourier series of t^2 converges to t^2 at every point. Therefore it only remains for us to deduce from the Fourier series of t^2 that (3.65) is valid. This is left as an exercise in the problems that follow.

PROBLEMS

1. Prove that (3.65) is true.
2. Obtain the Fourier series for the function

$$F(t) = \begin{cases} 0, & -\pi < t < 0, \\ t, & 0 \le t \le \pi. \end{cases}$$

   To what value does this series converge when $t = \pi$?
3. Obtain the Fourier series of the function

$$F(t) = \begin{cases} t^2, & \text{if } 0 < t < \pi, \\ 0, & \text{if } t = 0, \\ -t^2, & \text{if } -\pi < t < 0. \end{cases}$$

   Find the sum of this series for $-\pi \le t \le \pi$.
4. F is called an even function if $F(-t) = F(t)$ and an odd function if $F(-t) = -F(t)$. Show that if F is even $b_n = 0$ and if F is odd $a_n = 0$.
5. Solve the diffusion problem of this section assuming that both ends of the cylinder are closed by a semipermeable membrane, across which diffusion takes place. The boundary condition for a semipermeable membrane has the form $u_x(0,t) - c[u(0,t) - U_0] = 0$, where c is a constant and U_0 is the concentration outside the end of the cylinder. Note that in this problem you will have to solve the equation $\cot \lambda_n L = \lambda_n/c$ for the eigenvalues λ_n. Since this cannot be done in a closed form, you will have to leave the Fourier coefficients in integral form.
6. Solve the diffusion problem of this section assuming the diffusing gas dissociates at a rate proportional to the concentration of the gas at that point.

7. Solve Problem H_{11} for the case $L = \pi$ and the initial data $\varphi(x) = x^2$, $0 \leq x \leq \pi$, by continuing φ into $-\pi < x < 0$ so that it is an odd function. Use the theorem of this section to conclude that the Fourier series converges and comment on the value of this sum at $x = \pi$.

3.5 THE MAXIMUM-VALUE PRINCIPLE

In order to be able to prove some of the mathematical properties of the heat equation which we have thus far accepted without proof we need an additional tool, the maximum-value principle. The maximum-value principle, like the energy method, can be used to gather important information about solutions to a problem without requiring that a solution be known. In this section we present the maximum-value theorem with respect to the heat equation and in the process demonstrate the utility of the theorem.

Theorem 3.7

If the function u is continuous in the closed rectangle $0 \leq t \leq T$, $0 \leq x \leq L$ and satisfies the heat equation

$$u_t = a^2 u_{xx} \tag{3.70}$$

in the set $0 < x < L$, $0 < t \leq T$, then the maximum and minimum values of the function u occur either on the line segment, $0 \leq x \leq L$, $t = 0$, or on one of the segments, $0 \leq t \leq T$, $x = 0$, or $0 \leq t \leq T$, $x = L$.

A few preliminary remarks are in order before proving Theorem 3.7. Referring to Figure 3.1, the theorem simply states that u takes on its maximum and minimum values in the closed rectangle somewhere on the portion of the boundary denoted by the wavy line. It is essential to note that the theorem is stated for the closed rectangle of Figure 3.1. In fact we shall make use of one of the most important theorems of advanced calculus, which states that a function of x and t defined on the closed rectangle actually takes on its (absolute) maximum value and its (absolute)

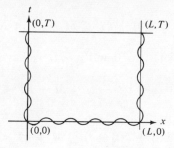

FIGURE 3.1

minimum value somewhere on this set. Finally, the physical significance of the theorem is simply that if the temperature at the boundary or at the initial time does not exceed a value M, then it is not possible to produce a temperature inside the rod in excess of M.

PROOF. Denote the maximum value of u on the three line segments defined in the statement of the theorem by M and suppose that at some point (x_0,t_0), $0 < x_0 < L$, $0 < t_0 \leq T$, u assumes the maximum value

$$u(x_0,t_0) = M + \epsilon, \quad \epsilon > 0.$$

The task is to prove that this supposition leads to a contradiction.

From the fact that u attains a maximum value at (x_0,t_0), it follows that

$$\frac{\partial u}{\partial x}(x_0,t_0) = 0, \quad \text{and} \quad \frac{\partial^2 u}{\partial x^2}(x_0,t_0) \leq 0, \tag{3.71}$$

for, if $u_{xx}(x_0,t_0)$ were positive at (x_0,t_0) the function $u(x_0,t_0)$ would have a minimum value on the segment $0 < x < L$ at the point x_0 and it would follow that (x_0,t_0) could not be a maximum point of u. We must also have

$$u_t(x_0,t_0) \geq 0, \tag{3.72}$$

for it follows that if $t_0 < T$ then $u_t(x_0,t_0) = 0$, but if $t_0 = T$ then $u_t(x_0,T) \geq 0$. Now if we knew that either $u_{xx}(x_0,t_0) < 0$ or $u_t(x_0,t_0) > 0$ or both, then the heat equation could not hold at (x_0,t_0) which is certainly a contradiction. Since it is possible that both $u_{xx}(x_0,t_0)$ and $u_t(x_0,t_0)$ are zero, we resort to the following device in order to reach a contradiction. Define the function

$$v(x,t) = u(x,t) + k(t_0 - t), \tag{3.73}$$

where $k > 0$ is to be properly chosen below. Clearly

$$v(x_0,t_0) = u(x_0,t_0) = M + \epsilon$$

and

$$k(t_0 - t) \leq kt_0 \leq kT, \quad \text{for } 0 < t \leq T.$$

Choose k so the $kT < \epsilon/2$, or $k < \epsilon/2T$. Then if $t = 0$ we have

$$v(x,0) = u(x,0) + kT \leq M + kT < M + \frac{\epsilon}{2}.$$

For the limiting conditions $x = 0$ and $x = L$ we have

$$v(0,t) = u(0,t) + k(t_0 - t) \leq M + kT < M + \frac{\epsilon}{2},$$

and

$$v(L,t) = u(L,t) + k(t_0 - t) \leq M + kT < M + \frac{\epsilon}{2}.$$

Consequently, $v(x,t)$ is less than $M + \epsilon/2$ on the wavy boundary of the closed rectangle and assumes the value $M + \epsilon > M + \epsilon/2$ at the point (x_0,t_0) inside. But v is continuous in the closed rectangle, since it is the sum of two continuous functions. It follows that v must have an absolute maximum in the closed rectangle at the point (x_1,t_1), and $x_1 \neq 0$, $x_1 \neq L$, and $t_1 \neq 0$. Now let us compute at the point (x_1,t_1). We have

$$v_{xx}(x_1,t_1) = u_{xx}(x_1,t_1) \leq 0$$

and

$$v_t(x_1,t_1) = u_t(x_1,t_1) - k \geq 0.$$

The last inequality implies u_t is positive at (x_1,t_1) while u_{xx} is nonpositive at (x_1,t_1). Consequently, Equation (3.70) cannot be satisfied by u at (x_1,t_1). This is the contradiction we have been seeking. It shows us that $u(x_0,t_0) \leq M$ and proves the theorem with respect to the maximum value.

Similar arguments may be used to prove the theorem with respect to the minimum value, but this is not really necessary since if u is a solution of (3.70) the function $\tilde{u} = -u$ is also a solution of (3.70) and has a maximum value where u has a minimum. Thus the theorem is completely proved.

3.6 UNIQUENESS IN THE HEAT-CONDUCTION PROBLEM

Let us start by invoking the maximum-value principle to prove that the solution to a particular class of heat-conduction problems is unique.

Theorem 3.8

If the two functions u_1 and u_2 are defined and continuous in the closed rectangle $0 \leq x \leq L$, $0 \leq t \leq T$ and satisfy the equation

$$u_t = a^2 u_{xx} + f(x,t), \quad \text{for} \quad 0 < x < L, \quad 0 < t \leq T, \tag{3.74}$$

together with the initial and boundary conditions

$$u_1(x,0) = u_2(x,0) = \varphi(x), \quad 0 \leq x \leq L,$$

$$u_1(0,t) = u_2(0,t) = l(t), \quad 0 < t \leq T,$$

and

$$u_1(L,t) = u_2(L,t) = r(t), \quad 0 \leq t \leq T,$$

then $u_1(x,t) = u_2(x,t)$, for all (x,t) in the closed rectangle.

PROOF. To prove this theorem we set

$$v(x,t) = u_2(x,t) - u_1(x,t),$$

noting that v is continuous wherever both u_1 and u_2 are continuous. Further,

$$v_t = \frac{\partial u_2}{\partial t} - \frac{\partial u_1}{\partial t} = a^2\left(\frac{\partial^2 u_2}{\partial x^2} - \frac{\partial^2 u_1}{\partial x^2}\right) = a^2 v_{xx},$$

so that v satisfies the homogeneous heat equation in $0 < x < L, 0 < t \leq T$. Finally we have

$$v(x,0) = 0, \quad v(0,t) = 0, \quad v(L,t) = 0$$

on the appropriate line segments.

Since v satisfies the conditions for the validity of the maximum-value principle, we may apply that result and conclude that v reaches its maximum and minimum values on one of the three line segments, at $t = 0, x = 0$ or $x = L$. It then follows that

$$v \equiv 0 \quad \text{for} \quad 0 \leq x \leq L, \quad 0 \leq t \leq T,$$

because v is zero on each of the three line segments. This proves that

$$u_1 \equiv u_2.$$

Another important consequence of the maximum-value principle may be stated as follows.

Theorem 3.9

Let u_1 and u_2 be continuous in $0 \leq x \leq L, 0 \leq t \leq T$ and be solutions of the heat equation

$$u_t = a^2 u_{xx}, \quad \text{for} \quad 0 < x < L, \quad 0 < t \leq T,$$

satisfying

$$u_1(x,0) \leq u_2(x,0), \quad 0 \leq x \leq L,$$
$$u_1(0,t) \leq u_2(0,t), \quad u_1(L,t) \leq u_2(L,t), \quad 0 \leq t \leq T.$$

Then

$$u_1(x,t) \leq u_2(x,t)$$

for all (x,t) such that $0 \leq x \leq L, 0 \leq t \leq T$.

PROOF. The difference $v = u_2 - u_1$ is continuous in $0 \leq x \leq L, 0 \leq t \leq T$ and is a solution of the heat equation in $0 < x < L, 0 < t \leq T$. Consequently, the maximum-value principle applies to v. Further, $v \geq 0$ on the line segments $t = 0, 0 \leq x \leq L, x = 0, 0 < t \leq T$, and $x = L, 0 < t \leq T$. It follows from the maximum-value principle that

$$v(x,t) \geq 0 \quad \text{for} \quad 0 \leq x \leq L, \quad 0 \leq t \leq T.$$

That is,

$$u_1(x,t) \leq u_2(x,t),$$

as was to be proved.

Corollary 3.1

If u, $\underline{u}$, and $\bar{u}$ are continuous in the closed rectangle $0 \leq x \leq L, 0 \leq t \leq T$ and satisfy

$$u_t = a^2 u_{xx} \quad \text{in} \quad 0 < x < L, \quad 0 < t \leq T,$$

together with the conditions

$$\underline{u}(x,0) \leq u(x,0) \leq \bar{u}(x,0), \quad 0 \leq x \leq L,$$

and $\underline{u}(0,t) \leq u(0,t) \leq \bar{u}(0,t), \underline{u}(L,t) \leq u(L,t), \leq \bar{u}(L,t), 0 < t \leq T$, then

$$\underline{u}(x,t) \leq u(x,t) \leq \bar{u}(x,t)$$

for $0 \leq x \leq L, 0 \leq t \leq T$.

PROOF. The corollary is proven by direct application of Theorem 3.9 to the pairs of functions $u, \bar{u}$ and $u, \underline{u}$.

Corollary 3.2

If u_1 and u_2 are continuous in the closed rectangle $0 \leq x \leq L, 0 \leq t \leq T$ and satisfy

$$u_t = a^2 u_{xx} \quad \text{in} \quad 0 < x < L, \quad 0 < t \leq T,$$

for

$$|u_1(x,0) - u_2(x,0)| \leq \epsilon, \quad 0 \leq x \leq L,$$

and

$$|u_1(0,t) - u_2(0,t)| \leq \epsilon, \quad |u_1(L,t) - u_2(L,t)| \leq \epsilon, \quad 0 < t \leq T,$$

then

$$|u_1(x,t) - u_2(x,t)| \leq \epsilon$$

for $0 \leq x \leq L, 0 \leq t \leq T$.

The proof of this corollary is left to the reader as an exercise.

The result of Corollary 3.2 is often spoken of as *continuous dependence* of the solution of the boundary-value problem on the initial and boundary data, or simply continuous dependence on the data. These considerations lead us to formulate the following criteria for a problem to be well-posed.

A problem is said to be well-posed if

(a) there exists a solution to the problem,
(b) the solution is unique, and
(c) the solution depends continuously upon the data.

This concept is general in that it applies to most of those problems that are formulated as differential equations, ordinary and partial, together with initial conditions, boundary conditions, or both. The criteria are appealing in that they are in accord with our physical intuition about the nature of the real world.

Of the three requirements that a problem be well-posed, the first requirement—that a solution exist—is usually the most difficult to verify. For the heat equation, or more precisely, for Problem H_{11}, we have yet to do so since we must first develop more working tools. That (b) and (c) are valid for Problem H_{11} is, of course, a consequence of our work in this section. Non-mathematicians are, in general, more concerned with (b) and (c) than with (a), since any problem at hand is derived from a physical system, hence it is reasonable to assume that "a" solution exists.

PROBLEMS

1. Prove Corollary 3.2.
2. Let $v(x,t)$ satisfy the conditions
   (i) $v(x,0) = 0$ for all real values of x, and
   (ii) $v(x,t) \leq 2M$ (M a constant) for all x and all $t \geq 0$,
   (iii) $v_t - a^2 v_{xx} = 0$ for $t > 0$ and all x.
   Prove that for $|x| \leq L$, and $t \geq 0$, the function
   $$V(x,t) = \frac{4M}{L^2}\left(\frac{x^2}{2} + a^2 t\right)$$
   satisfies
   $$-V(x,t) \leq v(x,t) \leq V(x,t).$$

3. Using the result of Problem 2 prove that the solution to the problem
   $$u_t - a^2 u_{xx} = 0, \quad -\infty < x < \infty, \quad t > 0,$$
   $$u(x,0) = f(x), \quad -\infty < x < \infty,$$
   $$|u(x,t)| \leq M, \quad -\infty < x < \infty, \quad t \geq 0$$
   is unique.

   Problems 4 through 9 are concerned with the application of the energy method expounded in Chapter 1 to the heat equation.

4. By multiplying the heat equation
   $$u_t - a^2 u_{xx} = 0$$
   by u_t and appropriate integrations, derive the identity

$$\int_0^L a^2 u_x^2(x,t)\, dx + 2\int_0^t \int_0^L u_t^2(x,t)\, dx\, dt$$
$$= \int_0^L a^2 u_x^2(x,0)\, dx + 2a^2 \int_0^t (u_x u_t)(L,\sigma)\, d\sigma - 2a^2 \int_0^t (u_x u_t)(0,\sigma)\, d\sigma.$$

5. By multiplying the heat equation
$$u_t - a^2 u_{xx} = 0$$
by u and appropriate integrations, derive the identity
$$\int_0^L u^2(x,t)\, dx + 2a^2 \int_0^t \int_0^L u_x^2(x,t)\, dx\, dt$$
$$= \int_0^L u^2(x,0)\, dx + 2a^2 \int_0^t (u u_x)(L,\sigma)\, d\sigma - 2a^2 \int_0^t (u u_x)(0,\sigma)\, d\sigma.$$

6. Assume that $u_x(L,t) = u_x(0,t) = 0$. Prove that
$$\int_0^L u^2(x,t)\, dx \le \int_0^L u^2(x,0)\, dx.$$

7. Assume that $u(0,t) = 0$, $u(L,t) + \alpha u_x(L,t) = 0$, $\alpha > 0$. Derive the inequalities
$$\int_0^L u^2(x,t)\, dx + 2a^2\alpha \int_0^t u_x^2(L,\sigma)\, d\sigma \le \int_0^L u^2(x,0)\, dx,$$
$$\int_0^L u_x^2(x,t)\, dx + \frac{1}{\alpha} u^2(L,t) \le \int_0^L u_x^2(x,0)\, dx + \frac{1}{\alpha} u^2(L,0).$$

8. Deduce that the solution of the boundary-value problem
$$u_t - a^2 u_{xx} = f(x,t), \quad 0 < x < L, \quad t > 0,$$
$$u_x(0,t) = f_1(t), \quad u_x(L,t) = f_2(t), \quad t > 0,$$
$$u(x,0) = g(x), \quad 0 < x < L,$$
is unique. [*Hint:* Use the inequality of Problem 6.]

9. Prove that the solution of the boundary-value problem
$$u_t - a^2 u_{xx} = f(x,t), \quad 0 < x < L, \quad t > 0,$$
$$u(0,t) = f_1(t), \quad u(L,t) + \alpha u_x(L,t) = f_2(t), \quad \alpha > 0, \quad t > 0,$$
$$u(x,0) = g(x), \quad 0 < x < L,$$
is unique. Can this result be deduced from the inequalities of Problem 7?

*10. Let $f(t)$ be an integrable function of t. Deduce from the inequality
$$\int_0^t (1 + \lambda |f(\sigma)|)^2\, d\sigma \ge 0$$
which is valid for arbitrary real λ, the inequality

$$\left(\int_0^t |f(\sigma)|\, d\sigma\right)^2 \le t \int_0^t [f(\sigma)]^2\, d\sigma.$$

This result is a special case of a theorem known as *Schwarz's inequality*.

11. Use the inequality obtained in Problem 10 to deduce the inequality

$$\int_0^L \left\{\frac{2}{t}[u(x,t) - u(x,0)]^2 + a^2 u_x^2(x,t)\right\} dx \le a^2 \int_0^L u_x^2(x,0)\, dx$$
$$+ 2a^2 \int_0^t (u_x u_t)(L,\sigma)\, d\sigma - 2a^2 \int_0^t (u_x u_t)(0,\sigma)\, d\sigma,$$

valid for $t > 0$, from the energy identity of Problem 4.

12. Prove that

$$2ab \le \epsilon a^2 + \frac{1}{\epsilon} b^2$$

for any $\epsilon > 0$.

13. Apply the inequality of Problem 12 to the result obtained in Problem 11 to deduce the inequality

$$\int_0^L \left[\frac{1}{t}u^2(x,t) + a^2 u_x^2(x,t)\right] dx \le \int_0^L \left[\frac{2}{t}u^2(x,0) + a^2 u_x^2(x,0)\right] dx$$
$$+ 2a^2 \left\{\int_0^t (u_x u_t)(L,\sigma)\, d\sigma - \int_0^t (u_x u_t)(0,\sigma)\, d\sigma\right\}.$$

3.7 THE NONHOMOGENEOUS HEAT EQUATION

Let us begin by considering a simple nonhomogeneous problem where we wish to find a function u satisfying

$$u_t = a^2 u_{xx} + F(x,t), \quad 0 < x < L, \quad t > 0, \tag{3.75}$$

the initial condition

$$u(x,0) = 0, \tag{3.76}$$

and the boundary conditions

$$u(0,t) = 0, \tag{3.77}$$
$$u(L,t) = 0.$$

Our work in Sections 3.2 and 3.3, and in particular Theorem 3.5 where we show that the sequence $\{\sin n\pi x/L\}_1^\infty$ is complete, makes it seem reasonable to search for a solution $u(x,t)$ having the form

$$u(x,t) = \sum_{n=1}^\infty u_n(t) \sin \frac{n\pi}{L} x. \tag{3.78}$$

In order to work with this function we shall have to assume that $F(x,t)$ can be represented in the same form, namely

$$F(x,t) = \sum_{n=1}^{\infty} F_n(t) \sin \frac{n\pi x}{L}. \tag{3.79}$$

Recalling the formal arguments we used in Section 3.2, this in turn requires that

$$F_n(t) = \frac{a}{L} \int_0^2 F(\xi,t) \sin \frac{n\pi \xi}{L} d\xi. \tag{3.80}$$

Thus $F_n(t)$ is the nth Fourier sine coefficient of the function $F(x,t)$. If $F(x,t)$ is continuous as a function of ξ for each fixed t, these coefficients are well-defined. The convergence of the series in (3.79) is, of course, another matter.

Continuing formally, we simply substitute (3.78) and (3.79) into (3.75). This gives

$$\sum_{n=0}^{\infty} \dot{u}_n(t) \sin \frac{n\pi x}{L} = -a^2 \sum_{n=0}^{\infty} \left(\frac{n\pi}{L}\right)^2 u_n(t) \sin \frac{n\pi x}{L} + \sum_{n=0}^{\infty} F_n(t) \sin \frac{n\pi x}{L}$$

or

$$\sum_{n=0}^{\infty} \sin \frac{n\pi x}{L} \left\{ \dot{u}_n(t) + \left(\frac{n\pi a}{L}\right)^2 u_n(t) - F_n(t) \right\} = 0.$$

Now on the interval $0 < x < L$ the functions $\sin n\pi x/L$ are linearly independent, and therefore the linear combination can be zero if and only if the coefficients are zero. Accordingly we obtain the infinite sequence of differential equations

$$\dot{u}_n(t) + \left(\frac{n\pi a}{L}\right)^2 u_n(t) = F_n(t), \qquad n = 1, 2, \ldots. \tag{3.81}$$

Next we make use of the initial condition (3.76). This yields

$$u_n(x,0) = \sum_{n=1}^{\infty} u_n(0) \sin \frac{n\pi x}{L} = 0,$$

where again because of the linear independence of the sine functions we obtain

$$u_n(0) = 0, \quad n = 1, 2, \ldots \tag{3.82}$$

Solving (3.81) subject to the initial conditions (3.82) yields

$$u_n(t) = \int_0^t e^{-\left(\frac{n\pi a}{L}\right)^2 (t-\tau)} F_n(\tau)\, d\tau. \tag{3.83}$$

Substituting (3.83) into (3.78) gives us

$$u(x,t) = \sum_{n=1}^{\infty} \left\{ \int_0^t e^{-\left(\frac{n\pi a}{L}\right)^2 (t-\tau)} F_n(\tau)\, d\tau \right\} \sin \frac{n\pi x}{L}. \tag{3.84}$$

Next substitute (3.80) into (3.84). This produces

$$u(x,t) = \sum_{n=1}^{\infty} \left\{ \frac{2}{L} \int_0^t e^{-\left(\frac{n\pi a}{L}\right)^2 (t-\tau)} \int_0^L F(\xi,\tau) \sin \frac{n\pi \xi}{L}\, d\xi\, d\tau \right\} \sin \frac{n\pi x}{L}.$$

An interchange of summation and integration yields

$$u(x,t) = \int_0^t \int_0^L \frac{2}{L} \left\{ \sum_{n=1}^{\infty} \sin \frac{n\pi \xi}{L} \sin \frac{n\pi x}{L} e^{-\left(\frac{n\pi a}{L}\right)^2 (t-\tau)} \right\} F(\xi,\tau)\, d\xi\, d\tau \tag{3.85}$$

$$= \int_0^t \int_0^L G(x,\xi, t-\tau) F(\xi,\tau)\, d\xi\, d\tau,$$

where $G(x,\xi, t-\tau)$ is the same Green's function we obtained in Section 3.2, evaluated now at $t-\tau$.

With this result at hand, let us finally produce a formal solution to the first general boundary-value problem for the heat equation, namely

$$u_t = a^2 u_{xx} + F(x,t) \tag{3.86}$$

with the additional conditions

$$u(x,0) = g(x), \tag{3.87}$$

$$u(0,t) = h_1(t),$$
$$u(L,t) = h_2(t). \tag{3.88}$$

To solve this problem we set

$$u(x,t) = w(x,t) + v(x,t) \tag{3.89}$$

and seek to determine $w(x,t)$ in such a way that

$$w(0,t) = h_1(t),\quad w(L,t) = h_2(t).$$

For this purpose let us suppose

$$w(x,t) = \alpha(t)x + \beta(t),$$

then

$$w(0,t) = \beta(t) = h_1(t),$$

and

so that
$$w(L,t) = \alpha(t)L + h_1(t) = h_2(t)$$
$$\alpha(t) = \frac{h_2(t) - h_1(t)}{L}.$$

Thus we have
$$w(x,t) = \frac{h_2(t) - h_1(t)}{L} x + h_1(t). \tag{3.90}$$

Assume for the present that h_1 and h_2 are differentiable functions of t for all $t \geq 0$. Then
$$u_t = [\dot{h}_2(t) - \dot{h}_1(t)] \frac{x}{L} + \dot{h}_1(t) + v_t,$$
$$u_{xx} = v_{xx},$$
so that
$$u_t - a^2 u_{xx} = F(x,t)$$
becomes
$$v_t - a^2 v_{xx} = F(x,t) - [(\dot{h}_2(t) - \dot{h}_1(t)] \frac{x}{L} - \dot{h}_1(t) = \tilde{F}(x,t). \tag{3.91}$$

The function v also satisfies
$$v(0,t) = 0,$$
$$v(L,t) = 0, \tag{3.92}$$
and
$$v(x,0) = g(x) - w(x,0) = g(x) - [h_2(0) - h_1(0)] \frac{x}{L} - h_1(0). \tag{3.93}$$

In order to solve the problem (3.91), (3.92), (3.93) we set
$$v(x,t) = v^*(x,t) + v^{**}(x,t) \tag{3.94}$$
where v^* satisfies the partial differential equation
$$v_t^* = a^2 v_{xx}^*,$$
subject to the conditions
$$v^*(0,t) = v^*(L,t) = 0,$$
$$v^*(x,0) = g(x) - [h_2(0) - h_1(0)] \frac{x}{L} - h_1(0),$$
and v^{**} satisfies
$$v_t^{**} = a^2 v_{xx}^{**} + F(x,t) - [\dot{h}_2(t) - \dot{h}_1(t)] \frac{x}{L} - \dot{h}_1(t),$$
and
$$v^{**}(0,t) = v^{**}(L,t) = 0,$$
$$v^{**}(x,0) = 0.$$

The first of these problems was solved above in Section 3.2 and the second in this section. Therefore, we obtain

$$v^*(x,t) = \int_0^L G(x,\xi,t) \left\{ g(\xi) - [h_2(0) - h_1(0)]\frac{\xi}{L} - h_1(0) \right\} d\xi, \quad (3.95)$$

and

$$v^{**}(x,t) = \int_0^t \int_0^L G(x,\xi, t-\tau) \\ \times \left\{ F(\xi,\tau) - [\dot{h}_2(\tau) - \dot{h}_1(\tau)]\frac{\xi}{L} - \dot{h}_1(\tau) \right\} d\xi\, d\tau. \quad (3.96)$$

Now combining (3.90), (3.95), and (3.96) into (3.89) yields the formula

$$u(x,t) = [h_2(t) - h_1(t)]\frac{x}{L} + h_1(t) + \int_0^L G(x,\xi,t) \left\{ g(\xi) \\ - [h_2(0) - h_1(0)]\frac{\xi}{L} - h_1(0) \right\} d\xi \\ + \int_0^t \int_0^L G(x,\xi, t-\tau) \left\{ F(\xi,\tau) - [\dot{h}_2(\tau) - \dot{h}_1(\tau)]\frac{\xi}{L} - \dot{h}_1(\tau) \right\} d\xi\, d\tau, \quad (3.97)$$

which is the solution for the entire problem.

It is possible to write (3.97) in a more elegant form as we shall now show. For this purpose note first that

$$G_{\xi\xi}(x,\xi,t-\tau) = -\frac{2}{L} \sum_{n=1}^{\infty} \left(\frac{n\pi}{L}\right)^2 \sin\frac{n\pi x}{L} \sin\frac{n\pi\xi}{L} e^{-\left(\frac{n\pi a}{L}\right)^2(t-\tau)},$$

and

$$G_\tau(x,\xi,t-\tau) = \frac{2}{L} \sum_{n=1}^{\infty} \left(\frac{n\pi a}{L}\right)^2 \sin\frac{n\pi x}{L} \sin\frac{n\pi\xi}{L} e^{-\left(\frac{n\pi a}{L}\right)^2(t-\tau)},$$

so that

$$G_\tau(x,\xi,t-\tau) = -a^2 G_{\xi\xi}(x,\xi,t-\tau).$$

We now apply this result to the double integral in (3.97) as follows:

$$\int_0^t \int_0^L G(x,\xi,t-\tau)[\dot{h}_2(\tau) - \dot{h}_1(\tau)]\xi\, d\xi\, d\tau \\ = \int_0^L \xi \int_0^t G(x,\xi,t-\tau)[\dot{h}_2(\tau) - \dot{h}_1(\tau)]\, d\tau\, d\xi \\ = \int_0^L \xi \left\{ [h_2(\tau) - h_1(\tau)]G(x,\xi,t-\tau)\Big|_0^t \right. \\ \left. - \int_0^t \frac{\partial G}{\partial \tau}(x,\xi,t-\tau)[h_2(\tau) - h_1(\tau)]\, d\tau \right\} d\xi$$

$$= [h_2(t) - h_1(t)] \int_0^L G(x,\xi,0)\xi \, d\xi - [h_2(0) - h_1(0)] \int_0^L G(x,\xi,t)\xi \, d\xi$$
$$+ a^2 \int_0^t [h_2(\tau) - h_1(\tau)] \int_0^L G_{\xi\xi}(x,\xi, t - \tau)\xi \, d\xi \, d\tau.$$

Further,
$$\int_0^L G_{\xi\xi}(x,\xi, t - \tau)\xi \, d\xi = G_\xi(x,\xi, t - \tau)\xi \Big|_0^L - \int_0^L G_\xi(x,\xi, t - \tau) \, d\xi$$
$$= LG_\xi(x,L, t - \tau)$$

since $G(x,L, t - \tau) = G(x,0, t - \tau) = 0$. Consider next the double integral

$$\int_0^t \int_0^L G(x,\xi, t - \tau)\dot{h}_1(\tau) \, d\xi \, d\tau = \int_0^L \int_0^t G(x,\xi, t - \tau)\dot{h}_1(\tau) \, d\tau \, d\xi$$
$$= \int_0^L \left\{ h_1(\tau)G(x,\xi, t - \tau) \Big|_0^t - \int_0^t h_1(\tau)G_\tau(x,\xi, y - \tau) \, d\tau \right\} d\xi$$
$$= h_1(t) \int_0^L G(x,\xi,0) \, d\xi - h_1(0) \int_0^L G(x,\xi,t) \, d\xi$$
$$+ a^2 \int_0^t h_1(\tau) \int_0^L G_{\xi\xi}(x,\xi, t - \tau) \, d\xi \, d\tau$$
$$= h_1(t) \int_0^L G(x,\xi,0) \, d\xi - h_1(0) \int_0^L G(x,\xi,t) \, d\xi$$
$$+ a^2 \int_0^t G_\xi(x,L, t - \tau)h_1(\tau) \, d\tau - a^2 \int_0^t G_\xi(x,0, t - \tau)h_1(\tau) \, d\tau.$$

Finally we observe that
$$x = \int_0^L G(x,\xi,0)\xi \, d\xi, \quad 1 = \int_0^L G(x,\xi,0) \, d\xi.$$

Inserting all of these results into (3.97) yields the formula

$$u(x,t) = \int_0^L G(x,\xi,t)g(\xi) \, d\xi - a^2 \int_0^t G_\xi(x,L, t - \tau)h_2(\tau) \, d\tau$$
$$+ a^2 \int_0^t G_\xi(x,0, t - \tau)h_1(\tau) \, d\tau + \int_0^t \int_0^L G(x,\xi, t - \tau)F(\xi,\tau) \, d\xi \, d\tau. \quad (3.98)$$

This remarkable formula shows that the first general boundary-value problem (3.86), (3.87), (3.88) for the heat equation can be completely solved with the help of the Green's function $G(x,\xi, t - \tau)$ for Problem H_{11}. Notice also that the derivatives $\dot{h}_1(\tau)$ and $\dot{h}_2(\tau)$ which appear in (3.97) do not appear in (3.98). Thus it seems plausible that we can discard the hypothesis that these functions are differentiable provided we can show that the Green's function is well-behaved.

We wish to emphasize once again that the most effective way to justify all of the formal manipulations made in Section 3.2 and in the present section is to verify directly that the integrals in formula (3.98) behave as specified. If we write (3.98) in the form

$$u(x,t) = \sum_{j=1}^4 u^{(j)}(x,t),$$

where u^j denotes the jth integral on the right-hand side, then the integrals solve the following problems:

(1) $u_t^{(1)} = a^2 u_{xx}^{(1)}$, $u^{(1)}(0,t) = u^{(1)}(L,t) = 0$, $u^{(1)}(x,0) = g(x)$,
(2) $u_t^{(2)} = a^2 u_{xx}^{(2)}$, $u^{(2)}(0,t) = 0$, $u^{(2)}(L,t) = h_2(t)$, $u^{(2)}(x,0) = 0$,
(3) $u_t^{(3)} = a^2 u_{xx}^{(3)}$, $u^{(3)}(0,t) = h_1(t)$, $u^{(3)}(L,t) = 0$, $u^{(3)}(x,0) = 0$,
(4) $u_t^{(4)} = a^2 u_{xx}^{(4)} + F(x,t)$, $u^{(4)}(0,t) = u^{(4)}(L,t) = 0$, $u^{(4)}(x,0) = 0$.

Finally, note that the representation (3.98) expressing u as a sum of integrals again suggests that conditions of integrability may suffice for the data g, h_1, h_2, and F. To obtain the most general conditions on these functions which will ensure the validity of the representation (3.98) is a fairly deep problem beyond the scope of this text. For example, it is not true that the double integral will solve Problem (4) if F is merely integrable. The student should be warned therefore that in unusual problems with discontinuities he should consult the mathematical literature before blindly applying such a formula.

PROBLEMS

1. Solve the boundary-value problem
$$u_t = a^2 u_{xx}, \quad 0 < x < L, \quad t > 0,$$
$$u(0,t) = 0, \quad u(L,t) = At, \quad 0 < L,$$
$$u(x,0) = 0, \quad 0 < x < L.$$

2. Solve the boundary-value problem
$$u_t = a^2 u_{xx} + F(x,t), \quad 0 < x < L, \quad 0 < t$$
$$u(0,t) = 0, \quad u_x(L,t) = g_1(t), \quad 0 < t,$$
$$u(x,0) = 0, \quad 0 < x < L.$$

3. In the text the assertion is made that
$$x = \int_0^L G(x,\xi,0)\xi \, d\xi, \quad 1 = \int_0^L G(x,\xi,0) \, d\xi.$$

Why must these formulas be valid?

The next few problems have to do with the solution of the boundary-value problem P defined by
$$u_t = a^2 u_{xx} + F(x,t), \quad 0 < x < L, \quad t > 0,$$
$$u_x(0,t) - hu(0,t) = g_1(t), \quad u_x(L,t) + hu(L,t) = g_2(t), \quad 0 < t,$$
$$u(x,0) = \varphi(x), \quad 0 < x < L.$$

4. Show that Problem P can be solved in the form
$$u(x,t) = v(x,t) + g(x,t)$$

where v is a solution of the Problem P' defined by

$$v_t = a^2 v_{xx} + F'(x,t), \quad 0 < x < L, \quad t > 0,$$
$$v_x(0,t) - hv(0,t) = 0, \quad v_x(L,t) + hv(L,t) = 0, \quad 0 < t,$$
$$v(x,0) = \varphi'(x).$$

Determine the functions $g(x,t)$, $F'(x,t)$, and $\varphi'(x)$ explicitly.

5. Determine the solution of Problem P' in the case where $F'(x,t) = 0$. Show that the eigenfunctions are

$$X_n(x) = \cos \lambda_n x + \frac{h}{\lambda_n} \sin \lambda_n x, \quad n = 1, 2, 3, \ldots,$$

where the λ_n, $n = 1, 2, 3, \ldots$ are the positive roots of the equation

$$\cot \frac{\lambda}{L} = \frac{1}{2}\left(\frac{\lambda}{hL^2} - \frac{hL^2}{\lambda}\right).$$

6. Show that Problem P' will be solved with the aid of the functions

$$v_n(t) = \int_0^t e^{-a^2 \lambda_n^2 (t-s)} \Theta_n(s) \, ds + a_n e^{-a^2 \lambda_n^2 t},$$

where

$$\Theta_n(t) = \frac{2\lambda_n^2}{(\lambda_n^2 + h^2)L + 2h} \int_0^L F'(\xi,t) X_n(\xi) \, d\xi,$$

$$a_n = \frac{2\lambda_n^2}{(\lambda_n^2 + h^2)L + 2h} \int_0^L g'(\xi) X_n(\xi) \, d\xi.$$

7. Write out the Green's function for Problem P. Express the solution in a form similar to formula (3.98) for the solution of the first boundary-value problem.

8. Solve the boundary-value problem

$$u_t = a^2 u_{xx} - Ku + F(x,t), \quad 0 < x < L, \quad 0 > t,$$
$$u_x(0,t) - hu(0,t) = g_1(t), \quad u_x(L,t) + hu(L,t) = g_2(t), \quad 0 < t,$$
$$u(x,0) = \varphi(x), \quad 0 < x < L.$$

9. Compute $\lim_{t \to +\infty} u(x,t)$ if u is the solution of the boundary-value problem

$$u_t = a^2 u_{xx}, \quad 0 < x < L, \quad 0 < t,$$
$$u(0,t) = 0, \quad u(L,t) = A \cos \omega t, \quad 0 < t,$$
$$u(x,0) = \varphi(x), \quad 0 < x < L.$$

10. Compute $\lim_{t \to +\infty} u(x,t)$ if u is the solution of the boundary-value problem

$$u_t = a^2 u_{xx}, \quad 0 < x < L, \quad 0 < t,$$
$$u(0,t) = 0, \quad u_x(L,t) = A e^{i\omega t}, \quad 0 < t,$$
$$u(x,0) = \varphi(x), \quad 0 < x < L.$$

11. Compute $\lim_{t \to +\infty} u(x,t)$ if u is the solution of the boundary-value problem

$$u_t = a^2 u_{xx}, \quad 0 < x < L, \quad 0 < t,$$
$$u(0,t) = 0, \quad u_x(L,t) - hu(L,t) = A e^{i\omega t}.$$

3.8 ASYMPTOTIC BEHAVIOR OF SOLUTIONS

By working the problems at the end of Section 3.5 the student had an opportunity to apply energy methods to the heat equation. Our aim in this section is to apply the same methods to a slightly more general equation in order to demonstrate how solutions to various boundary-value problems behave as $t \to \infty$.

Let us consider the equation

$$u_{xx} = c(x)(bu + u_t), \quad 0 < x < 1, \quad t > 0, \tag{3.99}$$

with the initial condition

$$u(x,0) = f(x), \quad 0 \leq x \leq 1. \tag{3.100}$$

We shall assume throughout this section that the function c is defined and has a continuous first derivative on the interval $[0,1]$ and, moreover, that there exist positive constants c_0 and c_1 such that

$$c_1 \geq c(x) \geq c_0 > 0 \quad \text{on } [0,1]. \tag{3.101}$$

We shall also assume that f has a continuous derivative on $[0,1]$ and $f(x) \not\equiv 0$. With these hypotheses on the functions c and f it is possible to show that every solution of (3.99), (3.100) satisfying any set of linear boundary conditions has a continuous mixed second partial derivative u_{xt} on the set $0 < x < 1, t > 0$. We shall accept this fact without proof. It will be needed later when we wish to interchange u_{xt} and u_{tx}. The constant b may be positive or negative.

The starting point of our analysis is to construct an energy identity by multiplying (3.99) by u. We have

$$uu_{xx} = c(x)(bu^2 + uu_t). \tag{3.102}$$

Note that the following identities may be formed using terms of (3.102).

$$uu_{xx} = (uu_x)_x - u_x^2$$

and

$$uu_t = \tfrac{1}{2}(u)_t^2.$$

With these identities (3.102) may be written as

$$(uu_x)_x - (u_x^2 + bc(x)u^2) = \tfrac{1}{2}(c(x)u^2)_t.$$

Let us set
$$h(t) = \int_0^1 c(x)u^2(x,t)\,dx. \tag{3.103}$$

Assume now that the boundary conditions are such that
$$u(0,t)u_x(0,t) = u(1,t)u_x(1,t) = 0. \tag{3.104}$$

Then integrating the revised form of (3.102) from 0 to 1 with respect to x yields
$$\dot h(t) = -2\int_0^1 [u_x^2(x,t) + bc(x)u^2(x,t)]\,dx. \tag{3.105}$$

The second step in our analysis starts by multiplying (3.99) by u_t to obtain
$$u_t u_{xx} = c(x)(buu_t + u_t^2).$$

Assuming $u_{xt} = u_{tx}$, we form
$$u_t u_{xx} = (u_t u_x)_x - \tfrac{1}{2}(u_x^2)_t,$$
and place it into the previous identity to obtain
$$(u_t u_x)_x - \tfrac{1}{2}(u_x^2 + bc(x)u^2)_t = c(x)u_t^2.$$

Assume next that the boundary conditions are such that
$$u_t(0,t)u_x(0,t) = u_t(1,t)u_x(1,t) = 0. \tag{3.106}$$

Then again integrating from 0 to 1 with respect to x, we obtain
$$-\frac{d}{dt}\int_0^1 [u_x^2(x,t) + bc(x)u^2(x,t)]\,dx = 2\int_0^1 c(x)u_t^2(x,t)\,dx. \tag{3.107}$$

The identities (3.105) and (3.107) are the basic energy identities associated with Equation (3.99). We shall use them in several ways.

Theorem 3.10

Suppose the functions f and c satisfy the conditions given above and that u is any solution of an initial boundary-value problem for Equation (3.99) with boundary conditions satisfying (3.104) and (3.106). Then there exist constants α and $\beta > 0$ such that
$$h(t) \geq \alpha t + \beta, \quad \text{for all } t \geq 0. \tag{3.108}$$

PROOF. From (3.105) and (3.107) we deduce at once that
$$\ddot h(t) = 4\int_0^1 c(x)u_t^2(x,t)\,dx.$$

Since $c(x) \geq c_0 > 0$ and noting that $u_t^2(x,t) > 0$ we have
$$\int_0^1 c(x) u_t^2(x,t)\, dx \geq c_0 \int_0^1 u_t^2(x,t)\, dx \geq 0.$$
It immediately follows that
$$\ddot{h}(t) \geq 0.$$
Integrating twice, we obtain
$$h(t) \geq \dot{h}(0) t + h(0).$$
Since $f \not\equiv 0$ on $[0,1]$ and f is continuous we have
$$h(0) = \int_0^1 c(x) f^2(x)\, dx > 0.$$
This proves the theorem with
$$\beta = h(0) \quad \text{and} \quad \alpha = \dot{h}(0).$$

Theorem 3.10 does not say anything about the behavior of $h(t)$ for large values of t when α is negative. Indeed we can see directly from its definition that $h(t)$ is always nonnegative while if $\alpha < 0$ the number $\alpha t + \beta$ becomes more and more negative as $t \to \infty$. Thus if $\alpha < 0$, Theorem 3.10 implies the positivity of $h(t)$ only for sufficiently small values of t. On the other hand, if $\alpha \geq 0$ (3.108) says much more about the behavior of $h(t)$. Let us pursue this problem further.

From (3.105) we have
$$\alpha = \dot{h}(0) = -2 \int_0^1 [f'^2(x) + bc(x) f^2(x)]\, dx.$$
It is slightly more suggestive to write this in the form
$$\alpha = -2 \left[\int_0^1 f'^2(x)\, dx + b \int_0^1 c(x) f^2(x)\, dx \right]$$
or
$$\alpha = -2 \left[\int_0^1 f'^2(x)\, dx + b h(0) \right]. \tag{3.109}$$

Theorem 3.11

Assume that the conditions of Theorem 3.10 are satisfied and, in addition, that
$$b < -\frac{1}{h(0)} \int_0^1 f'^2(x)\, dx, \tag{3.110}$$
then the function $u(x,t)$ does not remain bounded as $t \to \infty$.

PROOF. The inequality (3.110) implies, by virtue of (3.109), that $\alpha > 0$. It follows that $h(t)$ tends to $+\infty$ as $t \to \infty$. Suppose now that u remains bounded, that is, that there exists a constant $M > 0$ such that $|u(x,t)| \le M$ for all $0 < x < 1$, $t > 0$. Then we would have

$$h(t) \le C_1 M^2$$

by (3.103). But then (3.108) implies

$$C_1 M^2 \ge \alpha t + \beta, \quad \alpha > 0,$$

which is clearly a contradiction. Thus the hypothesis that u remains bounded leads to a contradiction, so u cannot remain bounded.

Before turning to some different results let us make two additional remarks on what we have done thus far. The first is that condition (3.110) clearly implies $b < 0$, and since it also involves the initial function f, for a given negative value of b one can give initial data for which the corresponding solution may remain bounded.

The second point to be made is that the boundary conditions for any of Problems H_{ij}, $i,j = 1, 2$, satisfy both of the conditions (3.104) and (3.106).

Now let us turn to the case where b is nonnegative. Since $c(x)$ is positive on $[0,1]$, it follows at once from (3.105) that $\dot{h}(t) \le 0$. Thus $h(t)$ is a nonincreasing function of t. But h is bounded below so, by Theorem 1.2,

$$\lim_{t \to \infty} h(t) = h_\infty$$

exists and satisfies $h_\infty \ge 0$. Next set

$$h(t) - h(0) = \int_0^t \dot{h}(\tau)\, d\tau.$$

Passing to the limit we obtain

$$h_\infty - h(0) = \lim_{t \to \infty} \int_0^t \dot{h}(\tau)\, d\tau. \tag{3.111}$$

In other words, $\int_0^\infty \dot{h}(\tau)\, d\tau$ exists. Previously we obtained

$$\ddot{h}(t) = 4 \int_0^1 c(x) u_t^2(x,t)\, dt \ge 0$$

from which we conclude that

$$\dot{h}(t_1) - \dot{h}(t_2) \ge 0, \quad \text{if} \quad t_1 > t_2.$$

Thus we have shown that $\dot{h}$ is a nondecreasing function of t. Since $\dot{h} \le 0$ it follows that

$$\lim_{t \to \infty} \dot{h}(t) = \dot{h}_\infty$$

exists and satisfies $\dot{h}_\infty \leq 0$. We now claim that $\dot{h}_\infty = 0$. In fact, for the purpose of creating a contradiction suppose that $\dot{h}_\infty < 0$. Since $\dot{h}_\infty < \dot{h}(t)$ for all t, we would then have

$$\lim_{t \to \infty} \int_0^t \dot{h}(\tau)\, d\tau \leq \lim_{t \to \infty} \dot{h}_\infty \int_0^t d\tau = \dot{h}_\infty \lim_{t \to \infty} t \to -\infty$$

because $\dot{h}_\infty < 0$. But this means that $\int_0^\infty \dot{h}(\tau)\, d\tau$ is a divergent integral contradicting the fact that this integral exists as established by (3.111). It follows that $\dot{h}_\infty = 0$. We summarize this argument with:

Lemma 3.1

Suppose that the hypotheses of Theorem 3.10 are satisfied and that $b \geq 0$. Then $\dot{h}(t)$ is nondecreasing for all $t > 0$ and

$$\lim_{t \to \infty} \dot{h}(t) = 0. \tag{3.112}$$

We shall exploit this lemma now to prove two additional theorems on asymptotic behavior.

Theorem 3.12

Suppose the hypotheses of Lemma 3.1 are satisfied and the boundary conditions are those of one of Problems H_{11}, H_{12}, or H_{21}. Then the solution u satisfies

$$\lim_{t \to \infty} u(x,t) = 0. \tag{3.113}$$

PROOF. The key point in the proof is the limitation on the boundary conditions. If we are dealing with Problem H_{11} for Equation (3.99) then u vanishes at $x = 0$ and at $x = 1$. In the case H_{12} u vanishes at $x = 0$ and for H_{21} u vanishes at $x = 1$. In other words, for each of these problems, u vanishes in the closed interval $[0,1]$ in at least one point for all values of t. Let x_0 be a point in $[0,1]$ such that $u(x_0,t) = 0$ for all t and write

$$u(x,t) = u(x_0,t) + \int_{x_0}^x u_\xi(\xi,t)\, d\xi$$
$$= \int_{x_0}^x u_\xi(\xi,t)\, d\xi.$$

By Lemma 1.2 of Chapter 1 this means that

$$|u(x,t)| = \left| \int_{x_0}^x u_\xi(\xi,t)\, d\xi \right|$$
$$\leq \left| \int_{x_0}^x |u_\xi(\xi,t)|\, d\xi \right|.$$

Note that we must use the absolute value of the integral since x_0 may be greater than x. Since the integrand is nonnegative we can assert that

$$|u(x,t)| \leq \int_0^1 |u_x(x,t)|\, dx.$$

(Here we have replaced the variable ξ by x because x no longer appears in the upper limit.) Therefore

$$|u(x,t)|^2 \leq \left(\int_0^1 |u_x(x,t)|\, dx\right)^2.$$

Now let us apply the inequality of Problem 10, Section 3.6 with $t = 1$ to the right-hand side. This gives us

$$|u(u,t)|^2 \leq \int_0^1 u_x^2(x,t)\, dx. \tag{3.114}$$

In order to complete the proof we return once again to (3.105) which we now write in the form

$$-\dot{h}(t) = 2\int_0^1 [u_x^2(x,t) + bc(x)u^2(x,t)]\, dx.$$

Since $b \geq 0$ and $c(x)$ is positive, we have

$$-\tfrac{1}{2}\dot{h}(t) \geq \int_0^1 u_x^2(x,t)\, dx.$$

Using (3.114) we may replace the right-hand side of this inequality to obtain

$$0 \leq |u(x,t)|^2 \leq -\tfrac{1}{2}\dot{h}(t).$$

By Lemma 3.1 the right-hand side tends to zero as $t \to \infty$. It follows that $|u|^2$, hence $|u|$ and u itself tend to zero as $t \to \infty$. This completes the proof of the theorem.

What happens to a solution of Problem H_{22}? The behavior of solutions to this problem turns out to be much more interesting than for the three problems covered by Theorem 3.12. We shall not attempt to cover all the possibilities here; instead we confine ourselves to two results of particular interest.

Basically the difficulty with Problem H_{22} is that $u(x_0,t)$ need not be equal to zero for a fixed x_0 and all t. Thus the proof of Theorem 3.12 fails for this case. The situation may be salvaged however by placing an additional condition on the problem. Let us assume that the initial data f satisfies

$$\int_0^1 c(x)f(x)\, dx = 0. \tag{3.115}$$

Theorem 3.13

Suppose the conditions of Lemma 3.1 are satisfied and the initial data satisfies (3.115). Then the solution u of Problem H_{22} tends to zero as $t \to \infty$.

PROOF. We begin by integrating Equation (3.99) with respect to x from 0 to 1. Applying the boundary conditions H_{22} we obtain

$$\int_0^1 c(x)[bu(x,t) + u_t(x,t)]\, dx = 0. \qquad (3.116)$$

Set $k(t) = \int_0^1 c(x)u(x,t)\, dx$. Then (3.116) can be written in the form

$$\frac{dk(t)}{dt} + bk(t) = 0.$$

This is equivalent to

$$\frac{d}{dt}[e^{bt}k(t)] = 0.$$

Hence

$$e^{bt}k(t) = k(0)$$

or

$$k(t) = k(0)e^{-bt}.$$

Now by (8.115),

$$k(0) = \int_0^1 c(x)f(x)\, dx = 0,$$

from which it follows that

$$\int_0^1 c(x)u(x,t)\, dx = 0 \qquad (3.117)$$

for all values of t.

Now let x,y be arbitrary points in $[0,1]$. We can write

$$u(x,t) = u(y,t) + \int_y^x u_\xi(\xi,t)\, d\xi$$

$$\leq u(y,t) + \int_0^1 |u_x(x,t)|\, dx$$

as in the proof of Theorem 3.12. Again we use the inequality of Problem 10 Section 3.6 to obtain

$$u(x,t) \leq u(y,t) + \left(\int_0^1 u_x^2(x,t)\, dx\right)^{1/2}.$$

Multiply both sides of this inequality by the positive function $c(y)$ to obtain

$$c(y)u(x,t) \leq c(y)u(y,t) + c(y)\left(\int_0^1 u_x^2(x,t)\, dx\right)^{1/2}.$$

If we now integrate with respect to y from 0 to 1, the first term on the right will drop out because of (3.117) and we are left with

$$u(x,t) \int_0^1 c(y)\, dy \le \left(\int_0^1 c(y)\, dy\right)\left(\int_0^1 u_x^2(x,t)\, dx\right)^{1/2}.$$

Since
$$\int_0^1 c(y)\, dy > 0,$$

we can divide both sides by this number and take absolute values to obtain

$$|u(x,t)| \le \left|\left(\int_0^1 u_x^2(x,t)\, dx\right)^{1/2}\right| = \left(\int_0^1 u_x^2(x,t)\, dx\right)^{1/2}. \quad (3.118)$$

Since both sides of (3.118) are positive, this inequality is equivalent to the inequality (3.114). Therefore the remainder of the proof of Theorem 3.12 applies without modification and we conclude that Theorem 3.13 is true.

The simple device used in the first portion of the proof of Theorem 3.13 can be used to extract some information from a considerably more general problem. We conclude this section by applying it to the problem

$$u_{xx} = c(x)[bu + u_t] - F(x,t), \quad 0 < x < 1, \quad t > 0, \quad (3.119)$$

$$u_x(0,t) = g_0(t), \quad t > 0,$$
$$u_x(1,t) = g_1(t), \quad t > 0, \quad (3.120)$$

$$u(0,x) = f(x), \quad 0 \le x \le 1. \quad (3.121)$$

Integrating (3.119) from 0 to 1 with respect to x, and applying (3.120), we have

$$\int_0^1 c(x)[bu + u_t]\, dx = g_1(t) - g_0(t) + \int_0^1 F(x,t)\, dx.$$

Let
$$k(t) = \int_0^1 c(x) u(x,t)\, dx, \quad \tilde{F}(t) = \int_0^1 F(x,t)\, dx, \quad (3.122)$$

then
$$\dot{k} = bk = g_1(t) - g_0(t) + \tilde{F}(t) \equiv \sigma(t).$$

This differential equation is equivalent to

$$\frac{d}{dt}[e^{bt} k(t)] = e^{bt}\sigma(t),$$

Integrating from 0 to t,

$$e^{bt} k(t) = k(0) + \int_0^t e^{b\tau} \sigma(\tau)\, d\tau$$

or

$$k(t) = k(0)e^{-bt} + \int_0^t e^{-b(t-\tau)} \sigma(\tau)\, d\tau.$$

Making the appropriate substitutions into the formula we obtain

$$\int_0^1 c(x)u(x,t)\,dx = e^{-bt}\int_0^1 c(x)f(x)\,dx$$
$$+ \int_0^t e^{-b(t-\tau)}[g_1(\tau) - g_0(\tau) + \hat{F}(\tau)]\,d\tau. \quad (3.123)$$

If $b > 0$ the first term on the right-hand side of (3.123) tends exponentially to zero. If $b < 0$, this term becomes unbounded as $t \to \infty$ while if $b = 0$ we have a constant term. To investigate the behavior of the second term let us first consider the case where $\hat{F} \equiv 0$, $g_1 \equiv \alpha$, $g_0 \equiv \beta$. Then we have

$$\int_0^t e^{-b(t-\tau)}(\alpha - \beta)\,d\tau = \frac{(\alpha - \beta)}{b}(1 - e^{-bt}) \to \frac{\alpha - \beta}{b} \quad \text{as} \quad t \to \infty,$$

where $b \geq 0$. Again when $b > 0$ this term becomes unbounded. If $b = 0$ we obtain

$$(\alpha - \beta)\int_0^t d\tau = (\alpha - \beta)t$$

which also becomes unbounded. Thus, if $F(x,t) \equiv 0$, and the constant values $g_0 \equiv \beta$, $g_1 \equiv \alpha$ are given, the weighted mean value $k(t)$ becomes unbounded if $b \leq 0$ and tends exponentially to

$$\frac{\alpha - \beta}{b}$$

if $b > 0$. Arguing as we did above in the proof of Theorem 3.11 it follows that u does not remain bounded as $t \to \infty$ if $b \leq 0$. We leave the consideration of other possibilities to the problems.

PROBLEMS

1. Suppose $g(t)$ is defined for all $t \geq t_0$ and $|g(t)| \to 0$ as $t \to \infty$. Prove that
$$\lim_{t \to \infty} g(t) = 0.$$

2. Suppose that $g(t)$ is defined for all $t \geq t_0$ and satisfies $g(t) > 0$. Prove that if $g^2(t) \to 0$ as $t \to \infty$, then
$$\lim_{t \to \infty} g(t) = 0.$$

[*Hint:* Use the fact that $\lim_{t \to \infty} \sqrt{g^2(t)} = \sqrt{\lim_{t \to \infty} g^2(t)}$.]

3. Assume that the conditions of Lemma 3.1 are satisfied, that $b = 0$, and that
$$\int_0^1 c(x)f(x)\,dx = \hat{f}. \quad (3.124)$$

Prove that in this case the solution u of the boundary-value problem H_{22} for (3.100) satisfies

$$\lim_{t \to \infty} u(x,t) = \hat{f}.$$

[Note that the condition (3.124) is not a restriction on the initial function f since the integral on the left-hand side is a constant.]

4. Prove that if $a \geq 0$, $b \geq 0$, then $a \leq b$ if and only if $a^2 \leq b^2$.
5. Investigate the behavior of the mean value of the solution of the problem (3.119), (3.120), (3.121) when $g_0 = g_1 \equiv 0$ and $F(x,t) = h(x)/(1-t)$.
6. Consider the boundary-value problem H_{13} for the differential equation (3.99). Derive energy identities similar to (3.105) and (3.107) for these boundary conditions.
7. Prove that Theorem 3.10 remains true for the boundary-value problem H_{13}.
8. Prove that an analog of Theorem 3.11 is true for the boundary-value problem H_{13}.
9. Show that Theorem 3.10 remains true for the boundary-value problems H_{31} and H_{33}.
10. Prove that Theorem 3.12 is true for the boundary-value problems H_{13} and H_{31}. Is Theorem 3.12 true for the boundary-value problem H_{33}?

3.9 A NONLINEAR PROBLEM

We shall conclude our work on the heat equation by examining a nonlinear problem. At the same time we will see that the techniques used in Section 3.8 have much wider applicability than indicated there. The emphasis here will not be on the various boundary conditions, so throughout this section we shall consider the differential equation

$$u^2 u_t = u_{xx} - bu - cu^3 \tag{3.125}$$

with the boundary conditions

$$u(0,t) = u_x(1,t) = 0, \tag{3.126}$$

and the initial condition

$$u(x,0) = g(x). \tag{3.127}$$

Here b is a constant which may be positive or negative and c is a constant which may be positive or zero.

As in Section 3.8, we shall consider solutions to this problem having continuous derivatives u_{xx}, u_t, and u_{xt} for $0 < x < 1$, $t > 0$. We suppose that g has a continuous derivative in [0,1] and that

$$\int_0^1 g^2(x)\,dx > 0. \tag{3.128}$$

The method is essentially the same as was employed in Section 3.8. Multiplying (3.128) by u leads to

$$\tfrac{1}{4}(u^4)_t = (uu_x)_x - u_x^2 - bu^2 - cu^4.$$

Integrating this result with respect to x from 0 to 1 yields

$$\tfrac{1}{4}\dot{k}(t) = -\int_0^1 [u_x^2(x,t) + bu^2(x,t)]\,dx - ck(t) \tag{3.129}$$

where

$$k(t) = \int_0^1 u^4(x,t)\,dx. \tag{3.130}$$

Now multiplying (3.125) by u_t and again integrating with respect to x from 0 to 1 yields

$$-\frac{1}{2}\frac{d}{dt}\int_0^1 [u_x^2(x,t) + bu^2(x,t)]\,dx - \frac{c}{4}\dot{k}(t) = \int_0^1 u^2(x,t)u_t^2(x,t)\,dx. \tag{3.131}$$

The identities (3.129) and (3.131) are analogous to those of Section 3.8.

Let us next set

$$p(t) = \int_0^1 [u_x^2(x,t) + bu^2(x,t)]\,dx \tag{3.132}$$

so that (3.129) becomes

$$\dot{k}(t) + 4ck(t) = -4p(t) \tag{3.133}$$

and (3.131) becomes

$$-4\dot{p}(t) = 2c\dot{k}(t) + 8\int_0^1 u^2(x,t)u_t^2(x,t)\,dx. \tag{3.134}$$

Combining (3.133) and (3.134) we arrive at the differential inequality

$$\ddot{k} + 4c\dot{k} \geq 2c\dot{k}$$

or

$$\ddot{k} + 2c\dot{k} \geq 0.$$

Integrating from 0 to t yields

$$\dot{k}(t) + 2ck(t) \geq \dot{k}(0) + 2ck(0).$$

This inequality is equivalent to the inequality

$$\frac{d}{dt}(k(t)e^{2ct}) \geq [\dot{k}(0) + 2ck(0)]e^{2ct}.$$

Hence, integrating again from 0 to t we obtain

$$k(t)e^{2ct} \geq k(0) + [\dot{k}(0) + 2ck(0)]\frac{1}{2c}(e^{2ct} - 1).$$

It follows that

$$k(t) \geq k(0) + \frac{\dot{k}(0)}{2c}(1 - e^{-2ct}). \tag{3.135}$$

Now, rearranging terms in (3.129) and evaluating the resulting equation at $t = 0$ yields

$$\dot{k}(0) = -4\int_0^1 [g'^2(x) + cg^4(x)]\,dx - 4b\int_0^1 g^2(x)\,dx,$$

hence we have:

Theorem 3.14

Assume u is a solution of the boundary-value problem (3.125), (3.126), (3.127) with u_{xt} continuous. Then if

$$b \leq \frac{-\int_0^1 [g'^2(x) + cg^4(x)]\,dx}{\int_0^1 g^2(x)\,dx} \tag{3.136}$$

$k(t)$ is bounded below by $k(0)$ for all t and the limit of $k(t)$ as $t \to \infty$, if it exists, is bounded below by $k(0) + \dot{k}(0)/2c$.

Recall that for the linear problem of Section 3.8 we were able to show that the solution becomes unbounded as $t \to \infty$ for b sufficiently negative. For the present problem it is clear that the nonlinear term cu^3 in Equation (3.125) has a considerable influence.

Theorem 3.14 can be complimented by a result showing that the solution to our boundary-value problem is also bounded above even if $b < 0$. For this purpose we use the inequality of Problem 10, Section 3.6. This yields

$$\left(\int_0^1 u^2(x,t)\,dx\right)^2 \leq \int_0^1 u^4(x,t)\,dx = k(t). \tag{3.137}$$

Now integrate (3.131) from 0 to t and rearrange terms to get

$$\int_0^1 \left[u_x^2(x,t) + bu^2(x,t) + \frac{c}{2}u^4(x,t)\right]dx + 2\int_0^t u^2(x,\tau)u_\tau^2(x,\tau)\,dx\,d\tau$$
$$= \int_0^1 \left[g'^2(x) + bg^2(x) + \frac{c}{2}g^4(x)\right]dx.$$

Now we note the obvious fact that the double integral on the left-hand side of the equation is nonnegative. Therefore we form the inequality

$$\int_0^1 u_x^2(x,t)\,dx + b\int_0^1 u^2(x,t)\,dx + \frac{c}{2}\left(\int_0^1 u^2(x,t)\,dx\right)^2 \le k, \quad (3.138)$$

where

$$k = \int_0^1 \left[g'^2(x) + bg^2(x) + \frac{c}{2}g^4(x)\right]dx.$$

Next rewrite (3.138) in the form

$$\int_0^1 u_x^2(x,t)\,dx + \left\{b + \frac{c}{2}\int_0^1 u^2(x,t)\,dx\right\}\int_0^1 u^2(x,t)\,dx \le k. \quad (3.139)$$

Two possibilities exist. First it may happen that for some values of t we have

$$b + \frac{c}{2}\int_0^1 u^2(x,t)\,dx \ge 1. \quad (3.140)$$

When (3.140) holds we deduce from (3.139) that

$$\int_0^1 u_x^2(x,t)\,dx + \int_0^1 u^2(x,t)\,dx \le k.$$

Since both terms on the left-hand side of this inequality are nonnegative it follows that

$$\int_0^1 u_x^2(x,t)\,dx \le k, \quad \int_0^1 u^2(x,t)\,dx \le k \quad (3.141)$$

both hold. The second possibility is that for some values of t we have

$$b + \frac{c}{2}\int_0^1 u^2(x,t)\,dx < 1$$

which implies in turn the inequality

$$\int_0^1 u^2(x,t)\,dx < \frac{2(1-b)}{c}. \quad (3.142)$$

Rewriting (3.139) in the form

$$\int_0^1 u_x^2(x,t)\,dx \le k - \left\{b + \frac{c}{2}\int_0^1 u^2(x,t)\,dx\right\}\int_0^1 u^2(x,t)\,dx,$$

we may then say that

$$\int_0^1 u_x^2(x,t)\,dx \le k + \left|b + \frac{c}{2}\int_0^1 u^2(x,t)\right|dx \int_0^1 u^2(x,t)\,dx.$$

Now, for the case at hand, the absolute value on the right-hand side is less than one and using (3.142) we have

$$\int_0^1 u_x^2(x,t)\, dx \leq k + \frac{2(1-b)}{c} = k'. \tag{3.143}$$

It follows from (3.141), (3.142), and (3.143) that $\int_0^1 u^2(x,t)\, dx$ and $\int_0^1 u_x^2(x,t)\, dx$ are bounded for all t.

We now may make use of inequality (3.114) which is valid whenever $u(x_0,t) = 0$ for a fixed value of x_0 and all $t > 0$. In our problem $u(0,t) = 0$, hence

$$|u(x,t)|^2 \leq \int_0^1 u_x^2(x,t)\, dx, \tag{3.144}$$

which means that u remains bounded for all x as $t \to \infty$. A further result implied by the foregoing analysis is:

Theorem 3.15

Assume u is a solution of the boundary-value problem (3.125), (3.126), (3.127) with u_{xt} continuous. Then if $b \leq 0$ there exists a constant $\tilde{k} > 0$ such that

$$|u(x,t)| \leq \tilde{k} \quad \text{for all} \quad 0 \leq x \leq 1, \quad t \geq 0.$$

PROOF. We have proved this theorem except for the statement that $|u(1,t)| \leq \tilde{k}$. To prove this statement we make use of the inequality presented in Problem 10, Section 3.6 to conclude that

$$\tilde{k} \geq \int_0^1 u_x^2(x,t)\, dx \geq \left(\int_0^1 u_x(x,t)\, dx \right)^2 = u^2(1,t),$$

since $u(0,t) = 0$. It follows that $|u(1,t)| \leq \sqrt{\tilde{k}}$. By making $\tilde{k}$ larger than unity, if necessary, we have $\tilde{k} \geq \sqrt{\tilde{k}}$. This proves the theorem.

Once again it is clear that the nonlinear term cu^3 plays a fundamental role in the proof of Theorem 3.15. We shall see below what happens when $c = 0$, but first some further results for the case $b \geq 0$, $c > 0$.

Theorem 3.16

Assume u is a solution of the boundary-value problem (3.125), (3.126), (3.127) and that $b \geq 0$. Then

$$\int_0^1 u^2(x,t)\, dx \leq \alpha e^{-\beta t} \tag{3.145}$$

for all $t > 0$, where $\alpha > 0$, $\beta > 0$ are constants.

176 PARTIAL DIFFERENTIAL EQUATIONS

PROOF. When $b \geq 0$ we deduce at once from (3.129) the inequality
$$\dot{k}(t) + 4ck(t) \leq 0.$$
This is equivalent to
$$\frac{dt}{d}[e^{4ct}k(t)] \leq 0,$$
hence
$$e^{4ct}k(t) \leq k(0)$$
or
$$k(t) \leq k(0)e^{-4ct}.$$
Using (3.137) we have
$$\left(\int_0^1 u^2(x,t)\,dx\right)^2 \leq k(0)e^{-4ct}$$
which implies (3.145) with $\alpha = \sqrt{k(0)}$ and $\beta = 2c$. This proves the theorem.

Note that the proof of Theorem 3.16 does not make use of the identity (3.131) and so does not require the hypothesis that u_{xt} be continuous. If we do use (3.131) we can conclude that u tends to zero as $t \to \infty$, but we do not obtain exponential convergence.

Theorem 3.17

Assume u is a solution of the boundary-value problem (3.125), (3.126), (3.127) with u_{xt} continuous. Then, if $b \geq 0$, $c \geq 0$,
$$|u(x,t)| \to 0 \quad \text{as} \quad t \to \infty.$$

PROOF. Since $b \geq 0$ and $c \geq 0$, from (3.129) we have $\dot{k} \leq 0$ and
$$\ddot{k} = -4\frac{dt}{d}\int_0^1 [u_x^2 + bu^2]\,dx - 4c\dot{k}(t)$$
$$= -4\frac{dt}{d}\int_0^1 [u_x^2 + bu^2]\,dx - 2c\dot{k}(t) - 2c\dot{k}(t).$$
Using (3.132) and (3.134) we then conclude that
$$\ddot{k} = -2c\dot{k}(t) + 8\int_0^1 u^2 u_t^2\,dx \geq 0.$$
Thus $\dot{k}$ is a nondecreasing function and k is nonincreasing. Therefore the argument used to prove Lemma 3.1 of this chapter can again be used here and we conclude that
$$\lim_{t \to \infty} \dot{k}(t) = 0.$$

But then we deduce from (3.129) that

$$\int_0^1 u_x^2(x,t)\, dx \le -\dot{k}(t)$$

and hence from (3.144) the estimate

$$|u(x,t)|^2 \le -\dot{k}(t) \to 0 \quad \text{as} \quad t \to \infty.$$

Thus the theorem is proved.

The next result we wish to look at concerns the case where $c = 0$ and $b < 0$. It shows once again the profound effect of the nonlinearity on the behavior of the system. The nonlinear term we refer to this time is, of course, the term $u^2 u_t$ in Equation (3.125). It follows at once from formulas (3.22) and (3.23) that every solution of the linear equation

$$u_t = u_{xx} - bu$$

grows exponentially with t if $b < 0$ unless it is independent of t. In Theorem 3.15 we have seen that the nonlinear term cu^3, $c \ge 0$, completely changes the behavior of solutions, causing all solutions to remain bounded. The content of our final theorem is that even when $c = 0$ solutions grow rather slowly—in fact, no faster than $\sqrt{t}$.

Theorem 3.18

Let u be a solution of the boundary-value problem (3.125), (3.126), (3.127) with u_{xt} continuous. Then if $c = 0$ and $b < 0$ there exist constants $\alpha > 0$ and $\beta > 0$ such that

$$|u(x,t)|^2 \le \alpha t + \beta \tag{3.146}$$

for all sufficiently large t.

PROOF. With $c = 0$ integrate (3.131) from 0 to t and rearrange to obtain

$$\int_0^1 [u_x^2(x,t) + bu^2(x,t)]\, dx + 2\int_0^t \int_0^1 u^2(x,\tau) u_t^2(x,\tau)\, dx\, d\tau$$

$$= \int_0^1 [g'^2(x) + bg^2(x)]\, dx \equiv k.$$

Discarding the integral of u_x^2 which is clearly nonnegative, we have

$$b\int_0^1 u^2(x,t)\, dx + 2\int_0^t \int_0^1 u^2(x,t) u_t^2(x,t)\, d\tau dx \le k. \tag{3.147}$$

Now by the inequality presented in Problem 10, Section 3.6 we have for each $t > 0$

$$\int_0^t u^2 u_\tau^2 \, d\tau \geq \frac{1}{t}\left(\int_0^t u u_\tau \, d\tau\right)^2$$
$$\geq \frac{1}{4t}\left(\int_0^t (u^2)_\tau \, d\tau\right)^2$$
$$\geq \frac{1}{4t}(u^2(x,t) - g^2(x))^2.$$

Using this estimate in (3.147) gives us the inequality

$$b\int_0^1 u^2(x,t) \, dx + \frac{1}{2t}\left\{\int_0^1 u^4(x,t) \, dx - 2\int_0^1 u^2(x,t)g^2(x) \, dx\right.$$
$$\left. + \int_0^1 g^4(x) \, dx\right\} \leq k. \quad (3.148)$$

Since g has a continuous derivative in $[0,1]$ it is bounded there, say by the number G. Then

$$\int_0^1 u^2(x,t) g^2(x) \, dx \leq G \int_0^1 u^2(x,t) \, dx$$

or

$$-\int_0^1 u^2 g^2 \, dx \geq -G \int_0^1 u^2 \, dx.$$

We next use this inequality and the estimate (3.137) in (3.148).

$$\left(\int_0^1 u^2(x,t) \, dx\right)^2 + (2tb - 2G)\int_0^1 u^2(x,t) \, dx \leq 2tk + k'. \quad (3.149)$$

Here $k' = \left|\int_0^1 g^4(x) \, dx\right| \leq G^4$. Now we factor the left-hand side of (3.149) and write it in the form

$$\left[\int_0^1 u^2(x,t) \, dx + 2tb - 2G\right]\int_0^1 u^2(x,t) \, dx \leq 2tk + k': \quad (3.150)$$

Suppose first that

$$\int_0^1 u^2(x,t) \, dx + 2tb - 2G \geq 1,$$

then (3.150) leads at once to

$$\int_0^1 u^2(x,t) \, dx \leq 2tk + k'.$$

In the contrary case,

$$\int_0^1 u^2(x,t) \, dx < -2tb + 2G \quad \text{(remember } b < 0\text{)}.$$

Thus in any case we have the estimate

$$\int_0^1 u^2(x,t) \, dx \leq \alpha' t + \beta', \quad \alpha' > 0, \quad \beta' > 0. \quad (3.151)$$

Now return to the identity immediately preceding (3.147) and discard the nonnegative double-integral term to derive the inequality

$$\int_0^1 u_x^2(x,t)\,dx \leq k - b \int_0^1 u^2(x,t)\,dx.$$

Then from (3.151) we obtain

$$\int_0^1 u_x^2(x,t)\,dx \leq k - b\alpha' t - b\beta' = \alpha t + \beta$$

where $\alpha = -b\alpha' > 0$ and $\beta = k - b\beta' > 0$. Finally, by once again invoking inequality (3.144), the theorem is proved.

PROBLEMS

1. In using energy methods in Sections 3.8 and 3.9 we have obtained in a natural way a variety of results concerning the behavior of the integrals

   $$\int_0^1 u^2(x,t)\,dx, \quad \int_0^1 u_x^2(x,t)\,dx, \quad \int_0^1 u^4(x,t)\,dx.$$

   For the linear problem

   $$\begin{aligned} u_t &= u_{xx} - bu, & 0 < x < 1, \ 0 < t, \\ u(0,t) &= 0 = u_x(0,t), & 0 < t, \\ u(x,0) &= g(x), & 0 < x < 1, \end{aligned}$$

   discuss the behavior of these integrals as $t \to \infty$ in the case where $b < 0$ and also for the case where $b > 0$. Compare your results with those of Section 3.9.

2. In the case where $c = 0$ and $b < 0$ and satisfies

   $$\int_0^1 [g'^2(x) + bg^2(x)]\,dx = \gamma < 0,$$

   prove that a solution of the boundary-value problem (3.125), (3.126), (3.127) with u_{xt} continuous satisfies

   $$\int_0^1 u^4(x,t)\,dx \geq -\gamma t + \delta.$$

   [*Hint:* Use the identity (3.131) to show that $-\int_0^1 [u_x^2(x,t) + bu^2(x,t)]\,dx \geq -\gamma$ for all t and hence deduce the result from (3.129).]

3. If the differential equation (3.125) is replaced by

   $$u^4 u_t = u_{xx} - bu + cu^3,$$

   which results of Section 3.9 remain valid?

4. Find results similar to those of Section 3.9 for the equation

$$(1 + u^2)u_t = u_{xx} - bu$$

with the boundary conditions (3.126) replaced by

$$u_x(0,t) - hu(0,t) = 0 = u(1,t).$$

5. If the differential equation (3.125) is replaced by

$$u_t = u_{xx} - bu + cu^{1/3}, \quad c \geq 0,$$

which results of Section 3.9 remain valid?

6. Show that the function $u(x,t) = \sin \pi x$ is a solution of the problem

$$u_t = u_{xx} - bu,$$

$u(0,t) = 0 = u(1,t)$, $u(x,0) = \sin \pi x$, when $b = -\pi^2$. Note that this solution is independent of t and so does not grow exponentially as $t \to \infty$.

Chapter 4

OTHER TIME-DEPENDENT PARTIAL DIFFERENTIAL EQUATIONS

4.1 LONGITUDINAL VIBRATION OF A PRISMATIC ROD

Consider a prismatic rod of length L with one end situated at $x = 0$ and the other at $x = L$. It is possible to create a motion in this rod such that each particle moves in such a way as to cause stresses which are oriented in the axial direction. Naturally, due to Poisson's ratio, there will simultaneously occur transverse motions. We will assume such motions to be negligible.

The longitudinal vibrations of the rod may be described by a function u of x and t whose values $u(x,t)$ represent the displacement of a particle having coordinate x in its equilibrium position at the time t. Here the geometric variable x, Lagrange's variable, is used in the sense that each particle of a given cross section of the rod is characterized by the same geometric coordinate x during the entire motion. Thus a particle occupying the position x at the initial moment (in the equilibrium state) is found at the subsequent time t at the position whose coordinate is $x = x + u(x,t)$.

The segment of the rod which initially occupies the open interval $(x, x + \Delta x)$ will be elongated (or shortened) at time t and occupy the interval

$$(x + u(x,t), x + \Delta x + u(x + \Delta x, t)).$$

Assuming that u is a differentiable function of x for $0 < x < L$ and continuous for $0 \leq x \leq L$, we may apply the mean-value theorem for fixed t and obtain

$$u(x + \Delta x, t) - u(x,t) = u_x(x + \theta \Delta x, t)\Delta x,$$

or
$$u_x(x + \theta\Delta x, t) = \frac{u(x + \Delta x, t) - u(x,t)}{\Delta x}, \quad (4.1)$$

where $0 < \theta < 1$. The quotient on the right-hand side of (4.1) is defined as the strain caused by the elongation of the segment $(x, x + \Delta x)$.

Passing to the limit in (4.1) as $\Delta x \to 0$ shows that the strain at the position x is equal to $u_x(x,t)$. If the vibrations of the rod are sufficiently small, Hooke's law, which states that the stress at any position in the rod is linearly proportional to the strain, is applicable. That is,

$$\sigma(x,t) = E(x)u_x(x,t), \quad (4.2)$$

where $E(x)$ is Young's modulus at the position x and σ is the longitudinal stress. The function $\sigma(x,t)$ denotes the longitudinal stress at position x at the time t.

Now consider a typical segment (x_1,x_2) of the rod and formulate the change in momentum (mass times velocity) from time t_1 to time t_2.

$$\int_{x_1}^{x_2} \rho A[u_t(\xi,t_2) - u_t(\xi,t_1)]\,d\xi, \quad (4.3)$$

where ρ is the density and A the cross-sectional area of the rod. The change in momentum could equally well be defined in terms of the stresses at the ends of the segment during the interval (t_1,t_2) and any external forces acting on the segment. Consequently, we deduce from (4.2) and (4.3)

$$\rho A \int_{x_1}^{x_2} [u_t(\xi,t_2) - u_t(\xi,t_1)]\,d\xi$$
$$= A \int_{t_1}^{t_2} [E(x_2)u_x(x_2,\tau) - E(x_1)u_x(x,\tau)]\,d\tau + \int_{x_1}^{x_2}\int_{t_1}^{t_2} f(\xi,t)\,d\xi\,d\tau \quad (4.4)$$

where f is the external force per unit length acting on the rod.

Making the assumption that the function u has continuous second derivatives in x and t for $0 < x < L$ and for all t we can apply the mean value theorems of differential and integral calculus in the same way as was done in Section 3.1 in the case of heat conduction. This results in the differential equation for the longitudinal vibrations of the rod

$$[E(x)u_x]_x = \rho u_{tt} - \frac{f}{A}(x,t). \quad (4.5)$$

For the case of a homogeneous rod equation (4.5) can be written as

$$u_{tt} = a^2 u_{xx} + g(x,t), \quad (4.6)$$

where

$$a = \sqrt{\frac{E}{\rho}} \quad \text{and} \quad g = \frac{f}{\rho A}.$$

This equation is usually called the (nonhomogeneous) *wave equation*. Note that the parameter a in (4.6) has the dimensions of velocity. It will be shown that the parameter a is the velocity of propagation of waves along the rod.

As in the case of heat conduction we may expect that the longitudinal vibrations of the rod will be influenced by its contact with its surrounding media. Specifically, the vibrations will be influenced by the conditions prevailing at the ends of the rod, and also by the rod's initial configuration. The wave equation has an indicated second derivative with respect to t, hence we may expect to require two initial conditions.

Now that we have a basic understanding of the origin of the homogeneous wave equation

$$u_{tt} = a^2 u_{xx}, \tag{4.7}$$

or, more generally, for

$$[E(x)u_x]_x = \rho u_{tt} \tag{4.8}$$

let us formulate a few of the more fundamental boundary-value problems associated with wave equations.

PROBLEM W_{11}

Find a function u satisfying (4.7) or (4.8) for $0 < x < L$, $0 \leq t$ together with the initial conditions

$$\begin{aligned} u(x,0) &= \varphi(x), \quad 0 < x < L, \\ u_t(x,0) &= \psi(x), \quad 0 < x < L, \end{aligned} \tag{4.9}$$

and the boundary conditions

$$u(0,t) = 0 = u(L,t), \quad 0 \leq t. \tag{4.10}$$

PROBLEM W_{12}

Replace the conditions (4.10) by

$$u(0,t) = 0, \quad 0 \leq t, \quad u_x(L,t) = 0, \quad 0 \leq t. \tag{4.11}$$

PROBLEM W_{32}

Replace the boundary conditions (4.10) of W_{11} by

$$u_x(0,t) + \alpha u(0,t) = 0, \quad u_x(L,t) = 0. \tag{4.12}$$

Of course, we have a permutation of the combinations of the foregoing boundary conditions for the vibration problem similar to that for the heat conduction problem.

Limiting problems can be formulated in exactly the same manner as was done in Section 3.2. This leads us to the *Cauchy problem for the wave equation*:

Find a solution of (4.7) for all $-\infty < x < \infty$ and all $t \geq 0$ satisfying the initial conditions (4.9) for all x.

The Cauchy problem is treated in detail in the next section.

PROBLEMS

1. A heavy string is displaced from its vertical position of equilibrium. The upper end at $x = L$ is fixed and the lower end at $x = 0$ is free. Show that the small transverse deflections $u(x,t)$ satisfy

$$u_{tt} = g(xu_x)_x, \quad 0 < x < L, \quad 0 < t,$$
$$u(0,t) \text{ is bounded}, \quad u(L,t) = 0, \quad 0 < t,$$
$$u(x,0) = f(x), \quad u_t(x,0) = g(x),$$

where g is the acceleration due to gravity.

2. If the string of Problem 1 rotates with an angular velocity $\omega = $ constant with respect to the vertical equilibrium position, show that the differential equation becomes

$$u_{tt} = \frac{\omega^2}{2}(x^2 u_x)_x, \quad 0 < x < L, \quad 0 < t,$$

and the boundary and initial conditions remain the same.

3. A homogeneous rod is hanging vertically. The upper end at $x = L$ is rigidly fixed and a load Q is attached at the lower end $x = 0$. At $t = 0$ a support is removed from under the load and the rod begins longitudinal vibrations. Neglecting the action of gravity on the particles of the rod, show that the longitudinal displacements satisfy

$$u_{tt} = a^2 u_{xx}, \quad 0 < x < L, \quad 0 < t,$$
$$u(L,t) = 0, \quad \frac{Q}{g} u_{tt}(0,t) = hu_x(0,t) + Q, \quad 0 < t,$$
$$u(x,0) = 0, \quad u_t(x,0) = 0, \quad 0 < x < L.$$

4. A flexible homogeneous solid cylinder is displaced from equilibrium by giving each cross section a small angular displacement in a plane at right angles to the axis of the cylinder. Show that if the cylinder has length L, and if the end $x = 0$ is held fixed and the end $x = L$ is flexibly attached, the equations satisfied by the deflection $\theta(x,t)$ of the cross section at x are

$$\theta_{tt} = a^2 \theta_{xx}, \quad 0 < x < L, \quad 0 < t,$$
$$\theta(0,t) = 0, \quad \theta_x(L,t) + h\theta(L,t) = 0, \quad 0 < t,$$
$$\theta(x,0) = f(x), \quad \theta_t(x,0) = g(x), \quad 0 < x < L.$$

5. Consider a long thin conductor of length L. Suppose the resistance R, capacitance C, self-inductance L, and leakage conductance G are continuously distributed. An emf $E(t)$ is applied at the end $x = 0$ and the end $x = L$ is grounded. The initial current $i(x,0) = f(x)$ and the initial potential $v(x,0) = g(x)$ are given. Show that the potential v satisfies

$$v_{xx} = CLv_{tt} + (CR + GL)v_t + GRv, \quad 0 < x < L, \quad 0 < t,$$
$$v(0,t) = E(t), \quad v(L,t) = 0, \quad 0 < t,$$
$$v(x,0) = g(x), \quad v_t(x,0) = \frac{Gg(x) - f'(x)}{C}, \quad 0 < x < L.$$

6. Suppose the conductor of Problem 5 has negligibly small resistance and loss and that the ends are grounded. This is accomplished at the end $x = 0$ through a lumped resistance R_0, and the end $x = L$ through a lumped capacitance C_0. Show that if v is the voltage and i the current, then

$$v_x + Li_t = 0, \quad i_x + Cv_t = 0, \quad 0 < x < L, \quad 0 < t,$$
$$-v(0,t) = R_0 i(0,t), \quad C_0 v_t(L,t) = i(l,t), \quad 0 < t,$$
$$v(x,0) = f(x), \quad i(x,0) = g(x), \quad 0 < x < L.$$

7. Using boundary-value Problem 6 derive the individual boundary-value problems for v and i.

8. Suppose the conductor of Problem 5 is grounded at each end through a lumped resistance and a lumped self-inductance connected in series. Show that v and i satisfy

$$v_x + Li_t + Ri = 0, \quad i_x + Cv_t + Gv = 0, \quad 0 < x < L, \quad 0 < t,$$
$$-v(0,t) = L_{01}i_t(0,t) + R_{01}i(0,t), \quad 0 < t,$$
$$v(L,t) = L_{02}i_t(L,t) + R_{02}i(L,t), \quad 0 < t,$$
$$v(x,0) = f(x), \quad i(x,0) = g(x), \quad 0 < x < L.$$

Derive the boundary-value problems for v and i from these conditions.

9. Set up the boundary-value problem for the conductor of Problem 5 when one end is grounded through a lumped self-inductance and an emf $e(t)$ is applied through a lumped self-inductance at the other end.

10. Suppose two semi-infinite conductors are joined through a lumped capacitance. Set up the boundary-value problem for determining the strength of the current if there are no losses.

11. Suppose the conductor of Problem 5 is grounded through a lumped resistance and a lumped self-inductance connected in parallel at one end and through a lumped capacitance and a lumped self-inductance in parallel at the other end. Set up the boundary-value problem.

12. Formulate the problem of the torsional vibrations of a solid cylinder

186 OTHER TIME-DEPENDENT PARTIAL DIFFERENTIAL EQUATIONS

which is similar to Problem 6 on electrical vibrations. Do this for the case of a voltage problem and for the case of a current-strength problem.

13. Consider two semi-infinite homogeneous flexible rods of identical cross section joined at their ends to form one infinite rod. Show that if initial longitudinal displacements and velocities are imparted to cross sections of the rods, one obtains the boundary-value problem

$$\frac{\partial^2 u_1}{\partial t^2} = a_1^2 \frac{\partial^2 u_1}{\partial x^2}, \quad -\infty < x < 0, \quad 0 < t,$$

$$\frac{\partial^2 u_2}{\partial t^2} = a_2^2 \frac{\partial^2 u_2}{\partial x^2}, \quad 0 < x < \infty, \quad 0 < t,$$

$$u_1(0,t) = u_2(0,t), \quad h_1 \frac{\partial u_1(0,t)}{\partial x} = h_2 \frac{\partial u_2(0,t)}{\partial x}, \quad 0 < t,$$

$$u_1(x,0) = f(x), \quad \frac{\partial u_1}{\partial t}(x,0) = g(x), \quad -\infty < x < 0,$$

$$u_2(x,0) = f(x), \quad \frac{\partial u_2}{\partial t}(x,0) = g(x), \quad 0 < x < \infty.$$

14. Suppose the rods of Problem 13 are joined by a weight of negligibly small thickness and mass M. Show that the conditions at $x = 0$ of Problem 13 are replaced by

$$u_1(0,t) = u_2(0,t),$$

$$M \frac{\partial^2 u_1(0,t)}{\partial t^2} - M \frac{\partial^2 u_2(0,t)}{\partial t^2} = h_1 \frac{\partial u_2(0,t)}{\partial x} - h_1 \frac{\partial u_1(0,t)}{\partial x},$$

and all other conditions remain the same.

15. Two cylinders of length L are attached to pulleys as indicated in Figure 4.1. The moments of inertia of the pulleys about the axis of the

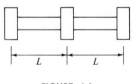

FIGURE 4.1

shaft are known. Show that the boundary-value problem for the small torsional vibrations of this system has the form

$$\frac{\partial^2 \theta_1}{\partial t^2} = a_1^2 \frac{\partial^2 \theta_1}{\partial x^2}, \quad 0 < x < L, \quad 0 < t,$$

$$\frac{\partial^2 \theta_2}{\partial t^2} = a_2^2 \frac{\partial^2 \theta_2}{\partial x^2}, \quad L < x < 2L, \quad 0 < t,$$

$$k_1 \frac{\partial^2 \theta_1(0,t)}{\partial t^2} = h_1 \frac{\partial \theta_1(0,t)}{\partial x}, \quad k_3 \frac{\partial^2 \theta_2(2L,t)}{\partial t^2} = h_2 \frac{\partial \theta_2(2L,t)}{\partial x}, \quad 0 < t,$$

$$k_2 \frac{\partial^2 \theta_1}{\partial t^2}(L,t) - k_2 \frac{\partial^2 \theta_2}{\partial t^2}(L,t) = h_2 \frac{\partial \theta_2}{\partial x}(L,t) - h_1 \frac{\partial \theta_1}{\partial x}(L,t), \quad 0 < t,$$

$$\theta_1(x,0) = f(x), \quad \frac{\partial \theta_1}{\partial t}(x,0) = g(x), \quad 0 < x < L,$$

$$\theta_2(x,0) = f(x), \quad \frac{\partial \theta_2}{\partial t}(x,0) = g(x), \quad L < x < 2L.$$

4.2 THE CAUCHY PROBLEM FOR THE WAVE EQUATION

In this section we consider the Cauchy problem, which consists of finding the solution to

$$u_{tt} - a^2 u_{xx} = 0, \tag{4.13}$$

subject to the initial conditions

$$u(x,0) = \varphi(x), \quad \text{and} \quad u_t(x,0) = \psi(x). \tag{4.14}$$

By introducing the new variables

$$\xi = x + at, \quad \text{and} \quad \eta = x - at, \tag{4.15}$$

Equation (4.13) can be reduced to a very simple form. In fact, using the chain rule we obtain

$$u_x = u_\xi + u_\eta, \quad u_{xx} = u_{\xi\xi} + 2u_{\xi\eta} + u_{\eta\eta},$$
$$u_t = a(u_\xi - u_\eta), \quad \text{and} \quad u_{tt} = a^2(u_{\xi\xi} - 2u_{\xi\eta} + u_{\eta\eta}),$$

and placing these quantities into (4.13) results in

$$u_{\xi\eta} = 0. \tag{4.16}$$

The foregoing equation implies that

$$u_\eta(\xi,\eta) = h(\eta)$$

where h is a function of the variable η only. For fixed ξ we integrate this equation with respect to η to obtain

$$u(\xi,\eta) = \int h(\eta)\, d\eta + f(\xi) \tag{4.17}$$
$$= f(\xi) + g(\eta),$$

where f is a function of ξ only and g a function of η only.

If f and g are any differentiable functions than the function u determined by (4.17) must be a solution of (4.16). That is, (4.17) yields the general solution of (4.16) which in turn yields

$$u(x,t) = f(x + at) + g(x - at) \tag{4.18}$$

as the general solution of (4.13).

If a solution to the Cauchy problem exists it is given by (4.18) subject to appropriate choices of the functions f and g. Therefore, the task of solving the problem is now reduced to choosing f and g so as to satisfy the initial conditions. Performing the required operations on (4.18) the initial conditions (4.14) become

$$u(x,0) = f(x) + g(x) = \varphi(x), \tag{4.19}$$

and

$$u_t(x,0) = af'(x) - ag'(x) = \psi(x), \tag{4.20}$$

where the primes denote derivatives of f and g by their respective arguments. For example, using the chain rule for the case where the argument of f is $w = x + at$ we get

$$\frac{df}{dt} = \frac{df}{dw}\frac{dw}{dt} = \frac{df}{dw}a = af'.$$

Integration of (4.20) leads to

$$f(x) - g(x) = \frac{1}{a}\int_{x_0}^{x}\psi(y)\,dy + c, \tag{4.21}$$

where x_0 and c are constants. Adding (4.19) and (4.21) gives

$$f(x) = \frac{1}{2}\varphi(x) + \frac{1}{2a}\int_{x_0}^{x}\psi(y)\,dy + \frac{c}{2}, \tag{4.22}$$

and subtraction of (4.21) from (4.19) gives

$$g(x) = \frac{1}{2}\varphi(x) - \frac{1}{2a}\int_{x_0}^{x}\psi(y)\,dy - \frac{c}{2}. \tag{4.23}$$

Equations (4.22) and (4.23) are valid for any values of their arguments. Substituting (4.22) and (4.23) into (4.18) we get

$$\begin{aligned}u(x,t) &= \frac{\varphi(x+at) + \varphi(x-at)}{2} + \frac{1}{2a}\left\{\int_{x_0}^{x+at}\psi(y)\,dy - \int_{x_0}^{x-at}\psi(y)\,dy\right\} \\ &= \frac{\varphi(x+at) + \varphi(x-at)}{2} + \frac{1}{2a}\int_{x-at}^{x+at}\psi(y)\,dy. \end{aligned} \tag{4.24}$$

Equation (4.24) is called *D'Alembert's formula*. It still remains to be shown that the solution to the Cauchy problem is unique. This will be done in Section 4.3. For the present let us verify the existence of the solution by showing that (4.24), which has been obtained by formal manipulations, actually solves the problem.

Theorem 4.1

Let $\varphi(y)$ be two times continuously differentiable for all y and let $\psi(y)$ be continuously differentiable for all y, then the function given by (4.24) is a solution of the Cauchy problem for the wave equation in the entire x,t plane.

PROOF. By direct computation in (4.24) we have

$$u_x = \frac{\varphi'(x+at) + \varphi'(x-at)}{2} + \frac{1}{2a}[\psi(x+at) - \psi(x-at)],$$

$$u_{xx} = \frac{\varphi''(x+at) + \varphi''(x-at)}{2} + \frac{1}{2a}[\psi'(x+at) - \psi'(x-at)], \quad (4.25)$$

$$u_t = \frac{a}{2}[\varphi'(x+at) - \varphi'(x-at)] + \frac{1}{2}[\psi(x+at) + \psi(x-at)],$$

$$u_{tt} = \frac{a^2}{2}[\varphi''(x+at) + \varphi''(x-at)] + \frac{a}{2}[\psi'(x+at) - \psi'(x-at)]. \quad (4.26)$$

In the foregoing the primes denote derivatives of φ and ψ with respect to their arguments. Substituting (4.25) and (4.26) into (4.13) shows us that $a^2 u_{xx} - u_{tt} = 0$, for all x and t. Also, when $t = 0$ (4.24) yields

$$u(x,0) = \varphi(x) + \frac{1}{2a}\int_x^x \psi(y)\,dy = \varphi(x),$$

and

$$u_t(x,0) = \psi(x).$$

This completes the proof of the existence of the solution of the Cauchy problem for smooth initial (Cauchy) data.

It still remains to be shown that (4.13) is indeed a wave equation. More specifically, we will show that the solution of (4.13) represents the motion of traveling waves. Recalling that in the general solution (4.18) f and g are arbitrary functions we will focus our attention on the special, but reasonable case when $f \equiv 0$. This reduces (4.18) to $u = g(x - at)$. If we now plotted u versus x for different values of t the result would be a family of curves in the u,x plane. Selecting a particular ordinate on any of these curves, say $t = 0$ and $x = x_1$ we have $u_1 = g(x_1)$, or more specifically

$$x - at = x_1. \quad (4.27)$$

If we now let x and t vary subject to the restriction (4.27) we see that in order to maintain $u = u_1 =$ constant the point x must travel forward along the x axis as t increases. Differentiating (4.27) with respect to time we obtain

$$x_t = a. \quad (4.28)$$

Clearly, from (4.28) we see that the parameter a is a velocity. Further, since every ordinate in the u,x plot can be located by some x_i as in (4.27) this requires that the entire curve travels forward along the x-axis with the velocity a and maintains its shape somewhat as a wave travels in water. In the same way we can argue that $u = f(x + at)$ is a wave traveling backwards along the rod. Thus the actual state for a position in the rod is composed of the sum of the states created by each wave.

Let us now turn to the analysis of a wave problem which does not fulfill all the requirements of Theorem 4.1. Suppose that $\psi(x) \equiv 0$ for all x and

$$\varphi(x) = \begin{cases} 1, & \text{for } -c \leq x \leq c \\ 0, & \text{otherwise.} \end{cases}$$

It is clear that the function ψ violates the hypothesis of Theorem 4.1. Nevertheless the function

$$u(x,t) = \frac{\varphi(x + at) + \varphi(x - at)}{2} \tag{4.29}$$

is defined and two times continuously differentiable for all x and t except when

$$x + at = \pm c, \quad \text{and} \quad x - at = \pm c. \tag{4.30}$$

Consequently the calculations involved in proving Theorem 4.1 are valid everywhere in the x,t plane except on the straight lines defined by (4.30). It then follows that the function given by (4.29) is a solution of the Cauchy problem except along these same lines. Figure 4.2 shows this information plotted in the x,t plane. The function $\varphi(x + at) = 1$ when $-c \leq x + at \leq c$ and is zero otherwise. This region is marked in Figure 4.2 with horizontal lines. The region defined by the function $\varphi(x - at) = 1$ when $-c \leq x -$

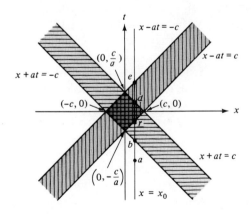

FIGURE 4.2

$at \leq c$ and zero otherwise is marked by vertical lines. By the given information u has the value 1 in the cross-hatched region, the value $\frac{1}{2}$ in the regions marked with vertical or horizontal lines only, and the value 0 elsewhere.

The lines parallel to $x - at = 0$ and $x + at = 0$ are called characteristics of the differential equation $a^2 u_{xx} - u_{tt} = 0$. More specifically, we have two families of characteristics, one consisting of all straight lines in the x,t plane parallel to the line $x - at = 0$ and the other consisting of all straight lines parallel to $x + at = 0$.

Suppose now that we consider the sequence of events which occur at a point x_0 in the interval $[0,c]$. For sufficiently large negative t, say the position a, there is no disturbance at x_0. Then at the time t determined by the intersection of the line $x = x_0$ with the characteristic $x + at = -c$, the position b, the value of u jumps to $\frac{1}{2}$ and remains at this value until the time t determined by the intersection of $x = x_0$ with the characteristic line $x - at = c$, the position r, when it jumps again to 1. Similarly, u remains at the value 1 until $t = (c - x_0)/a$, the position d, when it jumps back to the value $\frac{1}{2}$ where it remains until time $t = (c + x_0)/a$, the position e, after which $u = 0$. Thus the disturbance at x_0 is composed of a wave traveling to the left with discontinuous front and back and represented by the function $\varphi(x + at)$, and a wave traveling to the right with discontinuous front and back and represented by the function $\varphi(x - at)$. The characteristic line $x + at = c$ is the wavefront for the wave $\varphi(x - at)$ traveling to the left, and shows diagrammatically how the discontinuity in the Cauchy data at $x = -c$ propagates forward and backward in time due to the wave traveling to the left. The other three distinguished characteristics have similar interpretations.

It is clear from Figure 4.2 that the solution u at any point x_0 is composed of a left-traveling wave and a right-traveling wave. If $|x_0| > c$ these waves do not overlap at the points x_0. For example if $x_0 > c$ the left-traveling wave appears in the past and the right-traveling wave in the future.

Because of the foregoing simple, but important, geometric facts, the solution of the wave equation using D'Alembert's formula is often called the *method of traveling waves*.

PROBLEMS

1. An infinite string is given the initial deflection shown in Figure 4.3. Plot the position of the string at the times $t_k = k/a$ for $k = \frac{1}{2}, 1, 2, 5$.
2. An infinite string is given the initial deflection $f(x) = h(1 - (x/c)^2)$ on the interval $[-c,c]$, and zero elsewhere. Find the profile of the string for $t > 0$ and the law of motion of an arbitrary point x of the string for $t > 0$.

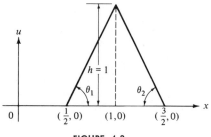

FIGURE 4.3

3. A transverse initial velocity v_0 = constant is given an infinite string on the interval $[0,c]$. Outside this interval the initial velocity is zero. Find the law of motion of points of the string for $t > 0$. Plot the positions of the string at times $t_k = k/a$ for $k = c/2,\, c,\, 2c,\, 4c$.

4. At time $t = 0$ an infinite string is given a transverse blow at the point x_0 transmitting an impulse I to the string. If the initial displacements and velocities of the other points of the string are zero, find the deflection $u(x,t)$.

5. A semi-infinite string is fixed at one end and excited by the initial deflection shown in the Figure 4.4. Plot the shape of the string at times $t_k = k/a$ for $k = 0,\, 1,\, 2,\, 4,\, 7$.

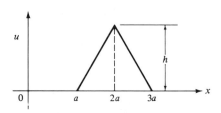

FIGURE 4.4

6. A semi-infinite rod with the end $x = 0$ free is given an initial velocity v_0 = constant on the segment $[a, 2a]$. The initial velocity is zero elsewhere and the initial displacement zero everywhere. Plot the longitudinal displacement $u(x,t)$ at the times

$$t_k = k/a, \quad k = 0,\, \tfrac{1}{2},\, 2,\, 4,\, 7.$$

7. Show how to solve the boundary-value problem

$$u_{tt} - u_{xx} = 0, \quad 0 < x < \infty, \quad 0 < t,$$
$$u(0,t) = 0,$$
$$u(x,0) = f(x), \quad u_t(x,0) = 0, \quad 0 < x < \infty,$$

by the method of traveling waves. This requires a suitable modification of D'Alembert's formula. [*Hint*: Extend the semi-infinite string to an infinite string and put appropriate initial data on the half $-\infty < x < 0$.]

8. Show how to solve
$$u_{tt} - u_{xx} = 0, \quad 0 < x < \infty, \quad 0 < t,$$
$$u_x(0,t) = 0,$$
$$u(x,0) = f(x), \quad u_t(x,0) = 0, \quad 0 < x < \infty,$$
by the method of traveling waves.

9. Show how to solve Problem 7 if $u(x,0) = 0$, $u_t(x,0) = y(x)$, and $0 < x < \infty$ by the method of traveling waves.

10. Show how to solve Problem 8 if $u(x,0) = 0$, $u_t(x,0) = g(x)$, and $0 < x < \infty$ by the method of traveling waves.

11. Find the solution to the boundary-value problem
$$u_{tt} = u_{xx}, \quad 0 < x < \infty, \quad 0 < t,$$
$$u_x(0,t) + \alpha u_t(0,t) = 0, \quad 0 < t,$$
$$u(x,0) = 0, \quad 0 < x < \infty,$$
$$u_t(x,0) = \omega, \quad 0 < x < \infty.$$

12. Find the solution to the boundary-value problem
$$u_{tt} - u_{xx} = 0, \quad 0 < x < \infty, \quad 0 < t,$$
$$u(0,t) = a(t), \quad 0 < t < \infty,$$
$$u(x,0) = u_t(x,0) = 0, \quad 0 < x < \infty.$$

13. Find the solution to the boundary-value problem
$$u_{tt} - u_{xx} = 0, \quad 0 < x < \infty, \quad 0 < t,$$
$$u_x(0,t) = -b(t), \quad 0 < t,$$
$$u(x,0) = u_t(x,0) = 0, \quad 0 < x < \infty.$$

4.3 THE ENERGY METHOD FOR THE WAVE EQUATION: UNIQUENESS

The technique of associating energy integrals with ordinary differential equations was used successfully in Chapters 1 and 2 to establish a variety of results. The technique was extended in Chapter 3 to the heat equation, a partial differential equation. Hence we expect that the same method can be applied to the wave equation. Accordingly we form the product

$$u_t u_{tt} - a^2 u_t u_{xx} = 0 \tag{4.31}$$

and observe the identities

$$u_t u_{tt} = \tfrac{1}{2}(u_t^2)_t,$$

where the notation $(u_t^2)_t$ stands for $(\partial/\partial t)(u_t^2)$, and

$$u_t u_{xx} = (u_t u_x)_x - u_{tx} u_x$$
$$= (u_t u_x)_x - \tfrac{1}{2}(u_x^2)_t.$$

Thus (4.31) can be written in the form

$$(u_t^2 + a^2 u_x^2)_t = 2a^2(u_x u_t)_x. \tag{4.32}$$

Equation (4.32) is the basic differential identity which we shall operate on to prove the uniqueness of the solution of the Cauchy problem and the uniqueness of certain boundary-value problems.

In order to show that the solution of the Cauchy problem is unique let us suppose there are two possibly distinct functions V and W defined for all x and t, satisfying the wave equation (except possibly along a finite number of characteristic lines) and having the same Cauchy data on the line $t = 0$.

It then follows that

$$V_{tt} - a^2 V_{xx} = 0, \quad W_{tt} - a^2 W_{xx} = 0,$$

and

$$V(x,0) = W(x,0), \quad V_t(x,0) = W_t(x,0).$$

Consider the function $u = V - W$. The foregoing systems then reduce to

$$u_{tt} - a^2 u_{xx} = 0$$

and

$$u(x,0) = u_t(x,0) = 0.$$

The task now becomes one of showing that $u(x,t) = 0$ at each point in the (x,t)-plane. For this purpose let (x_0, t_0) $t_0 > 0$ be a fixed point in the upper half-plane. Then the characteristic lines $x - x_0 - a(t - t_0) = 0$ and $x - x_0 + a(t - t_0) = 0$ pass through the point (x_0, t_0). These lines intersect the line $t = 0$ at the points $x = x_0 - at_0$ and $x = x_0 + at_0$ respectively, as shown in Figure 4.5. Integrating the identity (4.32) over the triangle whose

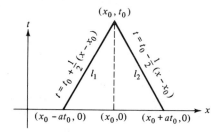

FIGURE 4.5

vertices are (x_0, t_0), $(x_0 - at_0, 0)$, and $(x_0 + at_0, 0)$, we obtain

$$0 = \int_0^{t_0} \int_{x_0-a(t-t_0)}^{x_0+a(t-t_0)} [(u_t^2 + a^2 u_x^2)_t - 2a^2(u_x u_t)_x] \, dx \, dt.$$

By appropriately interchanging orders of integration we can convert these double integrals into single integrals as follows:

$$0 = \int_{x_0-at_0}^{x_0} \int_0^{t_0+1/a(x-x_0)} (u_t^2 + a^2 u_x^2)_t \, dt \, dx - 2 \int_0^{t_0} \int_{x_0-a(t-t_0)}^{x_0+a(t-t_0)} a^2(u_x u_t)_x \, dx \, dt$$

$$+ \int_{x_0}^{x_0+at_0} \int_0^{t_0-1/a(x-x_0)} (u_t^2 + a^2 u_x^2)_t \, dt \, dx$$

$$= \int_{x_0-at_0}^{x_0} (u_t^2 + a^2 u_x^2)\left(x, t_0 + \frac{1}{a}(x - x_0)\right) dx$$

$$- \int_{x_0-at_0}^{x_0} (u_t^2 + a^2 u_x^2)(x, 0) \, dx$$

$$+ 2 \int_0^{t_0} a^2(u_x u_t)(x_0 - a(t - t_0), t) \, dt$$

$$+ \int_{x_0}^{x_0+at_0} (u_t^2 + a^2 u_x^2)\left(x, t_0 - \frac{1}{a}(x - x_0)\right) dx$$

$$- \int_{x_0}^{x_0+at_0} (u_t^2 + a^2 u_x^2)(x, 0) \, dx$$

$$- 2 \int_0^{t_0} a^2(u_x u_t)(x_0 + a(t - t_0), t) \, dt.$$

The second and sixth integrals involve u_t and u_x calculated on the base of the triangle. But on this segment $u_t = u = 0$, hence $u_x = 0$, and these two integrals have the value zero. The remaining four integrals are evaluated along the two characteristic lines and we can write the equality in the form

$$0 = \int_0^{t_0} (u_t + au_x)^2 (x_0 + a(t - t_0), t) \, dt$$

$$+ \int_0^{t_0} (u_t - au_x)^2 (x_0 - a(t - t_0), t) \, dt. \quad (4.33)$$

Observe that the integrands in these integrals are nonnegative, hence the integrals themselves are nonnegative. Since by (4.33) the sum is equal to zero each integral must itself be zero. If the derivatives u_x and u_t are continuous on the characteristic lines, it follows that the integrands must vanish. This yields $u_t + au_x = 0$ on the line segment l_1 and $u_t - aux = 0$ on the line segment l_2. On the line segment l_1 the directional derivative of u is $u_t + au_x$ and on l_2 the directional derivative of u is $u_t - au_x$. Since these derivatives are zero, u has a constant value on the line segments l_1 and l_2. But the value of u at the points where the characteristic lines intersect the x-axis is zero. Thus we conclude that the constant value of u on the segment l_1 equals the constant value of u on the segment l_2 equals zero. Therefore $u(x_0,t_0) = 0$. We have now proved that u equals zero at every point (x_0,t_0) in the upper half-plane. Clearly an entirely similar argument can be used to prove $u(x_0,t_0) = 0$ for $t_0 < 0$. The uniqueness of the wave equation may be stated as follows.

Theorem 4.2

Let u satisfy the wave equation

$$u_{tt} - a^2 u_{xx} = 0$$

for all (x,t) and suppose $u(x,0) = u_t(x,0) = 0$ for all x, then $u = 0$ everywhere in the (x,t)-plane.

Corollary 4.1

The solution to the Cauchy problem for the wave equation is unique.

The triangle with vertices (x_0,t_0), $(x_0 - at_0, 0)$, and $(x_0 + at_0, 0)$ is called the *characteristic triangle* with vertex (x_0,t_0). In order that u vanish at (x_0,t_0) it is clearly sufficient, on the basis of the foregoing proof, that the Cauchy data be zero on the closed interval $[x_0 - at_0, x_0 + at_0]$ of the x-axis. Another way of looking at this argument is to observe that the value of u at (x_0,t_0) is only affected by the data along the base of the characteristic triangle with vertex at (x_0,t_0). For this reason the interval $[x_0 - at_0, x_0 + at_0]$ is called the *domain of dependence* for the solution at (x_0,t_0).

Let us now use the identity (4.32) to prove that the solution to certain boundary-value problems for the wave equation is unique. The existence of solutions to these problems can be established by methods similar to that used in Section 3.2 where we formally obtained the solution to a boundary-value problem for the heat equation. We will apply this method to the wave equation in considerable detail in Section 4.4.

Consider a rod of length l with ends at $x = 0$ and $x = l$. Let (x_0,t_0) be any point with $t_0 > 0$ and integrate (4.32) over the rectangle $0 \le x \le l$, $0 \le t \le t_0$. Thus if u satisfies

$$u_{tt} - a^2 u_{xx} = 0,$$

we obtain from (4.32)

$$\int_0^{t_0} \int_0^l [(u_t^2 + a^2 u_x^2)_t - 2a^2 (u_x u_t)_x] \, dx \, dt = 0.$$

Carrying out the integrations as far as possible gives us the identity.

$$\int_0^l (u_t^2 + a^2 u_x^2)(x,t_0)\, dx - \int_0^l (u_t^2 + a^2 u_x^2)(x,0)\, dx$$
$$= 2a^2 \left\{ \int_0^{t_0} (u_x u_t)(l,t)\, dt - \int_0^{t_0} (u_x u_t)(0,t)\, dt \right\}. \quad (4.34)$$

Now suppose that $u(x,0) = u_t(x,0) = 0$ on $[0,l]$, then $u_x(x,0) = 0$ on $[0,l]$ and the second integral on the left-hand side of (4.34) has the value zero. On the other hand, consider any of the other possible homogeneous boundary conditions mentioned in Section 4.1. If $u = 0$ on the line $x = 0$ or $x = l$, then $u_t = 0$ on the line in question and the corresponding integral on the right-hand side of (4.34) vanishes. Similarly if $u_x = 0$ on $x = 0$ or on $x = l$ the corresponding integral on the right-hand side of (4.34) vanishes. As an example consider the boundary condition $u + \alpha u_x = 0$ on $x = 0$ for $\alpha \neq 0$. If this condition holds we have

$$\int_0^{t_0} (u_x u_t)(0,t)\, dt = \int_0^{t_0} \frac{1}{\alpha}(u u_t)(0,t)\, dt$$
$$= \frac{1}{2\alpha} \int_0^{t_0} \left[\frac{\partial}{\partial t}(u^2) \right](0,t)\, dt$$
$$= \frac{1}{2\alpha} u^2(0,t_0) - \frac{1}{2\alpha} u^2(0,0).$$

If $\alpha > 0$, we can then write (4.34) in the form

$$\int_0^l (u_t^2 + a^2 u_x^2)(x,t_0)\, dx + \left(\frac{a}{\beta}\right)^2 u^2(0,t_0)$$
$$= \int_0^l (u_t^2 + a^2 u_x^2)(x,0)\, dx + \left(\frac{a}{\beta}\right)^2 u^2(0,0)$$
$$+ 2a^2 \int_0^{t_0} (u_x u_t)(0,t)\, dt, \quad (4.35)$$

where $\beta = \sqrt{\alpha}$. Now suppose that u or u_x is zero on $x = l$, then (4.35) reduces to the identity

$$\int_0^l (u_t^2 + a^2 u_x^2)(x,t_0)\, dx + \left(\frac{a}{\beta}\right)^2 u^2(0,t_0)$$
$$= \int_0^l (u_t^2 + a^2 u_x^2)(x,0)\, dx + \left(\frac{a}{\beta}\right)^2 u^2(0,0). \quad (4.36)$$

Let us introduce the function

$$\mathcal{E}_{31}(t) = \int_0^l (u_t^2 + a^2 u_x^2)(x,t)\, dx + \left(\frac{a}{\beta}\right)^2 u^2(0,t). \quad (4.37)$$

Then we may write (4.36) in the form

$$\mathcal{E}_{31}(t_0) = \mathcal{E}_{31}(0) \quad \text{for all } t_0. \quad (4.38)$$

Except for a constant factor, the function $\mathcal{E}_{31}(t)$ represents the total energy function associated with the homogeneous boundary value problem denoted by W_{31} in Section 4.1. Thus (4.38) or (4.36) expresses the fact that Problem W_{31} is conservative.

Similarly we can introduce the function

$$\mathcal{E}_{11}(t) = \int_0^l (u_t^2 + a^2 u_x^2)(x,t)\, dx \tag{4.39}$$

and write (4.34) in the form

$$\mathcal{E}_{11}(t_0) = \mathcal{E}_{11}(0) \quad \text{for all } t_0 \tag{4.40}$$

when $u(0,t) = u(l,t) = 0$. This shows that Problem W_{11} represents a conservative system.

Clearly we can carry out the same analysis for each of the homogeneous problems W_{ij}, $i,j = 1, 2, 3$ of Section 4.1. Thus the following result has been proven.

Theorem 4.3

Each of the homogeneous problems W_{ij}, $i,j = 1, 2, 3$ of Section 4.1 for the wave equation has associated with it an energy function $\mathcal{E}_{ij}$ defined for all t and satisfying the conservation formula

$$\mathcal{E}_{ij}(t) = \mathcal{E}_{ij}(0) \quad \text{for all } t. \tag{4.41}$$

Moreover, with the exception of $\mathcal{E}_{22}$ each of the functions $\mathcal{E}_{ij}$ is positive definite.

This is simply stated as

$$\mathcal{E}_{ij}(t) > 0 \quad \text{unless } u(x,t) = 0 \quad \text{for fixed } t, \tag{4.42}$$

provided that for the boundary conditions

$$(u + \alpha u_x)(0,t) = 0 \quad \text{or} \quad (u + \alpha u_x)(l,t) = 0,$$

the constant α is positive or negative, respectively.

We shall call a homogeneous boundary-value problem *conservative* if the corresponding energy function $\mathcal{E}$ is positive definite. When we speak of the positive definiteness of $\mathcal{E}$ we are thinking of $\mathcal{E}$ as a function of u and its derivatives. $\mathcal{E}$ is only a function of t because u is a function of t. When we write the identity

$$\mathcal{E}(t) = \mathcal{E}(0) \tag{4.43}$$

which expresses the conservation law we purposely emphasize the apparent dependence of $\mathcal{E}$ upon t. This is because the various boundary-value prob-

Theorem 4.4

The solution to every conservative boundary-value problem for the wave equation is unique.

PROOF. From our previous uniqueness proofs we know that for a linear problem it is sufficient to show that the problem with zero initial conditions and zero boundary conditions for a homogeneous equation has the solution $u = 0$. On the other hand, for a conservative problem (4.43) holds with $\mathcal{E}$ positive definite and we will have

$$\mathcal{E}(0) = 0$$

because of the zero initial conditions. It follows from (4.43) that

$$\mathcal{E}(t) = 0,$$

for all t. Now positive definiteness means that $\mathcal{E}(t) = 0$ implies $u \equiv 0$. This proves the uniqueness of the solution to conservative boundary value problems.

PROBLEMS

1. In the statement of Theorem 4.3 we asserted that the energy functions $\mathcal{E}_{ij}$ are positive definite, except for the function $\mathcal{E}_{22}$. Explain on physical grounds why this function is an exception.
2. Prove, by modification of the methods used to prove Theorem 4.2, that the solution u of the boundary-value problem

$$u_{tt} - a^2 u_{xx} = f(x,t), \quad 0 \le x \le \infty, \quad \text{all } t,$$
$$u(0,t) = g(t), \quad \text{all } t,$$
$$u(x,0) = h_1(x), \quad u_t(x,0) = h_2(x), \quad 0 < x < \infty$$

is unique. [*Hint:* Integrate over the domain shown in Figure 4.6.]

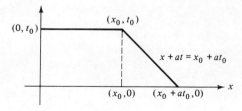

FIGURE 4.6

3. Prove that the solution u of the boundary-value problem
$$u_{tt} - a^2 u_{xx} = f(x,t), \quad 0 < x < \infty, \quad \text{all } t,$$
$$u(0,t) + \alpha u_x(0,t) = g(t), \quad \text{all } t,$$
$$u(x,0) = h_1(x), \quad u_t(x,0) = h_2(x), \quad 0 \leq x < \infty.$$
is unique if $\alpha < 0$.

4. Derive an energy identity for the solution of the equation
$$b(x)u_{tt} - (a^2(x)u_x)_x = 0, \quad b \geq 0, \quad 0 < x < l, \quad \text{all } t.$$

5. Prove that the solution of the boundary-value problem
$$u_{tt} - a^2 u_{xx} = 0, \quad 0 < x < l, \quad \text{all } t,$$
$$u(0,t) = f_1(t), \quad u_x(l,t) + \alpha u_t(l,t) = f_2(t), \quad \text{all } t, \quad \alpha > 0,$$
$$u(x,0) = g_1(x), \quad u_t(x,0) = g_2(x), \quad 0 < x \leq l,$$
is unique.

6. Show that the solution of the boundary-value problem
$$u_{tt} - \frac{a^2}{2}(x^2 u_x)_x = 0, \quad 0 < x < l, \quad \text{all } t,$$
$$u(0,t) \text{ is bounded for all } t,$$
$$u(l,t) = f(t), \quad \text{all } t,$$
$$u(x,0) = g_1(x), \quad u_t(x,0) = g_2(x), \quad 0 \leq x < l,$$
is unique.

7. Prove that the solution of the boundary-value problem
$$u_{tt} - a^2 u_{xx} = f(x,t), \quad 0 < x < l, \quad \text{all } t,$$
$$u(0,t) = g_1(t), \quad u_{tt}(l,t) + \alpha u_x(l,t) = g_2(t), \quad \alpha > 0, \quad \text{all } t,$$
$$u(x,0) = h_1(x), \quad u_t(x,0) = h_2(x), \quad 0 < x \leq l,$$
is unique.

8. Prove that the solution of the boundary-value Problem 7 is unique if the condition at $x = l$ is replaced by
$$u_{tt}(l,t) + \alpha u_x(l,t) - \beta u(l,t) = g_2(t), \quad \alpha > 0, \quad \beta > 0, \quad \text{all } t.$$

9. Discuss the conditions on the constants α, β, γ for which the solution of the boundary-value problem
$$u_{tt} - a^2 u_{xx} = f(x,t), \quad 0 < x < +\infty, \quad \text{all } t,$$
$$u_{tt}(0,t) + \alpha u_x(0,t) + \beta u_t(0,t) + \gamma u(0,t) = g(t), \quad \text{all } t,$$
$$u(x,0) = 0, \quad u_t(x,0) = 0, \quad 0 \leq x < \infty,$$
is unique.

10. Prove that the solution of the problem

$$u_{tt} - a^2 u_{xx} = 0, \quad \text{if } -\infty < x < 0, \quad \text{all } t,$$
$$v_{tt} - a^2 v_{xx} = 0, \quad \text{if } 0 < x < +\infty, \quad \text{all } t,$$
$$u(0,t) = v(0,t), \quad \text{all } t,$$
$$u_{tt}(0,t) - v_{tt}(0,t) = \alpha(v_x(0,t) - u_x(0,t)) - \beta u(0,t), \quad \text{all } t,$$
$$u(x,0) = f(x), \quad u_t(x,0) = 0, \quad -\infty < x < 0,$$
$$v(x,0) = f(x), \quad v_t(x,0) = 0, \quad 0 < x < \infty,$$

is unique.

11. Show that if

$$u_{tt} - a^2 u_{xx} - b u_x - c u = 0,$$

a, b, c constant, the function v defined by

$$u = v \exp(\mu x + \lambda t)$$

satisfies

$$v_{tt} - a^2 v_{xx} - c'v = 0$$

for proper choice of the constants μ and λ.

4.4 THE METHOD OF SEPARATION OF VARIABLES

In Section 3.7 we saw how to solve the first general boundary-value problem for the heat equation. The corresponding problem for the wave equation is given by the differential equation

$$u_{tt} = a^2 u_{xx} + b u_x + c u_t + d u + F(x,t), \quad 0 < x < L, \quad 0 < t, \quad (4.44)$$

together with the boundary conditions

$$u(0,t) = h_1(t), \quad u(L,t) = h_2(t), \quad 0 < t, \quad (4.45)$$

and the initial conditions

$$u(x,0) = g_1(x), \quad u_t(x,0) = g_2(x), \quad 0 < x < L. \quad (4.46)$$

In Equation (4.44) a, b, c, and d are constants. For this reason we refer to it as an equation with constant coefficients. Equations with variable coefficients will be examined to some extent in Chapter 5.

The method used in Chapter 3 to solve problems with nonhomogeneous boundary conditions consisted in setting

$$u(x,t) = v(x,t) + \varphi_1(t)x + \varphi_2(t)(x - L)$$

and then choosing φ_1 and φ_2 so that u satisfied the boundary conditions. In the present case this simply results in setting $\varphi_1(t) = (1/L)h_2(t)$ and

$\varphi_2(t) = -(1/L)h_1(t)$. Of course we will obtain a new boundary-value problem for the function V, but it again has the form (4.44), (4.45), (4.46) with $h_1(t) = h_2(t) = 0$. Accordingly, we shall confine our attention herein to this boundary-value problem with homogeneous boundary conditions.

Next let us reduce the differential equation (4.44) to a simpler form. For this purpose we can set

$$u(x,t) = e^{\alpha x + \beta t} v(x,t). \tag{4.47}$$

Differentiating and substituting into (4.44) yields

$$e^{\alpha x+\beta t}\{v_{tt} + 2\beta v_t + \beta^2 v\} = e^{\alpha x+\beta t}\{a^2(v_{xx} + 2\alpha v_x + \alpha^2 v) \\ + b(v_x + \alpha v) + c(v_t + \beta v) + dv\} + F(x,t).$$

Reorganizing the foregoing equation we obtain

$$v_{tt} = a^2 v_{xx} + (2\alpha a^2 + b)v_x + (c - 2\beta)v_t \\ + (a^2\alpha^2 + b\alpha + c\beta - \beta^2 + d)v + e^{-\alpha x - \beta t} F(x,t).$$

Letting $c = 2\beta$ and $\alpha = -b/2a^2$, we find that v satisfies

$$v_{tt} = a^2 v_{xx} + \left(d + \frac{c^2}{4} - \frac{b^2}{4a^2}\right)v + e^{\frac{b}{2a^2}x - \frac{c}{2}t} F(x,t). \tag{4.48}$$

This is an equation of the same form as (4.44) without first-derivative terms. Thus we have demonstrated a technique for reducing the complex system (4.44) to a simpler one and we shall therefore confine our attention to the reduced problem

$$\text{(P)} \begin{cases} u_{tt} = a^2 u_{xx} + du + F(x,t), & 0 < x < L, \quad 0 < t \\ u(0,t) = u(L,t) = 0, \\ u(x,0) = g_1(x), \quad u_t(x,0) = g_2(x). \end{cases}$$

Denote by P_0 Problem P with $F(x,t) = 0$. This, of course, corresponds to the homogeneous problem W_{11}. To solve this problem we first seek solutions having the form

$$u(x,t) = X(x)\,T(t). \tag{4.49}$$

Substituting into the homogeneous differential equation yields

$$X\ddot{T} = a^2 X''T + dXT,$$

where we have used a dot to denote differentiation with respect to t and a prime to denote differentiation with respect to x. Assuming, for the moment, that the product XT is different from zero, division by XT yields

$$\frac{1}{a^2}\frac{\ddot{T}}{T} = \frac{X''}{X} + \frac{d}{a^2}. \tag{4.50}$$

THE METHOD OF SEPARATION OF VARIABLES 203

The significant aspect of (4.50) is that the left-hand side of this equation is a function of t only and the right-hand side a function of x only. Let us write it in the form

$$M(t) = N(x) \qquad (4.51)$$

and note that the equality must hold for $0 < x < L$, $0 < t$.

Assume that M is a differentiable function of t, then from (4.51) we derive

$$\dot{M}(t) = 0, \quad \text{for } 0 < t.$$

It follows at once that $M(t)$ must equal a constant, say

$$M(t) = \mu. \qquad (4.52)$$

Similarly, if N is a differentiable function of x, we find that $N'(x) = 0$ for $0 < x < L$. Therefore, we must have

$$N(x) = \nu,$$

where ν is a constant. But from (4.51), $\mu = \nu$. Let us call this common value μ, so that in addition to (4.52) we also have

$$N(x) = \mu. \qquad (4.53)$$

We have therefore obtained the differential equations

$$\ddot{T} - a^2 \mu T = 0, \qquad (4.54)$$

and

$$X'' + \left(\frac{d}{a^2} - \mu\right) X = 0, \qquad (4.55)$$

which result from (4.52) and (4.53). Now we observe that if (4.54) and (4.55) are satisfied, the product XT satisfies the homogeneous partial differential equation. This is true even if $XT = 0$. To prove this we have only to substitute (4.54) and (4.55) into the equation following (4.49) to obtain

$$X\ddot{T} = a^2 \mu XT = (a^2 \mu - d)XT + dXT,$$

verifying the result.

Now to satisfy the boundary conditions we must have

$$X(0)T(t) = 0, \quad X(L)T(t) = 0, \quad 0 < t.$$

Hence either $T(t) = 0$ for all $t > 0$, or

$$X(0) = X(L) = 0. \qquad (4.56)$$

Now the choice $T(t) = 0$ implies from (4.49) that $u(x,t) = 0$. Obviously we are interested in nontrivial solutions to the differential equation. Consequently we seek nontrivial solutions of (4.55) satisfying (4.56).

The general solution of Equation (4.55) was obtained in Chapter 1 in the form

$$X(x) = c_1 e^{\sqrt{d/a^2 - \mu}\, x} + c_2 e^{-\sqrt{d/a^2 - \mu}\, x}, \quad \text{if} \quad \frac{d}{a^2} - \mu \neq 0.$$

An inspection of the exponential terms reveals that there are three cases to be considered,

$$\frac{d}{a^2} - \mu > 0, \quad \frac{d}{a^2} - \mu < 0, \quad \text{and} \quad \frac{d}{a^2} = \mu.$$

In the second case if we set $d/a^2 - \mu = -\tau^2$ the solution becomes

$$X(x) = c_1 e^{\tau x} + c_2 e^{-\tau x}.$$

Placing the boundary conditions (4.56) into this function yields

$$0 = c_1 + c_2,$$

and

$$0 = \tau c_1 e^{\tau L} - \tau c_2 e^{-\tau L}$$

which is a homogeneous system for the determination of c_1 and c_2. Therefore it can have a nontrivial solution if and only if the determinant is zero for some value of τ. But, since by hypothesis $\tau \neq 0$, this determinant is

$$\begin{vmatrix} 1 & 1 \\ \tau e^{\tau L} & -\tau e^{-\tau L} \end{vmatrix} = -\tau(d^{-\tau L} + e^{\tau L}) \neq 0.$$

Consider next the possibility that $\mu = d/a^2$, in which case (4.55) has the solution

$$X(x) = c_1 + c_2 x.$$

To satisfy the boundary conditions (4.56), we must have

$$c_1 = 0$$
$$c_2 L = 0,$$

implying that $X(x) \equiv 0$ on $0 \leq x \leq L$. Therefore the possibility that $\mu = d/a^2$ is eliminated.

Consequently, we may assume that

$$\mu < \frac{d}{a^2} \tag{4.57}$$

and write $d/a^2 - \mu = \lambda^2$ and

$$X(x) = c_1 \cos \lambda x + c_2 \sin \lambda x.$$

Again, from (4.56) we obtain

$$0 = c_1,$$

and

$$0 = c_2 \sin \lambda L,$$

where again, by hypothesis, $\lambda \neq 0$. Therefore we must choose λ so that

$$\sin \lambda L = 0.$$

From this we deduce that we must set

$$\lambda_n = \frac{n\pi}{L}, \quad n = 1, 2, \ldots$$

which yields

$$\mu_n = \frac{d}{a^2} - \left(\frac{n\pi}{L}\right)^2, \quad n = 1, 2, \ldots, \quad (4.58)$$

as the eigenvalues for our problem. Neglecting a constant multiplier the corresponding eigenfunctions are

$$X_n(x) = \sin \frac{n\pi x}{L}, \quad n = 1, 2, \ldots. \quad (4.59)$$

Now we observe that Equation (4.54) leads to the infinite set of equations

$$\ddot{T}_n - \left[d - \left(\frac{n\pi a}{L}\right)^2\right] T_n = 0, \quad n = 1, 2, \ldots. \quad (4.60)$$

We shall assume that

$$d < \left(\frac{a\pi}{L}\right)^2 \quad (4.61)$$

so that all of the numbers $d - (n\pi a/L)^2$ are negative. When this is not the case some of these numbers will be positive or zero and the corresponding equations (4.60) will have exponential rather than trigonometric solutions. Of course, for any fixed value of d, $d - (n\pi a/L)^2$ will be negative for sufficiently large n. With (4.61) in force we have

$$T_n(t) = A_n \cos a \sqrt{|\mu_n|} \, t + B_n \sin a \sqrt{|\mu_n|} \, t.$$

At this point we may proceed precisely as we did in the case of the heat equation. Therefore we will attempt to satisfy the initial conditions by forming the infinite series

$$u(x,t) = \sum_{n=1}^{\infty} [A_n \cos a \sqrt{|\mu_n|} \, t + B_n \sin a \sqrt{|\mu_n|} \, t] \sin \frac{n\pi x}{L}. \quad (4.62)$$

Placing the initial conditions into (4.62) yields

$$u(x,0) = g_1(x) = \sum_{n=1}^{\infty} A_n \sin \frac{n\pi x}{L},$$

and

$$u_t(x,0) = g_2(x) = \sum_{n=1}^{\infty} a \sqrt{\mu_n}\, B_n \sin \frac{n\pi x}{L}.$$

To find the numbers A_n we multiply the first of these equations by $\sin m\pi x/L$ and integrate from 0 to L. This gives

$$\int_0^L g_1(x) \sin \frac{m\pi x}{L}\, dx = \sum_{n=1}^{\infty} A_n \int_0^L \sin \frac{m\pi x}{L} \sin \frac{n\pi x}{L}\, dx.$$

Now we have seen in Chapter 3 that

$$\int_0^L \sin \frac{m\pi x}{L} \sin \frac{n\pi x}{L}\, dx = \begin{cases} 0, & \text{if } m \neq n, \\ \frac{L}{2}, & \text{if } m = n. \end{cases}$$

We therefore obtain

$$A_n = \frac{2}{L} \int_0^L g_1(x) \sin \frac{n\pi x}{L}\, dx. \tag{4.63}$$

In exactly the same way we obtain

$$B_n = \frac{2}{aL \sqrt{|\mu_n|}} \int_0^L g_2(x) \sin \frac{n\pi x}{L}\, dx. \tag{4.64}$$

Substitution of these results into (4.62) yields

$$u(x,t) = \sum_{n=1}^{\infty} \int_0^L \frac{2}{L} \sin \frac{n\pi \xi}{L} \sin \frac{n\pi x}{L} \cos a \sqrt{|\mu_n|}\, t\, g_1(\xi)\, d\xi$$

$$+ \sum_{n=1}^{\infty} \int_0^L \frac{2}{aL \sqrt{|\mu_n|}} \sin \frac{n\pi \xi}{L} \sin \frac{n\pi x}{L} \sin a \sqrt{|\mu_n|}\, t\, g_2(\xi)\, d\xi.$$

Now let us introduce the function

$$G(x,\xi,t) = \sum_{n=1}^{\infty} \frac{2}{aL \sqrt{|\mu_n|}} \sin \frac{n\pi x}{L} \sin \frac{n\pi \xi}{L} \sin a \sqrt{|\mu_n|}\, t. \tag{4.65}$$

Then we have

$$u(x,t) = \int_0^L \frac{\partial G}{\partial t}(x,\xi,t)g_1(\xi)\,d\xi + \int_0^L G(x,\xi,t)g_2(\xi)\,d\xi \qquad (4.66)$$

as the solution of the homogeneous problem P_0.

The function $G(x,\xi,t)$ given by formula (4.65) is called the *Green's function* for Problem W_{11} and formula (4.66) is entirely analogous to the solutions found for the homogeneous equations studied in Chapter 1. It is again worth calling the reader's attention to the fact that we have obtained (4.66) by purely formal manipulations, exactly as we obtained many corresponding results in Chapter 3.

Our next task is to solve the nonhomogeneous differential equation with homogeneous initial and boundary conditions. For this purpose it suits us to work again by analogy to our results of Section 3.7 and assume that $F(x,t)$ can be written in the form

$$F(x,t) = \sum_{n=1}^{\infty} F_n(t) \sin \frac{n\pi x}{L} \qquad (4.67)$$

where

$$F_n(t) = \frac{2}{L} \int_0^L F(\xi,t) \sin \frac{n\pi \xi}{L}\,d\xi. \qquad (4.68)$$

Similarly we shall suppose

$$u(x,t) = \sum_{n=1}^{\infty} u_n(t) \sin \frac{n\pi x}{L} \qquad (4.69)$$

with the functions $u_n(t)$ to be determined. Substituting into the differential equation gives

$$\sum_{n=1}^{\infty} \left\{ \ddot{u}_n + \left[\left(\frac{n\pi a}{L}\right)^2 - d\right] u_n - F_n(t) \right\} \sin \frac{n\pi x}{L} = 0.$$

Since the functions X_n are linearly independent, this means that we must solve the differential equations

$$\ddot{u}_n + \left[\left(\frac{n\pi a}{L}\right)^2 - d\right] u_n = F_n(t)$$

subject to the initial conditions

$$u_n(0) = \dot{u}_n(0) = 0.$$

This problem was solved in Chapter 1, and thus we obtain

$$u_n(t) = \frac{1}{a\sqrt{|\mu_n|}} \int_0^t \sin a\sqrt{|\mu_n|}\,(t-s)F_n(s)\,ds. \tag{4.70}$$

Substituting (4.68) into (4.70) and substituting the resulting equation into (4.69) yields the formula

$$u(x,t) = \int_0^t \int_0^L G(x,\xi, t-\tau)F(\xi,\tau)\,d\xi\,d\tau \tag{4.71}$$

as the formal solution of the nonhomogeneous equation with homogeneous data.

Finally, using linear superposition we combine (4.66) and (4.71) to obtain

$$u(x,t) = \int_0^L \frac{\partial G}{\partial t}(x,\xi,t)g_1(\xi)\,d\xi + \int_0^L G(x,\xi,t)g_2(\xi)\,d\xi$$
$$+ \int_0^t \int_0^L G(x,\xi, t-\tau)F(\xi,\tau)\,d\xi\,d\tau \tag{4.72}$$

as the solution of Problem P. Once again, just as we observed in Chapter 3, we see that the entire solution is formulated in terms of certain integrals involving the Green's function and the given data of the problem. Obviously this formal solution is easily justified if we can analyze these integrals or, what is the same, if we can obtain enough information about the Green's function. We shall not attempt to do this here.

The method of separation of variables we have used here can be applied to a large number of problems. It is indeed one of the most fundamental methods of solution used to solve a continuous, linear system.

PROBLEMS

1. Find the solution of the boundary-value problem

$$u_{tt} = a^2 u_{xx}, \quad 0 < x < L, \quad 0 < t,$$
$$u(0,t) = u(L,t) = 0, \quad 0 < t,$$
$$u(x,0) = \begin{cases} \dfrac{h}{x_0}x, & 0 < x < x_0, \\ \dfrac{h(L-x)}{L-x_0}, & x_0 < x < L, \end{cases}$$
$$u_t(x,0) = 0, \quad 0 < x < L.$$

2. Find the longitudinal vibrations of a rod with free ends and arbitrary initial velocity and displacement. Therefore,

$$u_x(0,t) = u_x(L,t) = 0 \quad \text{and} \quad u(x,0) = g_1(x) \quad \text{and} \quad u_t(x,0) = g_2(x).$$

What are the eigenvalues and eigenfunctions for this problem? What is the Green's function for this problem?

3. Find the vibrations of the rod of Problem 2 if it received a longitudinal impulse I at one end at $t = 0$. [*Hint:* Assume

$$u_t(x,0) = \begin{cases} 0, & 0 \le x \le L - \delta \\ -\dfrac{I}{\xi p}, & L - \delta < x \le L, \end{cases}$$

solve Problem 2, and pass to the limit as $\delta \to 0$.]

4. Show that the solution of the boundary-value problem

$$u_{tt} = a^2 u_{xx}, \quad 0 < x < L, \quad \text{all } t,$$
$$u_x(0,t) - hu(0,t) = 0, \quad u_x(L,t) + hu(L,t) = 0, \quad \text{all } t,$$
$$u(x,0) = \varphi_1(x), \quad u_t(x,0) = \varphi_2(x), \quad 0 \le x \le L,$$

is

$$u(x,t) = \sum_{n=1}^{\infty} (a_n \cos a\lambda_n t + b_n \sin a\lambda_n t) \sin (\lambda_n x + \mu_n)$$

where λ_n is the nth root of

$$\cot \lambda L = \frac{1}{2}\left(\frac{\lambda}{h} - \frac{h}{\lambda}\right),$$
$$\mu_n = \tan^{-1}\frac{\lambda_n}{h}.$$

Write out the Green's function for this problem.

5. Solve the boundary-value problem

$$u_{xx} = \alpha u_{tt} + \beta u_t, \quad 0 < x < L, \quad \text{all } t, \quad \alpha > 0, \quad \beta > 0,$$
$$u(0,t) = u_x(L,t) = 0, \quad \text{all } t,$$
$$u(x,0) = u_0, \quad u_t(x,0) = 0, \quad 0 < x < L.$$

What are the normalized eigenfunctions, and the eigenvalues of this problem? Write out the Green's function.

6. Solve the boundary-value problem

$$u_{tt} = a^2 u_{xx} + g, \quad 0 < x < L, \quad \text{all } t, \quad g = \text{const.},$$
$$u(x,0) = 0, \quad u_t(x,0) = u_0, \quad 0 < x < L,$$
$$u(0,t) = u_x(L,t) = 0, \quad \text{all } t.$$

Solve the boundary-value problem

$$u_{tt} = a^2 u_{xx}, \quad 0 < x < L, \quad \text{all } t,$$
$$u(0,t) = 0, \quad u_x(L,t) = \alpha \quad \text{all } t,$$
$$u(x,0) = u_t(x,0) = 0, \quad 0 < x < L.$$

Find the solution of the boundary-value problem

$$u_{tt} = a^2 u_{xx} + g(x)t, \quad 0 < x < L, \quad \text{all } t,$$
$$u(0,t) = u(L,t) = 0, \quad \text{all } t,$$
$$u(x,0) = u_t(x,0) = 0, \quad 0 < x < L.$$

9. Find the solution of the boundary-value problem
$$u_{tt} = a^2 u_{xx}, \quad 0 < x < L, \quad \text{all } t,$$
$$u(0,t) = 0, \quad u_x(L,t) = \alpha t, \quad \text{all } t,$$
$$u(x,0) = 0, \quad u_t(x,0) = 0, \quad 0 < x < L.$$

10. Find the solution of the boundary-value problem
$$u_{tt} = a^2 u_{xx} + g(x)t^m, \quad 0 < x < L, \quad \text{all } t, \quad m > -1,$$
$$u(0,t) = u(L,t) = 0, \quad \text{all } t,$$
$$u(x,0) = u_t(x,0) = 0, \quad 0 < x < L.$$

11. Find the solution of the boundary-value problem
$$u_{tt} = a^2 u_{xx} - 2\alpha u_t + g(x)t, \quad 0 < x < L, \quad \text{all } t,$$
$$u(0,t) = u(L,t) = 0, \quad 0 < t,$$
$$u(x,0) = u_t(x,0) = 0, \quad 0 < x < L.$$

4.5 TRANSVERSE VIBRATIONS OF A ROD

In Section 4.1 we saw that the small longitudinal vibrations of a rod are described by the wave equation. We have also indicated that the wave equation is applicable when describing a variety of boundary value problems related to physical systems, such as small transverse vibrations of a string, the variation of current and voltage in a conductor, and the torsional vibrations of a solid cylinder. Another important field which we have not discussed where the wave equation also occurs is the study of the propagation of small disturbances in a fluid, or more simply, the theory of sound. In this section we shall examine the equation for the small transverse vibrations of a rod or beam. This equation is somewhat more special than the wave equation because its application is limited to problems in the theory of elasticity.

Let $u = u(x,t)$ denote the transverse displacement of the neutral axis of a beam at the point x at time t and consider the beam to be homogeneous and of uniform cross-sectional area. From elementary beam theory we have

$$\frac{EI}{\zeta} = -M(x,t), \tag{4.73}$$

where EI is the beam stiffness which is assumed constant, ζ is the radius of curvature of the neutral axis and M is the bending moment in the beam. Assuming small curvature, (4.73) reduces to

$$EIu_{xx} = -M. \tag{4.74}$$

Referring to the free-body diagram of Figure 4.7, where m represents the

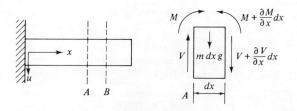

FIGURE 4.7

mass per unit length and V the shearing force, we may take moments and sum forces to obtain

$$V\,dx = dM \tag{4.75}$$

and

$$V_x + mg = mu_{tt}. \tag{4.76}$$

Combining (4.74), (4.75), and (4.76) we get

$$-EIu_{xxxx} + mg = mu_{tt}. \tag{4.77}$$

Neglecting the weight of the beam in comparison to the inertia force, (4.77) reduces to

$$u_{tt} + a^2 u_{xxxx} = 0 \tag{4.78}$$

which is the differential equation describing the transverse motion of the beam.

We shall assume that the beam has length L so that (4.78) is to be valid for $0 < x < L$ and $t > 0$. Since (4.78) is of second order with respect to t, we require the initial conditions

$$u(x,0) = F_1(x), \quad u_t(x,0) = F_2(x), \quad 0 < x < L. \tag{4.79}$$

In addition to the initial conditions, various sets of boundary conditions must be imposed. We mention here only some of these.

(i) If the beam is built in at one end, as in Figure 4.7, the boundary conditions are

$$u(0,t) = u_x(0,t) = 0.$$

(ii) If the beam is hinged (pinned) at $x = 0$, the boundary conditions are

$$u(0,t) = u_{xx}(0,t) = 0.$$

(iii) If the beam is free at $x = 0$, then

$$u_{xx}(0,t) = u_{xxx}(0,t) = 0.$$

A beam which is built in at one end and free at the other, such as the beam of Figure 4.7, is usually called a *cantilever beam* or simply a cantilever.

Obviously, many other boundary conditions in addition to those just mentioned are possible. We may obtain more insight into these by applying the energy method to Equation (4.78). Thus, multiply (4.78) by u_t to obtain

$$u_t u_{tt} + a^2 u_t u_{xxxx} = 0.$$

Now note that

$$u_t u_{tt} = \tfrac{1}{2}(u_t^2)_t,$$

and

$$u_t u_{xxxx} = (u_t u_{xxx} - u_{tx} u_{xx})_x + \tfrac{1}{2}(u_{xx}^2)_t.$$

Equation (4.78) multiplied by u_t now assumes the form

$$\tfrac{1}{2}(u_t^2 + a^2 u_{xx}^2)_t = a^2(u_{tx} u_{xx} - u_t u_{xxx})_x.$$

Integrating this equation with respect to x from 0 to L and with respect to t from 0 to t, we obtain

$$\int_0^L [u_t^2(x,t) + a^2 u_{xx}^2(x,t)]\, dx = \int_0^L [F_2^2(x) + a^2(F_1''(x))^2]\, dx$$
$$+ 2a^2 \int_0^t (u_{tx} u_{xx} - u_t u_{xxx})(L,s)\, ds$$
$$- 2a^2 \int_0^t (u_{tx} u_{xx} - u_t u_{xxx})(0,s)\, ds. \quad (4.80)$$

Now if we set

$$\int_0^L [u_t^2(x,t) + a^2 u_{xx}^2(x,t)]\, dx = \mathcal{E}_0(t,u),$$

then (4.80) becomes

$$\mathcal{E}_0(t,u) = \mathcal{E}_0(0,u) + 2a^2 \int_0^t [u_{tx}(L,s) u_{xx}(L,s) - u_t(L,s) u_{xxx}(L,s)]\, ds$$
$$- 2a^2 \int_0^t [u_{tx}(0,s) u_{xx}(0,s) - u_t(0,s) u_{xxx}(0,s)]\, ds. \quad (4.81)$$

We shall call the function $\mathcal{E}_0(t,u)$ the *basic energy function* for the beam equation. Note that $\mathcal{E}_0(t,u) > 0$ unless $u_t \equiv 0$ on $(0,L)$ and $u_{xx} \equiv 0$ on

$(0,L)$. This means that the only solutions $u = u(x,t)$ of the beam equation which permit $\mathcal{E}_0(t,u) = 0$ must have the form

$$u(x,t) = u_0 + u_1 x, \tag{4.82}$$

where u_0 and u_1 are constants.

Consider next what happens to the relation (4.81) in various special cases. For example, in the case of a cantilever satisfying

$$u(0,t) = u_x(0,t) = u_{xx}(L,t) = u_{xxx}(L,t) = 0,$$

we have $u_t(0,t) = u_{xt}(0,t) = 0$, and (4.81) reduces to

$$\mathcal{E}_0(t,u) = \mathcal{E}_0(0,u). \tag{4.83}$$

Equation (4.83) says that the basic energy function is a constant with respect to t. This means that we have conservation of energy for the cantilever beam. In the case of a beam pinned at both ends we have

$$u(0,t) = u_{xx}(0,t) = u(L,t) = u_{xx}(L,t) = 0.$$

In this case $u_t(0,t) = u_t(L,t) = 0$ and again (4.83) holds so that we have conservation of energy.

Let us look at a more interesting set of boundary conditions. Assume that a horizontal beam is built in at $x = 0$ and a load Q is attached at the end $x = L$. If we can neglect the moment of inertia of the load Q with respect to the x-axis, the boundary conditions for this problem are

$$u(0,t) = u_x(0,t) = u_{xx}(L,t) = 0,$$

$$\frac{Q}{\alpha} u_{tt}(L,t) = u_{xxx}(L,t), \quad \alpha > 0.$$

Therefore from (4.81) we obtain the equation

$$\mathcal{E}_0(t,u) = \mathcal{E}_0(0,u) - 2a^2 \frac{Q}{\alpha} \int_0^t u_t(L,s) u_{tt}(L,s)\, ds.$$

But

$$u_t u_{tt} = \tfrac{1}{2}(u_t^2)_t$$

so that we have

$$\mathcal{E}_0(t,u) = \mathcal{E}_0(0,u) - \frac{2a^2 Q}{\alpha} [u_t^2(L,t) - u_t^2(L,0)].$$

Thus we may define the energy function

$$\mathcal{E}_1(t,u) = \mathcal{E}_0(t,u) + \frac{2a^2 Q}{\alpha} u_t^2(L,t) \tag{4.84}$$

and (4.81) becomes

214   OTHER TIME-DEPENDENT PARTIAL DIFFERENTIAL EQUATIONS

$$\mathcal{E}_1(t,u) = \mathcal{E}_1(0,u) \tag{4.85}$$

which is again a conservation of energy formula, in terms of the energy function $\mathcal{E}_1(t,u)$ in place of $\mathcal{E}_0(t,u)$.

A general conclusion which can be formulated on the basis of the foregoing examples is that any set of boundary conditions which make the two integrals in (4.81) vanish or which convert these integrals into the difference of the squares of some derivative of u or of u itself with a negative coefficient at time t and positive coefficient at time 0 leads to the formulation of a conservation of energy problem for the beam equation. Obviously there are many such problems and we shall not attempt to formulate all of them here. However, we do wish to define exactly what we shall mean by a conservative problem for the beam equation (4.78).

Definition 4.1

A boundary-value problem for the beam equation will be called conservative if the boundary conditions are such that the basic energy relation (4.81) assumes the form

$$\mathcal{E}^*(t,u) = \mathcal{E}^*(0,u),$$

where the function $\mathcal{E}^*$ is such that the equation

$$\mathcal{E}^*(t,u) \equiv 0 \quad \text{in } (0,L)$$

implies that u satisfies (4.82).

As a further illustration of the use of the basic energy formula (4.81) let us consider a nonhomogeneous problem. Consider a cantilever beam fixed at the end $x = 0$ and acted upon at the free end $x = L$ by a transverse force, varying with time. We obtain the problem

$$\begin{aligned}
&u_{tt} + a^2 u_{xxxx} = 0, \quad 0 < x < L, \quad t > 0, \\
&u(x,0) = F_1(x), \quad u_t(x,0) = F_2(x), \quad 0 < x < L, \\
&u(0,t) = u_x(0,t) = u_{xx}(L,t) = 0, \quad u_{xxx}(L,t) = g(t), \quad 0 < t,
\end{aligned} \tag{4.86}$$

where g is a given function of time. Our aim is to prove that the solution of this problem is unique. To do this let us suppose, to the contrary, that the problem has two solutions v_1 and v_2. Let $u = v_1 - v_2$, then u satisfies

$$\begin{aligned}
&u_{tt} + a^2 u_{xxxx} = 0, \quad 0 < x < L, t > 0, \\
&u(x,0) = 0 = u_t(x,0), \quad 0 < x < L, \\
&u(0,t) = u_x(0,t) = u_{xx}(L,t) = u_{xxx}(L,t) = 0, \quad 0 < t.
\end{aligned}$$

These equations imply that $\mathcal{E}_0(0,u) = 0$ and also that $u_t(0,t) = u_{tx}(0,t) = 0$. It follows at once from (4.81) that

$$\mathcal{E}_0(t,u) \equiv 0 \quad \text{on } [0,L].$$

But then (4.82) implies that
$$u(x,t) = u_0 + u_1 x$$
where u_0 and u_1 are constants. Finally the initial conditions require that
$$u(x,0) = 0 = u_0 + u_1 x, \quad 0 < x < L,$$
which can only be satisfied if $u_0 = u_1 = 0$. This proves that $u \equiv 0$ for $0 \leq x \leq L$, $t \geq 0$. Hence $v_1 \equiv v_2$ which establishes the uniqueness.

4.6 A NONCONSERVATIVE PROBLEM

Let us consider a cantilever beam in a vertical position, built in at the end $x = 0$ and loaded at the free end $x = L$ by a force of magnitude P which is always directed along the axis of the beam at the free end. This situation which is generally termed a *fixed-free column*, is illustrated in Figure 4.8. The problem and variations of it have been widely studied in recent engineering literature where it is usually termed the *follower problem*.

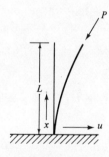

FIGURE 4.8

Using the same technique as in Section 4.5 we find that the differential equation for our problem is
$$a^2 u_{xxxx} + P u_{xx} + m u_{tt} = 0, \tag{4.87}$$
where m is the mass per unit length of the column. The term Pu_{xx} arises because of the axial load of magnitude P. The boundary conditions for this column are
$$u(0,t) = u_x(0,t) = u_{xx}(L,t) = u_{xxx}(L,t) = 0. \tag{4.88}$$
Let us assume the arbitrary initial conditions
$$u(x,0) = F_1(x), \quad u_t(x,0) = F_2(x). \tag{4.89}$$

We begin the analysis by showing that this problem is nonconservative. Multiplying (4.87) by u_t we obtain
$$\tfrac{1}{2}(mu_t^2 + a^2 u_{xx}^2 - P u_x^2)_t = \{a^2(u_{tx}u_{xx} - u_t u_{xxx}) - P u_x u_t\}_x.$$

216 OTHER TIME-DEPENDENT PARTIAL DIFFERENTIAL EQUATIONS

Integrating as before with respect to x and t yields

$$\tilde{\mathcal{E}}(t,u) = \tilde{\mathcal{E}}(0,u) + 2\int_0^t \{a^2[u_{tx}(L,s)u_{xx}(L,s) - u_t(L,s)u_{xxx}(L,s)]$$
$$- Pu_x(L,s)u_t(L,s)\} \, ds - 2\int_0^t \{a^2[u_{tx}(0,s)u_{xx}(0,s) - u_t(0,s)u_{xxx}(0,s)]$$
$$- Pu_x(0,s)u_t(0,s)\} \, ds, \quad (4.90)$$

where

$$\tilde{\mathcal{E}}(t,u) = \int_0^L [mu_t^2(x,t) + a^2 u_{xx}^2(x,t) - Pu_x^2(x,t)] \, dx. \quad (4.91)$$

Applying the boundary conditions (4.88), reduces (4.90) to

$$\tilde{\mathcal{E}}(t,u) = \tilde{\mathcal{E}}(0,u) - 2P\int_0^t u_x(L,s)u_t(L,s) \, ds. \quad (4.92)$$

From (4.92) we see at once that the problem will in general be non-conservative for $P \neq 0$. Indeed this will be the case unless $u_x(L,t) \equiv 0$ or $u(L,t) \equiv 0$.

Let us attempt to solve this follower problem by the method of separation of variables. We seek solutions having the form

$$u(x,t) = X(x)T(t). \quad (4.93)$$

Substituting into (4.87) we have

$$a^2 TX'''' + PTX'' + mX\ddot{T} = 0.$$

Proceeding formally yields

$$\frac{a^2 X''''}{X} + \frac{PX''}{X} = -m\frac{\ddot{T}}{T} = \lambda^2$$

where λ is a constant. (The choice of λ^2 will prove to be slightly more convenient in some of the calculations to follow below.) We therefore have to solve the two differential equations

$$\ddot{T} + \frac{\lambda^2}{m} T = 0. \quad (4.94)$$

and

$$X'''' + \frac{P}{a^2} X'' - \frac{\lambda^2}{a^2} X = 0. \quad (4.95)$$

From the boundary conditions we obtain

$$X(0) = X'(0) = X''(L) = X'''(L) = 0, \quad (4.96)$$

so that our task is to solve (4.95) subject to (4.96).

The fourth-order differential equation (4.95) can be approached in exactly the same way in which we analyzed the second-order differential equation for the RLC circuit in Chapter 1, namely by setting

$$X(x) = e^{\mu x} \tag{4.97}$$

and seeking values of μ which yield a solution. This gives rise to the characteristic equation

$$\mu^4 + \frac{P}{a^2}\mu^2 - \frac{\lambda^2}{a^2} = 0.$$

Solving for μ^2 we obtain

$$\mu^2 = -\frac{P}{2a^2} \pm \left[\left(\frac{P}{2a^2}\right)^2 + \left(\frac{\lambda}{a}\right)^2\right]^{1/2}.$$

For convenience let us set

$$r_1 = \left\{-\frac{P}{2a^2} + \sqrt{\left(\frac{P}{2a^2}\right)^2 + \left(\frac{\lambda}{a}\right)^2}\right\}^{1/2}$$

and $\tag{4.98}$

$$r_2 = \left\{\frac{P}{2a^2} + \sqrt{\left(\frac{P}{2a^2}\right)^2 + \left(\frac{\lambda}{a}\right)^2}\right\}^{1/2}.$$

Then as a general solution to (4.95) we have

$$X(x) = c_1 e^{r_1 x} + c_2 e^{r_1 x} + c_3 \sin r_2 x + c_4 \cos r_2 x.$$

Now, since we know that linear combinations of solutions of (4.95) are again solutions, we may put $X(x)$ into the more convenient form

$$X(x) = c_1 \sinh r_1 x + c_2 \cosh r_1 x + c_3 \sin r_2 x + c_4 \cos r_2 x. \tag{4.99}$$

It then follows that to satisfy the boundary conditions (4.96) we must have

$$c_2 + c_4 = 0,$$
$$r_1 c_1 + r_2 c_3 = 0,$$
$$r_1^2 c_1 \sinh r_1 L + r_1^2 c_2 \cosh r_1 L - r_2^2 (c_3 \sin r_2 L + c_4 \cos r_2 L) = 0,$$
$$r_1^3 (c_1 \cosh r_1 L + c_2 \sinh r_1 L) + r_2^3 (-c_3 \cos r_2 L + c_4 \sin r_2 L) = 0.$$

Substituting from the first two of these equations into the last two we arrive at

$$r_1(r_1 \sinh r_1 L + r_2 \sin r_2 L)c_1 + (r_1^2 \cosh r_1 L + r_2^2 \cos r_2 L)c_2 = 0,$$
$$r_1(r_1^2 \cosh r_1 L + r_2^2 \cos r_2 L)c_1 + (r_1^3 \sinh r_1 L - r_2^3 \sin r_2 L)c_2 = 0,$$

as a pair of homogeneous equations for the determination of c_1 and c_2. We can have a nontrivial solution if and only if the determinant of this system

218 OTHER TIME-DEPENDENT PARTIAL DIFFERENTIAL EQUATIONS

is different from zero. Let the determinant be $D(r_1,r_2)$. A little bit of manipulation yields

$$D(r_1,r_2) = -r_1[r_1^4 + r_2^4 + r_1r_2(r_2^2 - r_1^2) \sinh r_1L \sin r_2L \\ + 2r_1^2r_2^2 \cosh r_1L \cos r_2L] = 0.$$

One case to be considered is that of $r_1 = 0$. Then we must have $\lambda = 0$ regardless of the value of P. In this case formula (4.99) is no longer correct and we must solve the differential equation

$$X'''' + \frac{P}{a^2} X'' = 0.$$

Replace X'' by Y to obtain

$$Y'' + \frac{P}{a^2} Y = 0,$$

which from our earlier work we know has the general solution

$$Y(x) = c_1 \sin \sqrt{\frac{P}{a^2}} x + c_2 \cos \sqrt{\frac{P}{a^2}} x.$$

Now integrating twice yields

$$X(x) = c_1 \sin \sqrt{\frac{P}{a^2}} x + c_2 \cos \sqrt{\frac{P}{a^2}} x + c_3 x + c_4,$$

and applying the boundary conditions leads to

$$c_2 + c_4 = 0,$$

$$\sqrt{\frac{P}{a^2}} c_1 + c_3 = 0,$$

$$\frac{P}{a^2} \left(c_1 \sin \sqrt{\frac{P}{a^2}} L + c_2 \cos \sqrt{\frac{P}{a^2}} L \right) = 0,$$

$$\left(\frac{P}{a^2}\right)^{3/2} \left\{ c_1 \cos \sqrt{\frac{P}{a^2}} L - c_2 \sin \sqrt{\frac{P}{a^2}} L \right\} = 0.$$

The determinant of this homogeneous system is easily computed and one obtains the value

$$D = 1.$$

This means we cannot find a nontrivial solution of the boundary-value problem if $P > 0$ in the case when $\lambda = 0$.

Accordingly in order to have a nontrivial solution we must have

$$r_1^4 + r_2^4 + r_1r_2(r_2^2 - r_1^2) \sinh r_1L \sin r_2L + 2r_1^2r_2^2 \cosh r_1L \cos r_2L = 0.$$

Let us substitute the values of r_1 and r_2 as given in (4.98) into this formula. We find that

$$-\frac{p^2}{a}(1 + \lambda)\sinh r_1 L \sin r_2 L) = +2\lambda^2(1 + \cosh r_1 L \cos r_2 L). \quad (4.100)$$

Since r_1 and r_2 depend upon λ, our task is to find those values of λ, if any exist, for which the two sides of (4.100) are equal. Such values of λ will depend upon the magnitude P of the force applied to the column.

The characteristic equation (4.100) which we have found above is very complicated and we shall not seek to study it further in detail here. Instead we will simplify the problem in a manner familiar to engineers and indicate by studying the simplified problem how the column acted upon by a follower force may be expected to behave.

To this end consider Figure 4.9 in which a concentrated load of mass m is assumed to be located on the end of a weightless column. For small dis-

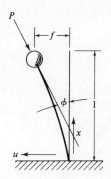

FIGURE 4.9

placements we may assume that the load only moves parallel to the y-axis so that the deflection of the upper end of the column is given by the function $f = f(t)$. As the system moves the upper end of the column is acted upon both by the follower force of magnitude P and the inertial force of magnitude.

$$R = -m\ddot{f} \quad (4.101)$$

If we take moments at the cross section with coordinate x and let $u = u(x,t)$ denote the deflection at this cross section, the differential equation for the motion is

$$EIu_{xx} = P(f - u) - (P\varphi - R)(1 - x), \quad (4.102)$$

where the column is now assumed to have unit length, E is the modulus of elasticity, and I the moment of inertia. We must regard the functions $f(t)$

and $\varphi(t)$ as unknown so that they as well as the deflection u are to be determined from the differential equation (4.102) and the boundary conditions

$$u(0,t) = u_x(0,t) = 0,$$
$$u(1,t) = f(t), \quad u_x(1,t) = \varphi(t). \tag{4.103}$$

We set

$$k^2 = \frac{P}{EI} \tag{4.104}$$

and write (4.103) in the form

$$u_{xx} + k^2 u = k^2 f - \left(k^2 \varphi - \frac{R}{EI}\right)(1-x). \tag{4.105}$$

In order to solve Equation (4.105) we will use the method of separation of variables, setting

$$u(x,t) = v(x)\sigma(t), \quad f(t) = f_0 \sigma(t), \quad \varphi(t) = \varphi_0 \sigma(t).$$

This yields

$$\sigma v'' + k^2 \sigma v = k^2 f_0 \sigma - k^2 \varphi_0 \sigma (1-x) - \frac{m}{EI} f_0 \ddot{\sigma}(1-x).$$

Dividing both sides by $f_0 \sigma(1-x)$ and rearranging, we have

$$\frac{v'' + k^2 v}{f_0(1-x)} - \frac{k^2}{1-x} + \frac{k^2 \varphi_0}{f_0} = -\frac{m}{EI}\frac{\ddot{\sigma}}{\sigma}.$$

Since the left-hand side is a function of x only and the right-hand side a function of t only, we must have

$$v'' + k^2 v = k^2 f_0 - k^2 \varphi_0 (1-x) + \omega^2 f_0 (1-x), \tag{4.106}$$

and

$$\ddot{\sigma} + \frac{EI}{m}\omega^2 \sigma = 0. \tag{4.107}$$

Equation (4.106) is to be solved subject to the boundary conditions

$$v(0) = v'(0) = 0, \quad v(1) = f_0, \quad v'(1) = \varphi_0. \tag{4.108}$$

The solution of (4.107) is found by the methods of Chapter 1 to be

$$v(x) = \alpha_1 \sin kx + \alpha_2 \cos kx + f_0 - \varphi_0(1-x) + \left(\frac{\omega}{k}\right)^2 f_0(1-x).$$

Substituting in the boundary conditions gives the four equations

$$\alpha_2 + \left(1 + \left(\frac{\omega}{k}\right)^2\right)f_0 - \varphi_0 = 0,$$

A NONCONSERVATIVE PROBLEM

$$\alpha_1 \sin k + \alpha_2 \cos k = 0,$$

$$k\alpha_1 - \left(\frac{\omega}{k}\right)^2 f_0 + \varphi_0 = 0,$$

$$k\alpha_1 \cos k - k\alpha_2 \sin k - \left(\frac{\omega}{k}\right)^2 f_0 = 0.$$

This system of homogeneous equations can be solved to determine the constants α_1, α_2, f_0, φ_0 if and only if the determinant

$$\begin{vmatrix} 0 & 1 & 1+\left(\frac{\omega}{k}\right)^2 & -1 \\ \sin k & \cos k & 0 & 0 \\ k & 0 & -\left(\frac{\omega}{k}\right)^2 & 1 \\ \cos k & -k \sin k & -\left(\frac{\omega}{k}\right)^2 & 0 \end{vmatrix} = 0.$$

Working this out leads to the equation

$$\left(\frac{\omega}{k}\right)^2 = \frac{1}{\sin k - k \cos k}. \qquad (4.109)$$

In order to understand the behavior of ω^2 we can graph $\sin k$ and $k \cos k$ simultaneously (Figure 4.10). We see that if $0 < k < \hat{k}$, ω^2 is positive.

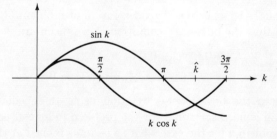

FIGURE 4.10

Therefore, for these values of k the solution of (4.107) is given by

$$\sigma(t) = \beta_1 \sin \sqrt{\frac{EI}{m}}\, \omega t + \beta_2 \cos \sqrt{\frac{EI}{m}}\, \omega t.$$

If k is in the interval $(\hat{k}, 3\pi/2)$, ω^2 is negative and the solution of (4.107) has the form

$$\sigma(t) = \beta_1 e^{-\sqrt{(EI/m)}|\omega|t} + \beta_2 e^{\sqrt{(EI/m)}|\omega|t}.$$

These formulas show that if k is in the interval $(0,\hat{k})$ the functions $f(t)$ and $\varphi(t)$ are periodic functions of t, but that when k is in the interval $(\hat{k}, 3\pi/2)$ the function f can become unbounded because of the presence of the term $\exp(\sqrt{(EI/m)}\,|\omega|t)$. In other words when k passes through the value $\hat{k}$ we pass from bounded vibrations to an unbounded regéme.

Now k is a function of P by (4.104). Hence we are saying that if the magnitude of the follower force P becomes larger then

$$P_{\text{crit}} = EI\hat{k}^2,$$

the column can have an unbounded deflection as a function of t. This critical magnitude for P is called the *buckling load* for the column and we may expect physically that for loads larger than P_{crit} a fixed-free column subjected to vibrations will buckle.

Note that when $k = \hat{k}$, that is, if $P = P_{\text{crit}}$, the determinant cannot be equal to zero. This value is thus one for which $\varphi_0 = f_0 = \alpha_1 = \alpha_2 = 0$.

Returning now to the characteristic equation (4.100) we merely report that it yields essentially similar results. For $P < P_{\text{crit}}$ all eigenvalues are real. As P increases to P_{crit} the smallest two eigenvalues become equal and as P increases beyond P_{crit} these two become complex with one of them having a negative imaginary part. This again leads to solutions which may become unbounded as t increases.

PROBLEMS

1. State exactly what is meant by a solution of problem (4.86).
2. Suppose that the boundary conditions for a beam are

    $$u(0,t) = u_x(0,t) = u_{xx}(L,t) = 0,$$
    $$u_{xxx}(L,t) + \alpha u_x(L,t) = 0.$$

    Investigate the basic energy relation using these conditions. What restriction is necessary on α for the problem to be conservative?

3. Work Problem 2 for the case where a beam is pinned at the end $x = 0$,

    $$u(0,t) = u_{xx}(0,t) \equiv 0,$$

    and satisfies

    $$u_{xxx}(L,t) = 0, \quad u_{xx}(L,t) + \alpha u_x(L,t) = 0.$$

4. Work Problem 2 if the boundary conditions are

    $$u_{xx}(0,t) - \alpha u_x(0,t) = 0, \quad u(0,t) = 0,$$
    $$u_{xx}(L,t) + \beta u_x(L,t) = 0, \quad u(L,t) = 0.$$

    Determine α and β so that the problem is conservative.

5. Work Problem 2 if the boundary conditions are

$$u_x(0,t) = u_{xxx}(0,t) = u_x(L,t) = u_{xxx}(L,t) = 0.$$

What physical meaning is attached to these boundary conditions?

6. Investigate the basic energy relation under the boundary conditions

$$u(0,t) = u_x(0,t) = u_{xx}(L,t) = 0,$$
$$u_{xxx}(L,t) - \alpha u_x(L,t) = 0.$$

Is this problem conservative?

7. Establish an energy identity similar to (4.81) for the beam equation

$$a^2 u_{xxxx} - b^2 u_{xx} + u_{tt} = 0.$$

What is the basic energy function for this equation? What must be true of any solution for which the basic energy function vanishes?

8. What modifications will result in your answers to Problem 7 if the differential equation is changed to

$$a^2 u_{xxxx} + b^2 u_{xx} + u_{tt} = 0?$$

9. Find the solution of the boundary-value problem

$$u_{tt} + a^2 u_{xxxx} = 0, \quad 0 \leq x \leq L, \quad 0 < t$$
$$u(x,0) = F(x), \quad u_t(x,0) = g(x), \quad 0 < x < L$$
$$u(0,t) = u_{xx}(0,t) = u(L,t) = u_{xx}(L,t) = 0, \quad 0 < t.$$

10. Solve Problem 9 if the boundary conditions are changed to

$$u(0,t) = u_x(0,t) = u(L,t) = u_x(L,t) = 0.$$

11. Solve Problem 9 for the boundary conditions

$$u_{xx}(0,t) = u_{xxx}(0,t) = u_{xx}(L,t) = u_{xxx}(L,t) = 0.$$

12. Formulate two boundary-value problems for the beam equation which are not homogeneous.

13. Show that the boundary-value problem

$$a^2 u_{xxxx} + P u_{xx} + u_{tt} = 0, \quad 0 < x < L, \quad 0 < t,$$
$$u(0,t) = u_x(0,t) = u_{xxx}(L,t) = 0, \quad u_{xx}(L,t) = \alpha, \quad 0 < t,$$
$$u(x,0) = F(x), \quad u_t(x,0) = g(x), \quad 0 < x < L$$

is nonconservative.

14. Carry out the solution of Problem 13 as far as you are able.
15. Derive Equation (4.87).
16. The simplified model of a column used in Section 4.6 does not take into account the inertia of rotation of the mass on the end. When this is

included, Equation (4.102) is replaced by the equation

$$EIu_{xx} = P(f - u) - (P\varphi - R)(1 - x) - m\rho^2\ddot{\varphi}, \qquad (4.110)$$

where ρ = constant is the radius of inertia of the mass. Show that one is led by this problem to the characteristic equation

$$\omega^4 - 2a\omega^2 + b = 0 \qquad (4.111)$$

in place of (4.109), where

$$a = \frac{kEI}{2m\rho^2} \frac{(1 + k^2\rho^2) \sin k - k \cos k}{2 - 2 \cos k - k \sin k},$$

and

$$b = \left(\frac{k^2 EI}{m\rho}\right)^2 \frac{1}{2 - 2 \cos k - k \sin k}.$$

17. Show that the motion of the column described in (4.109) is oscillatory in character only if both roots ω^2 of the biquadratic equation (4.111) are real and positive. Show also that this leads to the critical values given by the pair of equations

$$\left[\frac{1}{k\rho}(\sin k - k \cos k) + k\rho \sin k\right]^2 + 4(k \sin k + 2 \cos k - 2) = 0,$$

$$\frac{1}{k\rho}(\sin k - k \cos k) + k\rho \sin k = 0.$$

18. Plot graphs of the transcendental equations of Problem 17 in order to obtain an estimate of the smallest critical value for the column of Problem 16.

4.7 THE FOLLOWER PROBLEM

The nonconservative dynamic problem represented by the partial differential equation (4.87) and the boundary conditions (4.88) has been the source of a considerable amount of confusion to researchers. This is because of the many anomalies which have arisen when investigators tried to find a nontrivial solution ($u \neq 0$) to the problem by the static method. Recall from your elementary physics that a system is said to behave statically when it is at rest or moving at a constant velocity. Therefore by suppressing the inertia force mu_{tt} in (4.87) we are left with a static equation. The static method then asks for the solution of the differential equation

$$a^2 u'''' + Pu'' = 0 \qquad (4.112)$$

on the interval $(0,L)$ subject to the boundary conditions

$$u(0) = u'(0) = u''(L) = u'''(L) = 0. \qquad (4.113)$$

Note the difference between this static boundary-value problem and the dynamic boundary-value problem obtained by separation of variables in Section 4.6. There the differential equation has the form

$$a^2 u'''' + Pu'' - \lambda^2 u = 0$$

while the boundary conditions are, of course, the same.

Let us consider a slightly more general static follower problem. Suppose that we have a column with a variable cross section, then $a^2 = q(x)$ and the differential equation (4.112) is replaced by

$$(q(x)u'')'' + Pu'' = 0. \tag{4.114}$$

From the physics of the situation we know that the function q is positive and bounded on the closed interval $[0,L]$ and we may assume it is twice differentiable on the open interval $(0,L)$ with a continuous second derivative. Indeed we may assume that q satisfies

$$q_1 > q(x) > q_0 > 0 \quad \text{on} \quad [0,L]. \tag{4.115}$$

We shall now show that the only solution of the boundary-value problem (4.114), (4.113), which includes (4.112), (4.113) as a special case, is $u \equiv 0$ on $[0,L]$ for all positive values of P. Thus no information can be obtained from the static follower problem in contrast to the dynamic problem of the previous section.

Theorem 4.5

Suppose that q is positive and bounded on $[0,L]$ and twice continuously differentiable on $(0,L)$. Then the only solution of Equation (4.114) subject to the boundary conditions (4.113) is $u \equiv 0$ on $[0,L]$ if $p > 0$.

PROOF. We begin by defining the function y by setting

$$q(x)u'' = y. \tag{4.116}$$

In terms of the function y, Equation (4.114) becomes

$$y'' + \frac{P}{q(x)} y = 0$$

or

$$y'' + Pf(x)y = 0. \tag{4.117}$$

In view of the hypotheses on q embodied in (4.115), we have

$$f_1 \geq f(x) \geq f_0 > 0, \quad \text{on} \quad [0,L],$$

where $f_1 = 1/q_0$, $f_0 = 1/q_1$. The function y obviously must also satisfy the initial conditions

$$y(L) = y'(L) = 0. \tag{4.118}$$

Thus y is to be a solution of the differential equation (4.117) on $(0,L)$ satisfying the initial conditions (4.118) at $x = L$. By the uniqueness theorem (Theorem 1.3 proved in Chapter 1) it follows at once that $y \equiv 0$ on $[0,L]$. Since $q(x) \geq q_0 > 0$ on $[0,L]$, we deduce from (4.116) that

$$u'' \equiv 0 \quad \text{on} \quad [0,L].$$

Integrating twice yields

$$u(x) = bx + c.$$

Since $u(0) = 0$, we obtain $c = 0$ and, since $u'(0) = 0$, we obtain $b = 0$. Therefore $u \equiv 0$ on $[0,L]$, proving Theorem 4.5.

In view of Theorem 4.5 the static analysis of the follower problem does not yield any nontrivial solutions. At this point a logical question to be asked is why have researchers formulated the problem in this way? The reason is that the static method has been used successfully to solve a variety of stability problems. Of these, historically, the most famous stability problem and at the same time the problem most closely related to the follower problem is the Euler buckling problem.

The Euler problem asks for a solution of the differential equation (4.112) [or, more generally, of (4.114)] with the boundary conditions (4.113) replaced by

$$u(0) = u'(0) = u''(L) = 0, \quad a^2 u'''(L) + Pu'(L) = 0. \quad (4.119)$$

Physically the Euler problem differs from the follower problem in that the force with magnitude P at the free end of the beam $x = L$ is always directed parallel to the x-axis and is compressive (see Figure 4.8). This difference is accounted for in the last of conditions (4.119).

Now the general solution of the differential equation (4.112) was obtained in Section 4.6. It is

$$u(x) = c_1 \sin \frac{\sqrt{P}}{a} x + c_2 \cos \frac{\sqrt{P}}{a} x + c_3 x + c_4. \quad (4.120)$$

Substituting this solution into the boundary conditions (4.119), we obtain the equations

$$c_2 + c_4 = 0,$$
$$\frac{\sqrt{P}}{a} c_1 + c_3 = 0,$$
$$c_1 \sin \frac{\sqrt{P}}{a} l + c_2 \cos \frac{\sqrt{P}}{a} l = 0,$$
$$c_3 = 0.$$

Since $c_3 = 0$, the second equation yields $c_1 = 0$ and the third reduces to

$$c_2 \cos \frac{\sqrt{P}}{a} l = 0.$$

If we take $c_2 = 0$, then $c_4 = 0$ and we have only the trivial solution. It follows then that we must set $\cos (\sqrt{P/a})l = 0$ which leads to the values

$$\frac{\sqrt{P}}{a} l = (2k + 1) \frac{\pi}{2}, \quad k = 0, \pm 1, \ldots .$$

Thus for the values

$$P_k = \frac{(2k + 1)^2 \pi^2 a^2}{4l^2}$$

the Euler problem has nonzero solutions. The smallest of these is

$$P_0 = \frac{\pi^2 a^2}{4l^2}, \qquad (4.121)$$

and it is called the *Euler critical load* for the cantilever column.

From the physical point of view the structural engineer reasons as follows. If $0 < P < P_0$ the Euler problem has only the solution $u \equiv 0$ on $[0,L]$. When $P = P_0$ the column assumed the "buckled" shape obtained by setting $c_4 = -c_2$, allowing c_2 to remain arbitrary, and setting $P = \pi^2 c^2/4l^2$ in (4.120). Thus

$$u(x) = c_2 \left(\cos \frac{\pi}{2l} x - 1 \right) \qquad (4.122)$$

where $P = P_0$. Note that the magnitude of the deflection given by (4.122) is not determined by the solution to this static problem, but there can be a nonzero deflection of the column if $P = P_0$. The number P_0 is therefore interpreted as the smallest magnitude of a compressive force which will cause the column to buckle. Experimental results confirm that this value P_0 is indeed a very good approximation to the critical load for column buckling.

Mathematically the meaning of the analysis of the Euler problem is that the deflection u does not depend continuously upon the parameter P.

Why does the static method work for the Euler problem and fail for the follower problem? The answer is, at least in part, that the Euler problem is conservative and the follower problem is not. To see this difference as it applies to these two problems, multiply Equation (4.114) by a four times differentiable function v and integrate from 0 to L. We obtain after integrating by parts

$$\int_0^L [v(q(x)u'')'' + vPu''] \, dx = - \int_0^L [v'(q(x)u'')' + Pv'u'] \, dx$$
$$- v(0)(qu'')'(0) + v(L)(qu'')'(L) - Pv(0)u'(0) + Pv(L)u'(L).$$

We integrate the first term in the right-hand integral three more times by parts and the second term once more. This yields

$$\int_0^L v[(q(x)u'')'' + Pu''] \, dx - \int_0^L u[(q(x)v'')'' + Pv''] \, dx$$
$$= v(L)[(qu'')'(L) + Pu'(L)] - v(0)[(qu'')'(0) + Pu'(0)]$$
$$+ v'(0)q(0)u''(0) - v'(L)q(L)u''(L)$$
$$+ q(L)v''(L)u'(L) - q(0)v''(0)u'(0)$$
$$+ u(0)[(qv'')'(0) + Pv'(0)] - u(L)[(qv'')'(L) + Pv'(L)]. \quad (4.123)$$

Now suppose that u is a solution of (4.112) also satisfying the conditions (4.119). For this case (4.123) reduces to

$$-\int_0^L u[a^2 v'''' + Pv''] \, dx = -a^2 v(0)u'''(0) + a^2 v'(0)u''(0)$$
$$+ a^2 v''(L)u'(L) - u(L)[a^2 v'''(L) + Pv'(L)]. \quad (4.124)$$

We see at once from (4.124) that, if v is also a solution of (4.112) satisfying the boundary conditions (4.119), both sides of (4.124) are zero.

On the other hand, suppose u satisfies (4.112) and the boundary conditions (4.113). Then we derive from (4.123) the relation

$$-\int_0^L u[a^2 v'''' + Pv''] \, dx = -a^2 v(0)u'''(0) + u'(L)[Pv(L) + a^2 v''(L)]$$
$$+ a^2 v'(0)u''(0) - u(L)[a^2 v'''(L) + Pv'(L)]. \quad (4.125)$$

We see from this formula that if v is a solution of (4.112) satisfying the boundary conditions

$$v(0) = v'(0) = 0, \quad a^2 v''(L) + Pv(L) = 0,$$
$$a^2 v'''(L) + Pv'(L) = 0, \quad (4.126)$$

both sides are zero. This problem, namely (4.112), (4.126), is called the *adjoint problem* to the problem (4.112), (4.113).

Thus in the terminology of adjoint problems the Euler problem is *self-adjoint* while the follower problem is *nonself-adjoint*. For boundary value problems for ordinary differential equations we shall regard the distinction between self-adjoint and nonself-adjoint problems as equivalent to the distinction between conservative and nonconservative problems. As a general rule if a static problem is nonself-adjoint critical values cannot be found while if a problem is self-adjoint critical values can be found so that the Euler method is justified when applied to such problems. These remarks should not be interpreted as applying to all nonself-adjoint problems; they apply only to those which arise from the so-called *static-stability problems*.

PROBLEMS

1. Derive Equation (4.114) starting with a fixed-free column of variable stiffness $q(x)$.
2. Derive the last of the boundary conditions (4.119).
3. By using solution (4.120) of the differential equation (4.112) solve the static follower problem represented by (4.112) together with the boundary conditions (4.113) directly and hence show by this method that $u(x) \equiv 0$ on $[0,L]$ is the only solution.
4. Show directly that the only solution of the adjoint problem to the follower problem, that is, to Equation (4.112) with the boundary conditions (4.126), is $u(x) \equiv 0$ on $[0,L]$.
5. Let $h(x)$ be a continuous function on $[0,L]$. Establish the identity

$$\int_0^L \int_0^x h(\xi)\, d\xi\, dx = \int_0^L (L - x)h(x)\, dx. \qquad (4.127)$$

6. The adjoint problem to the general follower problem given by Equation (4.114) and the boundary conditions (4.113) can also be deduced from the identity (4.123). It is described by Equation (4.114) and the boundary conditions

$$u(0) = u'(0) = 0,$$
$$q(L)u''(L) + Pu(L) = 0,$$
$$(qu'')'(L) + Pu'(L) = 0.$$

By introducing a function w through the equation $q(x)u'' = w$ as was done in the proof of Theorem 4.5, show that the only solution of this boundary-value problem is $u \equiv 0$ on $[0,L]$.

[*Hint:* Show that the last two boundary conditions can be written in the form

$$q(L)u''(L) + P \int_0^L \int_0^x u''(\xi)\, d\xi\, dx = 0,$$
$$(qu'')'(L) + P \int_0^L u''(x)\, dx = 0,$$

and make use of the identity (4.127) at an appropriate point.]

7. A certain static follower problem leads to the differential equation

$$a^2 u'''' + Pu'' = 0,$$

subject to the boundary conditions

$$u(0) = u'(0) = u'''(L) = 0, \quad u''(L) = -Pe$$

where $e > 0$ is a constant. Note that with the boundary condition $u''(L) = -Pe$ the problem is not homogeneous so that we cannot expect the solution to be $u \equiv 0$ on $[0,L]$. By solving this problem directly show that one does not obtain any critical value of P at which the deflection u undergoes an abrupt change as in the Euler problem. In other words, the deflection u depends continuously upon P for $P > 0$.

8. The Euler problem and the static follower problem are both special cases of the problem of solving Equation (4.112) subject to the boundary conditions

$$u(0) = u'(0) = u''(L) = 0, \quad a^2 u'''(L) + (1 - k)Pu'(L) = 0.$$

The Euler problem occurs when $k = 0$ and the follower problem when $k = 1$. For the case where k is different both from 0 and from 1 make the substitution $u'' = W$ and formulate the boundary-value problem which W must satisfy. In doing this you will have need to refer to the hint at the end of Problem 6. Find the values of k for which this boundary-value problem can have nonzero solutions and the values of P which yield such nonzero solutions.

9. Consider the boundary-value problem defined by Equation (4.112) subject to the boundary conditions

$$u(0) = u'(0) = u''(L) = 0, \quad a^2 u'''(L) + Qu'(L) = 0.$$

Assume Q is different from zero and P, set $u'' = W$ and formulate the boundary value problem W must satisfy. Prove that this boundary-value problem can have nonzero solutions if P and Q satisfy

$$\frac{P}{Q} = 1 - \cos\frac{\sqrt{P}}{a}L. \tag{4.128}$$

10. Show that for a fixed Q there are at most a finite number of positive values of P which can satisfy (4.128). Investigate the situation both for positive and for negative values of Q. For what values of Q are there no positive values of P satisfying (4.128)? What is the solution u of the original problem formulated in Problem 9 for such values of Q?

11. Suppose that Problem 9 is changed and we concern ourselves with $P < 0$ (the compressive force on the column is changed to a tensile force). Rework Problems 9 and 11 for $P < 0$.

12. The problems we analyzed in the last two sections were all fixed-free columns. Set up the proper static boundary-value problem for both the Euler and follower cases for
    (a) a pinned-pinned column
    (b) a fixed-fixed column, and
    (c) a fixed-pinned column.

13. Solve the Euler boundary-value problems of Problem 12. Note that if one thinks in terms of an effective column length (point of inflection to point of inflection), having the solution to any one of the Euler problems immediately yields the solution for all of the others. Verify this by deducing the smallest critical load for the pinned-pinned column from your knowledge of the smallest critical load for the fixed-free column.
14. Discuss the solutions to the follower problems of Problem 12.

4.8 THE TIMOSHENKO BEAM EQUATION

In Section 4.5 we derived the equation of motion for the transverse vibrations of a beam using the assumption of pure bending. The resulting differential equation (4.78), the Bernoulli-Euler equation, was then studied under several conditions establishing some well known facts about the physical system it represents. In deriving (4.78) we neglected the effects of the inertia of rotation of the differential element, Figure 4.7, the deformation caused by the shearing forces, and the weight of the beam.

Recall that in Chapter 1 we made reference to the fact that the results obtained from studying the properties of an equation (or system of equations) were meaningful if, and only if, the equation truly represented the physical system from which it was derived. Thus the verification of a theoretical study is obtained when the theoretical results compare favorably with experimental results.

Unfortunately, the foregoing observations are somewhat oversimplified. There are cases where a simple mathematical representation of a system suffices to describe the behavior of the system for many situations, but breaks down for other situations. Such is the case with the Bernoulli-Euler beam. There are problems for which it is not sufficiently delicate to correctly describe the dynamic response of a beam. In such cases when we account for the effects of rotatory inertia, shear deformation, and internal damping of the beam we obtain results which agree with experimental data. The equation obtained when the effects of shear deformation and rotatory inertia are considered is called the *Timoshenko beam equation*.

In deriving the Timoshenko beam equation we will once again consider a differential element of the beam, Figure 4.11. Note that the D'Alembert representation for rotatory inertia, $I_p dx (\partial^2 u/\partial t^2)$, and translational inertia $m dx\ (\partial^2 u/\partial t^2)$ are included in the free-body diagram. The constant $I_p = mr^2$, where m is the mass per unit length, and r is the radius of gyration of the cross section about an axis perpendicular to the plane of motion and through the neutral axis. The angle φ in Figure 4.11 is the component of slope caused by the bending stress. Thus for small curvature we may write

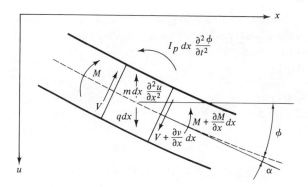

FIGURE 4.11

$$EI \frac{\partial \varphi}{\partial x} = -M(x,t). \tag{4.129}$$

The angle α is the component of slope caused by the shear deformation. From the foregoing definitions it follows that

$$\frac{\partial u}{\partial x} = \varphi + \alpha \tag{4.130}$$

and

$$V(x,t) = K\alpha = K\left(\frac{\partial u}{\partial x} - \varphi\right), \tag{4.131}$$

where K is a constant which depends on the shear modulus, the cross-sectional area and the shape of the cross section. Finally we note that q is the transverse load per unit length of beam and that the weight of the beam is neglected in comparison to the inertia force.

Assuming small deformations we may sum forces in the u direction to obtain

$$V_x + q = mu_{tt} \tag{4.132}$$

and take moments to obtain

$$V = M_x + I_p \varphi_{tt}. \tag{4.133}$$

Now replacing V in (4.132) using (4.131) we obtain

$$K(u_{xx} - \varphi_x) + q = mu_{tt}, \tag{4.134}$$

and replacing M and V in (4.133) using (4.129) and (4.131) yields

$$K(u_x - \varphi) = -EI\varphi_{xx} + I_p \varphi_{tt}. \tag{4.135}$$

Equations (4.134) and (4.135) are the basic equations for the motion of the beam. They are homogeneous if the external load q is zero. They must be solved subject to appropriate initial and boundary conditions. To be specific suppose (4.134) and (4.135) are to hold for $0 < x < 1$, $t > 0$, then we specify the initial conditions

$$u(x,0) = h_1(x), \quad \varphi(x,0) = f_1(x), \quad 0 < x < 1, \quad (4.136)$$
$$u_t(x,0) = h_2(x), \quad \varphi_t(x,0) = f_2(x).$$

A variety of boundary conditions are possible. For example,

$$u(0,t) = \varphi(0,t) = 0,$$
$$u(1,t) = \varphi(1,t) = 0, \quad 0 < t,$$

or

$$u_x(0,t) = \varphi(0,t) = 0,$$
$$u(1,t) = 0, \quad \varphi_x(1,t) + \alpha\varphi(1,t) = 0, \quad 0 < t.$$

It is clear that a large variety of such conditions can be formulated.

We now wish to eliminate φ between (4.134) and (4.135). Doing so we arrive at

$$EI \frac{\partial^4 u}{\partial x^4} + m \frac{\partial^2 u}{\partial t^2} - \left[\frac{EIm}{k} + I_p\right] \frac{\partial^4 u}{\partial x^2 \partial t^2} + \frac{I_p m}{k} \frac{\partial^4 u}{\partial t^4}$$
$$= q - \frac{EI}{K} \frac{\partial^2 q}{\partial x^2} + \frac{I_p}{K} \frac{\partial^2 q}{\partial t^2}, \quad (4.137)$$

which is the usual form of the Timoshenko beam equation.

We observe that the appropriate form for initial conditions for Equation (4.137) is that

$$u(x,0) = h_1(x), \quad u_t(x,0) = h_2(x), \quad u_{tt}(x,0) = h_3(x),$$
$$u_{ttt}(x,0) = h_4(x) \quad (4.138)$$

be given on $(0,1)$. Boundary conditions suitable to this equation are similar to those suitable to the Bernoulli-Euler equation.

Let us now consider the problem of the propagation of stress waves in an infinitely long beam. For the case of the Bernoulli-Euler beam we seek a solution to

$$u_{tt} + a^2 u_{xxxx} = 0.$$

From (4.77) and (4.78) we see that the constant $a^2 = EI/m$. Let $m = \rho A$, where ρ is the mass density of the beam and A is its cross-sectional area. Then we may write

$$a^2 = c_1^2 r^2, \quad (4.139)$$

where

$$c_1^2 = E/\rho \quad \text{and} \quad r^2 = I/A.$$

234 OTHER TIME-DEPENDENT PARTIAL DIFFERENTIAL EQUATIONS

Therefore the Bernoulli-Euler equation becomes

$$u_{tt} - c_1^2 r^2 u_{xxxx} = 0. \tag{4.140}$$

Drawing from our previous experience we may assume a wave solution to (4.140) of the form

$$u = u_0 \sin \frac{2\pi}{L}(x - ct), \tag{4.141}$$

where L is the wavelength and c is the wave velocity. Placing (4.141) into (4.140) we obtain

$$c = \pm c_1 r \frac{2\pi}{L}, \tag{4.142}$$

as the characteristic equation which c must satisfy. Studying (4.142) we see that in the limit as the wavelength approaches zero the wave velocity tends to infinity. A common type of beam loading which is characterized by extremely short wavelengths is a suddenly applied concentrated force. Thus (4.142) implies that the effect of a suddenly applied concentrated force on a beam will be felt almost instantaneously everywhere along the beam. This result does not agree with experimental observations.

Let us now investigate the same problem using the Timoshenko beam equation (4.137). Rewriting (4.137) in a more suggestive form and noting that for our case $q = 0$ we have

$$\frac{\partial^4 u}{\partial x^4} - \left[\frac{1}{c_1^2} + \frac{1}{c_2^2}\right]\frac{\partial^4 u}{\partial x^2 \partial t^2} + \frac{1}{c_1^2 c_2^2}\frac{\partial^4 u}{\partial t^4} + \frac{1}{c^2 r^2}\frac{\partial^2 u}{\partial t^2} = 0 \tag{4.143}$$

where

$$c_1^2 = \frac{E}{\rho}, \quad c_2^2 = \frac{K}{\rho A} \quad \text{and} \quad r^2 = \frac{I}{A}.$$

Again using the general solution (4.141) we arrive at the characteristic equation

$$1 - \left(\frac{c^2}{c_1^2} + \frac{c^2}{c_2^2}\right) + \frac{c^4}{c_1^2 c_2^2} - \frac{c^2}{c_1^2 r^2}\left(\frac{L}{2\pi}\right)^2 = 0. \tag{4.144}$$

An inspection of (4.144) shows that two roots exist for c/c_1 versus L. Hence two modes of motion are described. They correspond to two different shear to bending deflection ratios for the same wavelength.

It is interesting to note that we began with a continuous system, a beam, and made a series of approximations which reduced the system to what in one case we called the Bernoulli-Euler beam and in another case the Timoshenko beam. Where the wave length was large the Bernoulli-Euler beam yielded satisfactory results and where it was not we had to resort to the Timoshenko beam equation. The next logical question is, what are the limits of usefulness for the Timoshenko beam equation? To answer this one would have to resort to the elastic equations of motion of the system and determine an "exact" solution. The result of such a study is shown in Figure 4.12. The curves are for the special case of a circular beam of radius a. As determined above we see that the Bernoulli-Euler equation is valid for the case of large wavelengths and that the fundamental eigenvalue of the Timoshenko beam equation fits the exact theory reasonably well, but beyond that there exists a wide discrepancy.

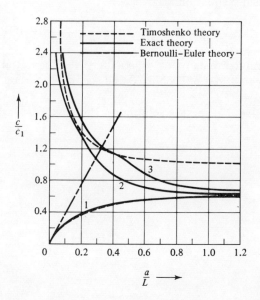

FIGURE 4.12 Adapted from H. N. Abramson, J. Acoust. Soc. Am., 29, 42, 1957.

A general question which arises from the foregoing study is one of the fruitfulness of studying a continuous model of a system (partial differential equations) versus studying a discrete analog of the same system (ordinary differential equations). This question is dealt with in the next chapter.

PROBLEMS

1. Starting with (4.134) and (4.135) derive (4.137).
2. If we suppress the effect of shear deformation in the Timoshenko beam equation the resulting equation is called the *Rayleigh equation*. Solve the infinite beam problem of this section using the Rayleigh equation. What conclusions can you state?

3. Consider a cantilever beam of length l traveling downward with a velocity v when the motion of the support is suddenly arrested. Using the Bernoulli-Euler equation formulate the initial boundary-value problem for this system.
4. Find a general solution for the system of Problem 3.
5. Using the solution found in Problem 4 determine the bending moment at the support $[u_{xx}(0,t)]$. Is the resulting series convergent? What conclusions can you state?
6. If we restate Problem 3 taking material damping into account the Bernoulli-Euler equation becomes

$$\frac{d_1}{\rho A}\frac{\partial u}{\partial t} + \frac{a^2 d_2}{E}\frac{\partial^5 u}{\partial t \partial x^4} + a^2 \frac{\partial^4 u}{\partial x^4} + \frac{\partial^2 u}{\partial t^2} = 0$$

where d_1 and d_2 are damping constants. Using the initial conditions and boundary conditions of Problem 3 find a general solution for the damped system.
7. Work Problem 5 using the solution to Problem 6.

Chapter 5

THE VARIOUS ASPECTS OF PHYSICAL PROBLEMS

5.1 INTRODUCTION

In Chapters 1 through 4 a variety of physical problems were introduced and examined from the point of view of mathematical analysis. Whereas a number of those problems were elementary in relation to many of the problems physicists and engineers are called upon to study in their day-to-day work, they do represent a formidable background of solution methods and solutions which enable the analyst to begin to attack more difficult problems. In this section we shall present a problem which can be quite complex; our objective being to examine this problem in the succeeding sections from several points of view thus enabling the reader to see a variety of ways in which such complex problems can be analyzed.

Let us consider the longitudinal vibrations of a flexible rod which occupies the segment $[0,l]$ and has its cross-sectional area $A(x)$, density $\rho(x)$ and modulus of elasticity $E(x)$, all variable. The rod is fixed at the end $x = 0$ and free at the end $x = l$. We will assume that deformations of any cross section are negligibly small.

The equation of motion for small longitudinal vibrations of the rod is derived precisely as was done in Section 4.1 except now we have the added complication of variable ρ, A, and E. Accordingly we obtain

$$\rho(x)A(x)u_{tt} - (E(x)A(x)u_x)_x = 0, \tag{5.1}$$

for $0 < x < l$ and $0 < t < +\infty$.

The boundary conditions are

$$u(0,t) = 0 \quad \text{and} \quad u_x(l,t) = 0, \quad \text{for} \quad 0 < t < \infty, \tag{5.2}$$

and the initial conditions are

$$u(x,0) = f(x) \quad \text{and} \quad u_t(x,0) = g(x), \quad \text{for} \quad 0 < x < l, \tag{5.3}$$

where f is the initial displacement of a point (hence cross section) in the rod and g the initial velocity of a point in the rod.

Equation (5.1) can also be written in the form

$$u_{tt} - a^2(x)u_{xx} - a(x)\frac{s'(x)}{s(x)}u_x = 0 \tag{5.4}$$

where s' is the derivative of s with respect to x and $a(x) = \sqrt{E(x)/\rho(x)}$. For (5.4) to be valid we must assume that s is differentiable on $(0,l)$ and does not vanish in this interval. From physical considerations all assumptions made thus far are reasonable.

Let us now formulate a couple of specific examples which are representative of problems whose mathematical models lead to the equation of motion (5.1). Consider a rod in the shape of a truncated right cone with the base at $x = 0$ of radius R and the base at $x = l$ of radius r. Letting ρ and E be constants, (5.1) becomes

$$\left(1 - \frac{H}{x}\right)^2 u_{tt} - a^2\left[\left(1 - \frac{x}{H}\right)^2 u_x\right]_x = 0 \tag{5.5}$$

where $a = \sqrt{E/\rho}$ as above and $H = lr/(R - r)$ is the altitude of the right cone. The boundary conditions (5.2) and the initial conditions (5.3) together with (5.5) fully define an initial boundary-value problem.

The problem formulated above is also appropriate for the transverse vibrations of a flexible string which is fixed at the end $x = 0$ and attached to a frictionless slide at the end $x = l$ such that the string is free to move transverse to the equilibrium position of the string. As an example of such a motion consider a string of variable density

$$\rho(x) = \frac{1}{(1 + x)^2}. \tag{5.6}$$

If A and E (which is now the tension in the string) are constant, Equation (5.1) becomes

$$\frac{1}{(1 + x)^2}u_{tt} - b^2 u_{xx} = 0, \tag{5.7}$$

where

$$b = \sqrt{E}.$$

Other examples suggested by the problems in Chapter 4 and in the present chapter indicate that a wide variety of physical problems can be reduced to Equations (5.1), (5.2), and (5.3).

In the next two sections we will indicate the extent to which the techniques introduced in Chapters 1 through 4 can be applied to the present general problem. The remainder of the chapter will be devoted to some fresh insights into the nature of our problem which can be derived from examining it in other forms.

5.2 THE ENERGY INTEGRAL

The energy method exploited in the previous chapters can be effectively used to analyze Equation (5.1). Accordingly, multiply (5.1) by u_t to obtain

$$\frac{1}{2}\rho(x)A(x)(u_t^2)_t - \frac{\partial}{\partial x}(E(x)A(x)u_x u_t) - E(x)A(x)(u_x^2)_t = 0,$$

or

$$\frac{\partial}{\partial t}[\rho(x)A(x)u_t^2 + E(x)A(x)u_x^2] = 2\frac{\partial}{\partial x}[E(x)A(x)u_x u_t]. \tag{5.8}$$

Integrating (5.8) over the rectangle $0 \leq x \leq l$, $0 \leq t \leq t_0$ for fixed $t_0 > 0$ we obtain the energy identity

$$\int_0^l [\rho(x)A(x)u_t^2(x,t_0) + E(x)A(x)u_x^2(x,t_0)]\,dx$$
$$= 2\int_0^{t_0} E(l)A(l)u_x u_t(l,t)\,dt - 2\int_0^{t_0} E(0)A(0)u_x u_t(0,t)\,dt$$
$$+ \int_0^l [\rho(x)A(x)u_t^2(x,0) + E(x)A(x)u_x^2(x,0)]\,dx. \tag{5.9}$$

Applying the boundary conditions (5.2) to the first two integrals on the right-hand side of (5.9) we see that these terms vanish. Therefore, concentrating on the remaining term we define the energy function

$$\mathcal{E}(t) = \int_0^l [\rho(x)A(x)u_t^2(x,t) + E(x)A(x)u_x^2(x,t)]\,dx. \tag{5.10}$$

Using the definition (5.10) the identity (5.9) reduces to the energy formulation

$$\mathcal{E}(t_0) = \mathcal{E}(0) \quad \text{for all } t_0 > 0. \tag{5.11}$$

By virtue of the fact that ρ, E and A are given positive functions the integral defining $\mathcal{E}$ will vanish only if $u_x = u_t = 0$. It follows that the energy function is positive unless $u = 0$, and that the system is a conservative system. Thus, as can be seen in (5.11), the system's energy level is controlled by the initial conditions. Recalling our previous work on the Cauchy problem for the wave equation, Section 4.2, we can conclude that the solution to the boundary-value problem (5.1), (5.2), and (5.3) is unique.

5.3 SEPARATION OF VARIABLES

The technique of separation of variables, discussed at length in Section 4.4, is also applicable to the present problem. Following the technique established there we seek solutions of Equation (5.1) having the form

$$u(x,t) = X(x)T(t). \tag{5.12}$$

Substituting (5.12) into (5.8) yields

$$\rho A X \ddot{T} - (EAX')' = 0. \tag{5.13}$$

Assuming, for the moment, that $X(x)T(t) \neq 0$, we multiply (5.13) by $(\rho AXT)^{-1}$ to obtain

$$\frac{\ddot{T}}{T} = \frac{1}{\rho(x)A(x)} \frac{(E(x)A(x)X')'}{X}. \tag{5.14}$$

We may now argue that the left-hand side of (5.14) is a function of t and the right-hand side a function of x. Hence, since $x \neq x(t)$ their common value must be a constant, say $-\lambda$, independent of x and t. Setting each side of (5.14) equal to $-\lambda$ we see that the functions X and T must satisfy the differential equations

$$(E(x)A(x)X')' + \lambda \rho(x)A(x)X = 0 \tag{5.15}$$

and

$$\ddot{T} + \lambda T = 0, \tag{5.16}$$

respectively. Conversely, once again we observe that if X and T satisfy these differential equations the product XT satisfies (5.1) whether or not $X(x)T(t) = 0$ for some pair (x,t). In order that (5.12) satisfy the homogeneous boundary conditions (5.2) we must have

$$X(0)T(t) = 0, \quad X'(l)T(t) = 0.$$

Consequently we may simply require

$$X(0) = X'(l) = 0 \tag{5.17}$$

in order that (5.12) be a nontrivial solution.

The function X may now be determined. It is the solution to the boundary-value problem defined by Equations (5.15) and (5.17). The solution depends upon the parameter λ. The values of λ, if any exist (for which a nontrivial solution exists) are the eigenvalues. The corresponding nontrivial functions X are the eigenfunctions.

The problem defined by Equations (5.15) and (5.17) is called a *Sturm-Liouville problem*. The study of the Sturm-Liouville theory leads to many interesting results, especially with reference to orthogonal functions. For the present let us observe that the presence of variable coefficients in (5.15) may make this equation difficult or impossible to solve explicitly. This is why we have gone to such great length to formulate the general theory of such problems. It enables us to determine important information concerning the eigenvalues and eigenfunctions even when they cannot be obtained explicitly, save by numerical methods.

As a particular example let us consider the string with variable density vibrating according to its equation of motion as expressed by (5.7). For this problem, Equations (5.15) and (5.16) become

$$X'' + \frac{\lambda}{(1+x)^2} X = 0, \tag{5.18}$$

and

$$\ddot{T} + \lambda T = 0, \tag{5.19}$$

respectively. We shall attempt to solve (5.18) by selecting the form

$$X(x) = (1+x)^\alpha. \tag{5.20}$$

This yields

$$(\alpha(\alpha - 1) + \lambda)(1 + x)^{\alpha-2} = 0,$$

which implies that

$$\alpha(\alpha - 1) + \lambda = 0$$

by virtue of the fact that $1 + x \neq 0$ for $x \geq 0$. Thus

$$\alpha = \tfrac{1}{2}(1 \pm \sqrt{1 - 4\lambda}).$$

The Wronski determinant of the functions

$$(1 + x)^{(1/2)(1+\sqrt{1-4\lambda})} \quad \text{and} \quad (1 + x)^{(1/2)(1-\sqrt{1-4\lambda})}$$

is $-\sqrt{1 - 4\lambda}$, which means that these functions are linearly independent for $x \geq 0$ except when $\lambda = \tfrac{1}{4}$. The reader can verify that the linear combination

$$X(x) = (1 + x)^{(1/2)(1+\sqrt{1-4\lambda})} - (1 + x)^{(1/2)(1-\sqrt{1-4\lambda})} \tag{5.21}$$

satisfies the condition $X(0) = 0$ for all λ. The condition $X'(l) = 0$ reduces to

$$\tfrac{1}{2}(1 + \sqrt{1 - 4\lambda})(1 + l)^{(1/2)(-1+\sqrt{1-4\lambda})}$$
$$- \tfrac{1}{2}(1 - \sqrt{1 - 4\lambda})(1 + l)^{-(1/2)(1+\sqrt{1-4})} = 0,$$

which further simplifies to

$$(1+l)^{\sqrt{1-4\lambda}} = \frac{1-\sqrt{1-4\lambda}}{1+\sqrt{1-4\lambda}}. \tag{5.22}$$

If $\lambda < \frac{1}{4}$, $\sqrt{1-4\lambda}$ is real and hence nonnegative. Consequently the left-hand side of (5.22) is greater than one. On the other hand, if $\lambda < \frac{1}{4}$, then the right-hand side of (5.22) is less than one. If $\lambda = \frac{1}{4}$, both sides of (5.22) equal one, but in this case we only have the one solution $(1+x)^{1/2}$ and $X(x)$ as given by (5.21) is identically equal to zero. However, it is easy to verify that $(1+x)^{1/2} \ln(1+x)$ is also a solution of (5.18) when $\lambda = \frac{1}{4}$. This solution has the value zero when $x = 0$, but its derivative is not zero at $x = l$. Consequently $\lambda = \frac{1}{4}$ is not an eigenvalue.

It remains for us to seek eigenvalues $\lambda > \frac{1}{4}$. In this case the square root becomes imaginary and we can write

$$(1+x)^{(1/2)(1+i\sqrt{4\lambda-1})} = (1+x)^{1/2} e^{(1/2)i\sqrt{4\lambda-1}\log(1+x)}$$
$$= (1+x)^{1/2} \{\cos[\sqrt{\lambda-\tfrac{1}{4}}\log(1+x)]$$
$$+ i\sin[\sqrt{\lambda-\tfrac{1}{4}}\log(1+x)]\}.$$

A direct verification shows that the functions

$$(1+x)^{1/2} \cos(\sqrt{\lambda-\tfrac{1}{4}}\log(1+x))$$

and

$$(1+x)^{1/2} \sin(\sqrt{\lambda-\tfrac{1}{4}}\log(1+x))$$

are solutions of (5.18). The Wronskian of these two functions is different from zero and hence they are linearly independent. Working with the function

$$(1+x)^{(1/2)(1-i\sqrt{4\lambda-1})}$$

leads to the same two functions. To satisfy the condition $X(0) = 0$, we set

$$X(x) = (1+x)^{1/2} \sin(\sqrt{\lambda-\tfrac{1}{4}}\log(1+x)).$$

Then

$$X'(l) = \tfrac{1}{2}(1+l)^{-1/2}\sin(\sqrt{\lambda-\tfrac{1}{4}}\log(1+l))$$
$$+ (1+l)^{-1/2}\sqrt{\lambda-\tfrac{1}{4}}\cos(\sqrt{\lambda-\tfrac{1}{4}}\log(1+l)).$$

Consequently the condition $X'(l) = 0$ yields the equation

$$\tan(\sqrt{\lambda-\tfrac{1}{4}}\log(1+l)) + \sqrt{4\lambda-1} = 0. \tag{5.23}$$

It is easy to see that Equation (5.23) has infinitely many positive solutions. Recall that $\tan y$ is periodic with period π and with vertical asymptotes at $y = (2k+1)(\pi/2)$, for $k = 0, 1, 2, \ldots$. Therefore the function

$$\tan(\sqrt{\lambda - \tfrac{1}{4}}(\log 1 + l))$$

is periodic in λ with vertical asymptotes given by

$$\sqrt{\lambda - \tfrac{1}{4}}\log(1+l) = (2k+1)\frac{\pi}{2}$$

or by

$$\tilde{\lambda}_k = \frac{1}{4} + \left[\frac{(2k+1)\dfrac{\pi}{2}}{\log(1+l)}\right]^2.$$

Let us plot the functions $\tan(\sqrt{\lambda - \tfrac{1}{4}}\log(1+l))$ and $-\sqrt{4\lambda - 1}$ on the same graph. This is done in Figure 5.1. It is clear from the figure that

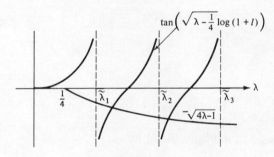

FIGURE 5.1

the eigenvalue λ_k is such that

$$\lambda_k > \tilde{\lambda}_k$$

for each $k = 1, 2, \ldots$, and that as k tends to ∞, λ_k tends to $\tilde{\lambda}_k$. That is,

$$\lim_{k \to \infty}(\lambda_k - \tilde{\lambda}_k) = 0.$$

We now have the eigenfunctions

$$X_n(x) = (1+x)^{1/2}\sin(\sqrt{\lambda_n - \tfrac{1}{4}}\log(1+x)) \qquad (5.24)$$

corresponding to the eigenvalues λ_n obtained from (5.23). Returning to (5.19), we obtain the solutions

$$T_n(t) = \alpha_n \sin \lambda_n t + \beta_n \cos \lambda_n t, \quad n = 1, 2, \ldots.$$

Thus the functions

$$u_n(x,t) = [\alpha_n \sin \lambda_n t + \beta_n \cos \lambda_n t](1+x)^{1/2}\sin(\sqrt{\lambda_n - \tfrac{1}{4}}\log(1+x))$$

satisfy (5.7) and the boundary conditions (5.2). A formal solution to the problem defined by (5.7), (5.2), and (5.3) can be obtained from the series

$$u(x,t) = \sum_{n=1}^{\infty} [\alpha_n \sin \lambda_n t + \beta_n \cos \lambda_n t](1 + x)^{1/2} \sin (\sqrt{\lambda_n - \tfrac{1}{4}} \log (1 + x)) \quad (5.25)$$

by choosing the coefficients α_n and β_n, $n = 1, 2, \ldots$, properly. For example, suppose $f(x) = 0$ on $(0,l)$, and $g \not\equiv 0$. Then we must have

$$g(x) = \sum_{n=1}^{\infty} \lambda_n \alpha_n (1 + x)^{1/2} \sin (\sqrt{\lambda_n - \tfrac{1}{4}} \log (1 + x)). \quad (5.26)$$

The reader may verify that for $n \neq m$

$$0 = \int_0^l \frac{dx}{(1 + x)^2} \{(1 + x)^{1/2} \sin (\sqrt{\lambda_n - \tfrac{1}{4}} \log (1 + x))(1 + x)^{1/2}$$
$$\sin (\sqrt{\lambda_m - \tfrac{1}{4}} \log (1 + x))\},$$

and that

$$\int_0^l \frac{dx}{(1 + x)^2} [(1 + x)^{1/2} \sin (\sqrt{\lambda_n - \tfrac{1}{4}} \log (1 + x))]^2 \neq 0.$$

Accordingly we can multiply (5.26) by

$$(1 + x)^{-3/2} \sin (\sqrt{\lambda_m - \tfrac{1}{4}} \log (1 + x)),$$

integrate both sides from 0 to l, interchange integration and summation on the right and obtain

$$\alpha_n = \frac{1}{\lambda_n} \frac{\int_0^l g(x)(1 + x)^{-3/2} \sin (\sqrt{\lambda_n - \tfrac{1}{4}} \log (1 + x)) \, dx}{\int_0^l (1 + x)^{-1} \sin^2 (\sqrt{\lambda_n - \tfrac{1}{4}} \log (1 + x)) \, dx}. \quad (5.27)$$

The same method can be used to determine the coefficients β_n.

It can be shown that the facts just established in the special case of Equations (5.18) and (5.7) in the presence of the boundary conditions (5.2) can be established for a great variety of problems. In the general case, however, these calculations are not so explicit as they are in the present case.

PROBLEMS

1. How many different fields of classical physics can you name in which some segment of that field gives rise to boundary-value problems whose differential equation is characterized by the wave equation? Elaborate on your answer by developing the appropriate equations.

2. Construct examples of boundary-value problems for the wave equation for which the total energy functions become unbounded as $t \to \infty$.
3. Use your answers to Problem 1 to aid you in giving physical interpretations of the boundary-value problems you created in Problem 2.
4. Show that the total energy associated with the boundary-value problem

$$u_{tt} - a^2 u_{xx} = 0, \quad 0 < x < l,$$
$$u(0,t) = u_x(l,t) = 0, \quad 0 < t,$$
$$u(x,0) = f_1(x), \quad u_t(x,0) = f_2(x), \quad 0 < x < l,$$

is proportional to

$$\mathcal{E}_{12}(t) = \int_0^l \{u_t^2(x,t) + a^2 u_x^2(x,t)\}\, dx.$$

5. Solve Problem 4 by separation of variables.
6. Prove that if Problem 4 is solved by separation of variables in the form

$$u(x,t) = \sum_{n=1}^\infty U_n(x,t) = \sum_{n=1}^\infty T_n(t) X_n(x),$$

then the energy $\mathcal{E}$ can be written in the form

$$\mathcal{E}_{12}(t) = \sum_{n=1}^\infty \mathcal{E}_{12}(n;t),$$

where

$$\mathcal{E}_{12}(n;t) = \int_0^l \left\{ \left(\frac{\partial U_n}{\partial t}\right)^2 (x,t) + a^2 \left(\frac{\partial U_n}{\partial x}\right)^2 (x,t) \right\} dx.$$

7. Show that the total energy $\mathcal{E}_{31}$ associated with the boundary-value problem

$$u_{tt} - a^2 u_{xx} = 0, \quad 0 < x < l, \quad 0 < t,$$
$$u_x(0,t) - hu(0,t) = 0, \quad u(l,t) = 0, \quad 0 < t,$$
$$u(x,0) = f_1(x), \quad u_t(x,0) = f_2(x),$$

is given by

$$\mathcal{E}_{31}(t) = \int_0^l \{u_t^2(x,t) + a^2 u_x^2(x,t)\}\, dx + hu^2(0,t).$$

8. Solve Problem 7 by separation of variables.
9. Prove that an addition formula similar to that for $\mathcal{E}_{12}$ in Problem 6 holds under separation of variables for Problem 7.
10. Explain on physical grounds (or mathematical if you can) why the infinite series of energy functions obtained in Problems 6 and 9 must be convergent for all $t > 0$.

5.4 A DIFFERENCE APPROXIMATION

Let us return to Equation (5.1) which was assumed to hold on the interval $(0,l)$. We shall subdivide this interval into N equal subintervals of length l/N. Accordingly, the points of subdivision will be denoted by $x_j = j(l/N)$, $j = 0, 1, \ldots, N$. For notational convenience set $h = l/N$. The requirement that the intervals be of equal length is not essential. It is used here to simplify the calculations which follow.

We begin by integrating (5.1) from $x_j - h/2$ to $x_j + h/2$ for $j = 1, \ldots, N - 1$. This yields

$$\int_{x_j-h/2}^{x_j+h/2} \rho(x)A(x)u_{tt}(x,t)\, dx - \int_{x_j-h/2}^{x_j+h/2} (E(x)A(x)u_x(x,t))_x\, dx = 0. \quad (5.28)$$

The second term of (5.28) can obviously be integrated directly to obtain

$$\int_{x_j-h/2}^{x_j+h/2} (E(x)A(x)u_x(x,t))_x\, dx = E\left(x_j + \frac{h}{2}\right) A\left(x_j + \frac{h}{2}\right) u_x\left(x_j + \frac{h}{2}, t\right)$$

$$- E\left(x_j - \frac{h}{2}\right) A\left(x_j - \frac{h}{2}\right) u_x\left(x_j - \frac{h}{2}, t\right). \quad (5.29)$$

Now we wish to replace the right-hand side of (5.29) by an approximation for the derivative u_x which we can justify in some mathematical sense. For this purpose let us hold t fixed and write Taylor's formula with remainder. First we have

$$u(x_j,t) = u\left(x_j + \frac{h}{2}, t\right) - u_x\left(x_j + \frac{h}{2}, t\right)\frac{h}{2}$$

$$+ u_{xx}\left(x_j + \frac{h}{2}, t\right)\frac{h^2}{8} - u_{xxx}(x_1,t)\frac{h^3}{48},$$

where $x_j < x_1 < x_j + h/2$. This formula is certainly valid for a thrice continuously differentiable function of x. Next we write

$$u(x_{j+1},t) = u\left(x_j + \frac{h}{2}, t\right) + u_x\left(x_j + \frac{h}{2}, t\right)\frac{h}{2}$$

$$+ u_{xx}\left(x_j + \frac{h}{2}, t\right)\frac{h^2}{4} + u_{xxx}(x_2,t)\frac{h^3}{8},$$

where $x_j + h/2 < x_2 < x_{j+1}$. Subtracting the first of these formulas from the second and rearranging the terms yields the result we seek, namely

$$u_x\left(x_j + \frac{h}{2}, t\right) = \frac{u(x_{j+1},t) - u(x_j,t)}{h} + [u_{xxx}(x_2,t) - u_{xxx}(x_1,t)]\frac{h^2}{48}. \quad (5.30)$$

The first term on the right-hand side of (5.30) is called the *central difference* of u at $x_j + h/2$. Mathematically it is a close approximation to $u_x(x_j + h/2, t)$ if the third derivative $u_{xxx}(x,t)$ is bounded on (x_j, x_{j+1}) and h is small, for then we have

$$\left| u_x\left(x_j + \frac{h}{2}, t\right) - \frac{u(x_{j+1},t) - u(x_j,t)}{h} \right| \leq Kh^2,$$

where $24K$ is a bound for $|u_{xxx}(x,t)|$. In particular, in the limit as $h \to 0$ the central difference tends to the partial derivative.

In exactly the same way as we derived (5.30) we can also show that the central difference

$$\frac{u(x_j,t) - u(x_{j-1},t)}{h}$$

is a close approximation of $u_x(x_j - h/2, t)$ for small h.

The order of the approximation of the central difference to the derivative is said to be h^2 because the second term on the right-hand side of (5.30) involves h^2. In general, the higher the order, the better the approximation is considered to be, because if $h < 1$, $h^{n+1} < h^n$, $n = 1, 2, \ldots$.

Again, for notational convenience we set

$$u(x_j,t) = u_j(t), \quad u(x_{j+1},t) = u_{j+1}(t),$$

$$E\left(x_j + \frac{h}{2}\right) = E_{j+1}, \quad E\left(x_j - \frac{h}{2}\right) = E_j, \quad \text{and so on.}$$

Using the foregoing development (5.29) is written as

$$\int_{x_j-h/2}^{x_j+h/2} (E(x)A(x)u_x(x,t))_x \, dx \sim E_{j+1}A_{j+1} \frac{u_{j+1}(t) - u_j(t)}{h}$$

$$- E_j A_j \frac{u_j(t) - u_{j-1}(t)}{h}. \quad (5.31)$$

Here the symbol $\sim$ can be read "is approximated by." We shall approximate the first term of (5.28) as follows:

$$\int_{x_j-h/2}^{x_j+h/2} \rho(x)A(x)u_{tt}(x,t) \, dx \sim h\rho(x_j)A(x_j)u_{tt}(x_j,t). \quad (5.32)$$

Thus the integral has been approximated by the product of the value of the integral at the midpoint of the interval of integration with the length of the interval.

In order to further compact the notation let us set

$$E_j A_j = S_j, \quad j = 1, \ldots, N, \quad \rho(x_j)A(x_j) = m_j, \quad j = 1, \ldots, N-1. \quad (5.33)$$

Observe that $u_{tt}(x_j,t)$ can equally well be written in the notation of Chapter

2 as $\ddot{u}_j(t)$. Using the representations (5.31) and (5.33) we have now replaced (5.28) by its discrete analog

$$hm_j\ddot{u}_j - s_{j+1}\frac{u_{j+1} - u_j}{h} + s_j\frac{u_j - u_{j-1}}{h} = 0, \quad j = 1, \ldots, N - 1,$$

or

$$u_j - \frac{s_{j+1}}{h^2 m_j}u_{j+1} + \frac{s_j + s_{j+1}}{h^2 m_j}u_j - \frac{s_j}{h^2 m_j}u_{j-1} = 0, \quad j = 1, \ldots, N - 1. \tag{5.34}$$

The equation corresponding to $j = 1$ has special significance, since for this case the term $u_{1-1} = u_0 = u(0,t)$ occurs. Thus we must deal with the boundary conditions. By the first of Equations (5.2) the function $u(0,t)$ is zero. Hence for $j = 1$, (5.34) reduces to the special equation

$$u_1 - \frac{s_2}{h^2 m_1}u_2 + \frac{s_1 + s_2}{h^2 m_1}u_1 = 0. \tag{5.35}$$

In the equation corresponding to $j = N - 1$ the function $u_N(t) = u(l,t)$ occurs. By the second condition of (5.2), $u_x(l,t)$ is zero.

We shall now replace $u_x(l,t)$ by a difference approximation. The point $x = l$ corresponds to $j = N$ so that we are required to calculate $u_x(x_N,t)$. The central difference used before is no longer appropriate because we do not have any information about the function u for $x > x_N$. However, we can again use Taylor's formula with remainder in the form

$$u(x_{N-1},t) = u(x_N,t) - u_x(x_N,t)h + u_{xx}(\hat{x},t)\frac{h^2}{2}, \quad x_{N-1} < \hat{x} < x_N.$$

It then follows that

$$u_x(x_N,t) = \frac{u(x_N,t) - u(x_{N-1},t)}{h} - u_{xx}(\hat{x},t)\frac{h}{2}. \tag{5.36}$$

The difference quotient on the right-hand side of (5.36) is called the backward difference of u at x_N. In the limit as $h \to 0$ it tends to the derivative from the left at $x = x_N$. The order of the approximation of the backward difference to the derivative is h.

Using the backward difference, the boundary condition $u_x(l,t) = 0$ is approximated by

$$\frac{u_N - u_{N-1}}{h} = 0$$

or by

$$u_N = u_{N-1}.$$

Therefore for $j = N - 1$ we have the special equation

$$\ddot{u}_{N-1} + \frac{s_{N-1}}{h^2 m_{N-1}} u_{N-1} - \frac{s_{N-1}}{h^2 m_{N-1}} u_{N-2} = 0. \tag{5.37}$$

Finally we wish to take account of the initial conditions (5.3). For this purpose we introduce the numbers

$$f_j = f(x_j), \quad g_j = g(x_j), \quad j = 1, \ldots, N - 1.$$

Then we require that

$$u_j(0) = f_j, \quad \dot{u}_j(0) = g_j, \quad j = 1, \ldots, N - 1. \tag{5.38}$$

At this point let us reflect what has been accomplished. The continuous problem represented by Equations (5.1) and (5.2) has been replaced by a discrete (lumped parameter) problem consisting of $N - 1$ differential equations, specifically, Equations (5.34) for $j = 2, \ldots, N - 2$, Equation (5.35) for $j = 1$, and Equation (5.37) for $j = N - 1$. These differential equations are of second order and accordingly are subject to the initial conditions (5.3) which in this development are given by (5.38). To this system of equations we can attempt to apply the methods of Chapter 2 for finding explicit solutions. More important this may be done whatever the form of the functions E, A, and ρ appearing in Equation (5.1). Of course, we do not know how well the functions $u_j(t)$, $j = 1, \ldots, N - 1$, will approximate the actual solution of the given problem along the lines $x = x_j$. This is a question which requires mathematical investigation, but which we will not pursue in this book. However, let us recall an earlier physical argument. If the results obtained from a mathematical model agree closely with experimental results then one can conclude that the model (for the given problem) is a good one. Experience will show the reader that analyses made using difference techniques will, in general, yield excellent results.

For the present, let us seek solutions to our system of differential equations in the exponential form

$$u_j(t) = v_j e^{\lambda t}, \quad j = 1, \ldots, N - 1, \tag{5.39}$$

where $v_1, \ldots, v_{N-1}$ are constants to be determined. We remind the reader that this technique was used in Chapter 2 to solve the double-pendulum problem and, it is worthwhile noting, is quite analogous to the use of the method of separation of variables in Section 5.3. Substituting (5.39) into (5.35) gives

$$\left(\lambda^2 + \frac{s_1 + s_2}{h^2 m_1}\right) v_1 - \frac{s_2}{h^2 m_1} v_2 = 0 \tag{5.40}$$

after the nonzero function $e^{\lambda t}$ has been divided out. Similarly we obtain from (5.34), after division by $e^{\lambda t}$, the equations

$$\left(\lambda^2 + \frac{s_j + s_{j+1}}{h^2 m_j}\right) v_j - \frac{s_{j+1}}{h^2 m_j} v_{j+1}$$
$$- \frac{s_j}{h^2 m_j} v_{j-1} = 0, \quad j = 2, \ldots, N-2, \quad (5.41)$$

and lastly from (5.37) we obtain

$$\left(\lambda^2 + \frac{s_{N-1}}{h^2 m_{N-1}}\right) v_{N-1} - \frac{s_{N-1}}{h^2 m_{N-1}} v_{N-2} = 0. \quad (5.42)$$

Equations (5.40), (5.41), and (5.42) constitute a system of $N-1$ linear homogeneous equations for the determination of the $N-1$ numbers v_1, ..., v_{N-1}. This system will have a nontrivial solution if and only if the determinant of the system is zero. Specifically, the determinant is

$$d(\lambda) = \begin{vmatrix} \lambda^2 + \frac{s_1 + s_2}{h^2 m_1} & -\frac{s_2}{h^2 m_1} & 0 & 0 & \cdots & 0 \\ -\frac{s_2}{h^2 m_2} & \lambda^2 + \frac{s_2 + s_3}{h^2 m_2} & -\frac{s_3}{h^2 m_2} & 0 & \cdots & 0 \\ 0 & -\frac{s_3}{h^2 m_3} & \lambda^2 + \frac{s_3 + s_4}{h^2 m_3} & -\frac{s_4}{h^2 m_3} & \cdots & 0 \\ \cdots & & & & & \cdots \\ 0 & \cdots & & & -\frac{s_{N-1}}{h^2 m_{N-1}} & \lambda^2 + \frac{s_{N-1}}{h^2 m_{N-1}} \end{vmatrix}.$$

(5.43)

Defining $\lambda^2 = -\omega$, Equation (5.43) becomes an equation of degree $N-1$ in ω (or of degree $2N-2$ in λ). That is, the left-hand side of the equation

$$d(\omega) = 0 \quad (5.44)$$

is a polynomial of degree $N-1$ in ω. Accordingly (5.44) will have $N-1$ zeros $\omega_1, \ldots, \omega_{N-1}$ which in turn give rise to the $2N-2$ numbers

$$\lambda_j^+ = i\sqrt{\omega_j}, \quad \lambda_j^- = -i\sqrt{\omega_j}, \quad j = 1, \ldots, N-1. \quad (5.45)$$

These are the eigenvalues of the system of differential equations (5.34), (5.35), and (5.37).

For each fixed value of λ one can then solve the corresponding system of homogeneous linear equations (5.40), (5.41), and (5.43) to determine a set of numbers $v_1, \ldots, v_{N-1}$. Once again drawing from our experience with the problems of Chapter 2 the set $v_1, \ldots, v_{N-1}$ corresponding to a fixed λ can only be determined up to a constant α. That is, if the set $v_1, \ldots, v_{N-1}$ corresponds to λ so does the set $\alpha v_1, \ldots, \alpha v_{N-1}$ for any α. A detailed treatment of the eigenvalues will be given in the next section.

In closing this section we call the reader's attention to the fact that the system of equations we have derived here is very similar in structure to the system of first-order equations of Section 2.3. In particular the terms not involving derivatives with respect to t are entirely similar to the corresponding terms in system (2.17).

PROBLEMS

1. Show that the central difference
$$\frac{u(x_j,t) - u(x_{j-1},t)}{h}$$
is a close approximation of $u_x(x_j - h/2, t)$ for small h.

2. Set up a difference approximation for the boundary-value problem
$$u_{tt} = a^2 u_{xx}, \qquad 0 < x < l, \quad 0 < t,$$
$$u(0,t) = u_x(l,t) = 0, \quad 0 < t < \infty,$$
$$u(x,0) = u_t(x,0) = 0, \quad 0 < x < l.$$

3. Work Problem 2 for the case where the initial conditions are
$$u(x,0) = f(x), \quad u_t(x,0) = g(x), \quad 0 < x < l.$$

4. Set up a difference approximation for the boundary-value problem
$$u_{tt} = a^2 u_{xx} - 2\nu u_t + g(x) \sin \omega t, \quad 0 < x < l, \quad 0 < t,$$
$$u(0,t) = u(l,t) = 0, \quad 0 < t,$$
$$u(x,0) = u_t(x,0) = 0, \quad 0 < x < l.$$

5. Find a mechanical and an electrical model which give rise to the system of differential equations you have obtained in Problem 4. Can you find a suitable model which is neither electrical or mechanical in origin?

6. Set up a difference approximation for the boundary-value problem
$$u_{tt} = a^2 u_{xx} - 2\nu u_t, \quad 0 < x < l, \quad 0 < t,$$
$$u(0,t) = 0, \quad u_x(l,t) = \frac{A}{ES} \sin \omega t, \quad 0 < t,$$
$$u(x,0) = u_t(x,0) = 0, \quad 0 < x < l.$$

7. Find a mechanical and an electrical model which give rise to the system of differential equations you have obtained in Problem 6.

8. Set up a difference approximation for the boundary-value problem
$$v_{xx} = CL v_{tt} + CR v_t, \quad 0 < x < l, \quad 0 < t,$$
$$v(0,t) = v_x(l,t) = 0, \quad 0 < t,$$
$$v(x,0) = v_0, \; v_t(x,0).$$

252   THE VARIOUS ASPECTS OF PHYSICAL PROBLEMS

9. Find a mechanical and an electrical model which give rise to the system of differential equations you obtained in Problem 8.
10. Assuming that $L > C^2R^2l^2/\pi^2$, show that the solution to Problem 8 is

$$v(x,t) = e^{-(R/2L)t} \sum_{n=1}^{\infty} a_n \sin \frac{(2n+1)\pi x}{2l} \sin(\omega_n t + \phi_n),$$

where

$$\omega_n = \frac{(2n+1)\pi}{2l\sqrt{CL}} \sqrt{1 - \frac{C^2R^2l^2}{L\pi^2(2n+1)^2}},$$

$$a_n = \frac{4\theta_0}{\pi(2N+1)\sin\phi_n}, \quad \tan\phi_n = 2\omega_n \frac{L}{R}.$$

What happens if $L \le C^2R^2l^2/\pi^2$?

11. Set up a difference approximation for the boundary-value problem

$$u_t = \alpha^2 u_{xx} - \beta u + g(x), \quad 0 < x < l, \quad 0 < t,$$
$$u(0,t) = 0, \quad u_x(l,t) - hu(l,t) = 0, \quad 0 < t,$$
$$u(x,0) = f(x), \quad 0 < x < l.$$

12. Find a mechanical and an electrical model which give rise to the system of (first-order) differential equations you have obtained in Problem 11.
13. Set up a difference approximation for the boundary-value problem

$$\left(1 - \frac{x}{L}\right)^2 u_t = a^2 \frac{\partial}{\partial x}\left[\left(1 - \frac{x}{L}\right)^2 u_x\right]$$
$$\qquad - k\left(1 - \frac{x}{L}\right) u, \quad 0 < x < l, \quad 0 < t,$$
$$u_x(0,t) = u_x(l,t) = 0, \quad 0 < t,$$
$$u(x,0) = U_0, \quad 0 < x < l.$$

5.5 SOME PROPERTIES OF THE EIGENVALUES

Let us prove that the numbers $\omega_1, \ldots, \omega_{N-1}$ are real, positive, and distinct. Recalling that $\lambda^2 = -\omega$ it will be useful to denote the determinant (5.43) by $d_p(\omega)$. Using the elements in the first p rows and columns of (5.43) we write

$$d_1(\omega) = \frac{s_1 + s_2}{h^2 m_1} - \omega,$$

$$d_2(\omega) = \begin{vmatrix} \frac{s_1 + s_2}{h^2 m_1} - \omega & -\frac{s_2}{h^2 m_1} \\ -\frac{s_2}{h^2 m_2} & \frac{s_2 + s_3}{h^2 m_2} - \omega \end{vmatrix},$$

and so forth. Set $d_0(\omega) \equiv 1$ and denote by $\omega_{1p}, \ldots, \omega_{pp}$ the zeros of the polynomial equation
$$d_p(\omega) = 0, \quad p = 1, \ldots, N-1.$$

We shall need the following preliminary result.

Lemma 5.1

$$d_p(0) = h^{-2p}(m_1 m_2 \ldots m_p)^{-1} s_1 s_2 \ldots s_{p+1} \left(\sum_{j=1}^{p+1} \frac{1}{s_j} \right), \quad p \leq N-2 \quad (5.46)$$

and

$$d_{N-1}(0) = h^{-2p}(m_1 m_2 \ldots m_{N-1})^{-1} s_1 s_2 \ldots s_{N-1}. \quad (5.47)$$

PROOF. The proof is by mathematical induction on p. For $p = 1$ we have

$$d_1(0) = \frac{s_1 + s_2}{h^2 m_1} = h^{-2} m_1^{-1}(s_1 s_2) \left(\frac{1}{s_1} + \frac{1}{s_2} \right),$$

so that formula (5.46) is valid if $p = 1$. In order to handle the inductive step we establish an identity. Expand $d_{p+1}(0)$ by the last row. This gives

$$d_{p+1}(0) = \frac{s_{p+1} + s_{p+2}}{h^2 m_{p+1}} d_p(0) + \frac{s_{p+1}}{h^2 m_{p+1}} \hat{d}_p(0),$$

where $\hat{d}_p(0)$ is the determinant obtained from $d_{p+1}(0)$ by deleting the pth row and $(p-1)$th column. Since the pth column of $\hat{d}_p(0)$ contains only one element (in the last row), we can expand $\hat{d}_p(0)$ relative to the last column to obtain

$$\hat{d}_p(0) = -\frac{s_{p+1}}{h^2 m_p} d_{p-1}(0).$$

Accordingly

$$d_{p+1}(0) = \frac{s_{p+1} + s_{p+2}}{h^2 m_{p+1}} d_p(0) - \frac{(s_{p+1})^2}{h^2 m_{p+1} m_p} d_{p-1}(0). \quad (5.48)$$

Now we make the inductive hypothesis that (5.46) holds up to order p. Then (5.48) gives

$$d_{p+1}(0) = \frac{s_{p+1} + s_{p+2}}{h^2 m_{p+1}} h^{-2p}(m_1 m_2 \ldots m_p)^{-1} s_1 s_2 \ldots s_{p+1} \left(\sum_{j=1}^{p+1} \frac{1}{s_j} \right)$$

$$- \frac{(s_{p+1})^2}{h^4 m_{p+1} m_p} h^{-2p+2}(m_1 m_2 \ldots m_{p-1})^{-1} s_1 s_2 \ldots s_p \left(\sum_{j=1}^{p} \frac{1}{s_j} \right)$$

$$= h^{-2p-2}(m_1 \ldots m_p m_{p+1})^{-1} s_1 s_2 \ldots (s_{p+1})^2 \left(\sum_{j=1}^{p} \frac{1}{s_j} \right)$$

$$+ h^{-2p-2}(m_1 \ldots m_{p+1})^{-1} s_1 \ldots s_{p+2} \left(\sum_{j=1}^{p+1} \frac{1}{s_j} \right)$$

$$- h^{-2p-2}(m_1 \ldots m_{p+1})^{-1} s_1 \ldots (s_{p+1})^2 \left(\sum_{j=1}^{p} \frac{1}{s_j} \right)$$

$$+ h^{-2p-2}(m_1 \ldots m_{p+1})^{-1} s_1 \ldots s_{p+1}$$

$$= h^{-2(p+1)}(m_1 \ldots m_{p+1})^{-1} s_1 \ldots s_{p+2} \left(\sum_{j=1}^{p+2} \frac{1}{s_j} \right).$$

This proves the inductive step and hence formula (5.46) for $p = N - 2$. For $p = N - 1$, formula (5.48) becomes

$$d_{N-1}(0) = \frac{s_{N-1}}{h^2 m_{N-1}} d_{N-2}(0) - \frac{(s_{N-1})^2}{h^2 m_{N-1} m_{N-2}} d_{N-3}(0).$$

Applying formula (5.46) for the cases $p = N - 2$ and $p = N - 3$ yields (5.47) immediately. Thus the lemma is proved.

Returning to the problem we set for ourselves at the start of this section let us now prove:

Theorem 5.1

The numbers $\omega_{1p}, \ldots, \omega_{pp}, \omega_{1p+1}, \ldots, \omega_{p+1\,p+1}$ can be so arranged that for each $p = 1, \ldots, N - 2$

$$0 < \omega_{1p+1} < \omega_{1p} < \omega_{2p+1} < \cdots < \omega_{pp} < \omega_{p+1\,p+1}. \quad (5.49)$$

PROOF. The proof is again by mathematical induction on p. Before beginning the induction, however, observe that Lemma 5.1 implies that each polynomial $d_p(\omega)$ is positive for $\omega = 0$, that is $d_p(0) > 0$, $p = 1, \ldots, N - 1$. Observe also that the identity (5.48) remains valid for all ω in the form

$$d_{p+1}(\omega) = \left(\frac{s_{p+1} + s_{p+2}}{h^2 m_{p+1}} - \omega \right) dp(\omega) - \frac{(s_{p+1})^2}{h^4 m_{p+1} m_p} d_{p-1}(\omega). \quad (5.50)$$

Now consider the case $p = 1$. From $d_1(\omega) = 0$ we deduce immediately that

$$\omega_{11} = \frac{s_1 + s_2}{h^2 m_1}.$$

Next observe that

$$d_2 \left(\frac{s_1 + s_2}{h^2 m_1} \right) = - \frac{s_2^2}{h^4 m_1 m_2} < 0,$$

SOME PROPERTIES OF THE EIGENVALUES

and that for sufficiently large ω, $d_2(\omega) > 0$ since the coefficient of ω^2 is $+1$. Thus, the polynomial $d_2(\omega)$ is positive if $\omega = 0$ and for sufficiently large ω, and negative if $\omega = \omega_{11}$. It follows immediately that (5.49) is valid for $p = 1$.

For the inductive step suppose that (5.49) is valid up to $p - 1$ so that the polynomial $d_p(\omega)$ has p distinct positive zeros, $0 < \omega_{1p} < \omega_{2p} < \cdots < \omega_{pp}$, and that these zeros interlace with those of $d_{p-1}(\omega)$ in the manner shown in (5.49) when $p - 1$ replaces p. Multiply (5.50) by $d_{p-1}(\omega)$ and evaluate the result at ω_{jp} for $j = 1, \ldots, p$. Since $d_p(\omega_{jp}) = 0$, this gives us

$$d_{p+1}(\omega_{jp}) \, d_{p-1}(\omega_{jp}) = - \frac{(s_{p+1})^2}{h^4 m_{p+1} m_p} (d_{p-1})^2(\omega_{jp}).$$

By the inductive hypothesis $d_{p-1}(\omega_{jp}) \neq 0$, hence we have

$$d_{p+1}(\omega_{jp}) \, d_{p-1}(\omega_{jp}) < 0, \quad j = 1, \ldots, p. \tag{5.51}$$

Now let us investigate how the polynomial $d_{p+1}(\omega)$ changes sign as ω increases from zero. By Lemma 5.1 $d_{p+1}(0) > 0$. Since $d_{p-1}(0) > 0$ and the smallest zero of $d_{p-1}(\omega) = 0$ is larger than ω_{1p}, we have $d_{p-1}(\omega_{1p}) > 0$. It follows for (5.51) that $d_{p+1}(\omega_{1p}) < 0$ so that $d_{p+1}(\omega)$ has a zero between $\omega = 0$ and $\omega = \omega_{1p}$. Call this ω_{1p+1}. Next observe that $d_{p-1}(\omega_{2p}) < 0$ because $\omega_{1p-1} < \omega_{2p}$, hence $d_{p+1}(\omega_{2p}) > 0$ by (5.51). Consequently d_{p+1} has a zero between ω_{1p} and ω_{2p}. The argument clearly continues step by step, with $d_{p-1}(\omega_{jp}) = (-1)^{j-1}$ times a positive number. Hence $d_{p+1}(\omega_{jp}) = (-1)^j$ times a positive number and finally the coefficient of ω^{p+1} in $d_{p+1}(\omega)$ is $(-1)^{p+1}$ so that there is the zero $\omega_{p+1 \, p+1}$ larger than ω_{pp}. This completes the inductive step and the proof of the theorem.

The significance of the theorem is easily spelled out. We have proved that the zeros of $d(\omega)$, the numbers $\omega_1, \ldots, \omega_{N-1}$, are real, positive, and distinct. According to (5.43) the eigenvalues of the system of differential equations are therefore purely imaginary. This means that the $2N - 2$ solution sets to our system are trigonometric in character. In fact, using (5.37) and (5.43) we can clearly write

$$u_{jk}^+(t) = v_{jk}^+(\cos \sqrt{\omega_k} t + i \sin \sqrt{\omega_k} t),$$
$$u_{jk}^-(t) = v_{jk}^-(\cos \sqrt{\omega_k} t - i \sin \sqrt{\omega_k} t), \quad k, j = 1, \ldots, N - 1.$$

In the field of vibrations the numbers $\sqrt{\omega_1}, \ldots, \sqrt{\omega_{N-1}}$ are of primary importance. These positive numbers correspond to the possible frequencies of vibration of the system. They are also analogous to some of the eigenvalues obtained in the method of separation of variables of Section 5.3. One hopes that for a sufficiently large value of N the first few values $\sqrt{\omega_1}, \ldots, \sqrt{\omega_p}$ will be very close to the first few eigenvalues of problems

(5.15) and (5.17). This is the assumption engineers and physicists commonly use as a working hypothesis when they approximate continuous problems by discrete systems as we did in Section 5.4.

5.6 A DISCRETE SYSTEM

The philosophy expressed in Section 5.4 which resulted in the replacement of a complex partial differential equation problem by a seemingly less complex system of ordinary differential equations can be given a simple mechanical (or electrical) interpretation. The fact of the matter is that the *finite-difference technique* of Section 5.4 represents a more rigorous mathematical approach than engineers and physicists normally use in formulating a mathematical model. Here we explain how an analyst might replace a mechanical system by a series of masses connected by springs, and show that this technique is equivalent to the more formal method of Section 5.4.

Consider a beam fixed at one end, free at the other, and resting on a smooth surface. Let us approximate this beam by a sequence of masses $m_1, m_2, \ldots, m_{N-1}$ and springs with stiffness $k_1, k_2, \ldots, k_{N-1}$ as shown in Figure 5.2.

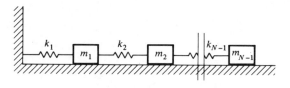

FIGURE 5.2

The motivation for the foregoing representation is easily seen when one considers the natural transition from the continuous stress-strain representation (Hooke's law) $\sigma = E\epsilon$ to its discrete analog, the force-displacement representation $\mathbf{F} = k\mathbf{r}$. Thus it follows that the modulus of elasticity E of the continuous model transforms to the spring force per unit length k in the discrete model.

If the masses in Figure 5.2 are constrained to slide on a frictionless horizontal glide the entire system represents a discrete physical approximation to the longitudinal vibrations of a bar with one end fixed and the other end free.

In order to derive the equations of motion for the system we introduce the independent coordinates $x_1, \ldots, x_{N-1}$ in such a way that each variable x_j represents the horizontal displacement of the mass m_j from the equilibrium position. To this end we could use Lagrange's equations or simply invoke Newton's second law directly. Let us use the latter. For this purpose the critical initial step is to formulate a free-body diagram from the mathematical model. That is, we select a specific mass, say m_j, isolated from the remainder of the system and subject to all the forces acting on the mass while in some state away from static equilibrium. To select the static equilibrium position would, of course, result in a trivial solution. Figure 5.3 is such a free-body diagram. Without loss of generality assume

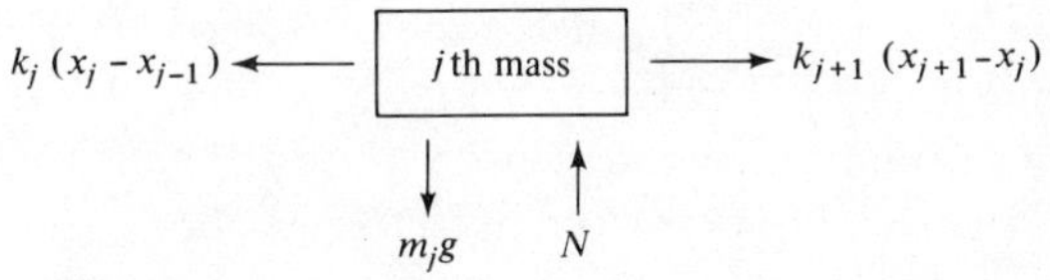

$k_j(x_j - x_{j-1})$ ← [jth mass] → $k_{j+1}(x_{j+1}-x_j)$

↓ ↑
$m_j g$ N

FIGURE 5.3

that $x_j > x_{j-1}$. Thus all the springs are in tension and positive acceleration is to the right. Since there is no motion in the vertical direction, summing those forces (with $\ddot{y}_j \equiv 0$) yields $m_j g \equiv N$. Summing forces in the horizontal direction we obtain $(F = ma)$

$$-k_j(x_j - x_{j-1}) + k_{j+1}(x_{j+1} - x_j) = m\ddot{x}_j. \tag{5.52}$$

Thus for the two extremes, $j = 1$ and $j = N - 1$, (5.52) reduces to

$$-k_1 x_1 + k_2(x_2 - x_1) = m_1 \ddot{x}_1, \tag{5.53}$$

and

$$-k_{N-1}(x_{N-1} - x_{N-2}) = m_{N-1}\ddot{x}_{N-1},$$

respectively.

The system of differential equations (5.52) may be solved subject to some initial conditions, say, the initial displacements

$$x_j(0) = a_j, \quad j = 1, \ldots, N - 1, \tag{5.54}$$

and initial velocities

$$\dot{x}_j(0) = b_j, \quad j = 1, \ldots, N - 1. \tag{5.55}$$

Dividing each Equation (5.52) by its related mass term and rearranging terms we may write

$$\ddot{x}_1 + \frac{k_1 + k_2}{m_1} x_1 - \frac{k_2}{m_1} x_2 = 0,$$

$$\ddot{x}_j - \frac{k_j}{m_j} x_{j-1} + \frac{k_j + k_{j+1}}{m_j} x_j - \frac{k_{j+1}}{m_j} x_{j+1} = 0, \quad j = 2, \ldots, N - 2 \tag{5.56}$$

$$\ddot{x}_{N-1} - \frac{k_{N-1}}{m_{N-1}} x_{N-2} + \frac{k_{N-1}}{m_{N-1}} x_{N-1} = 0.$$

With the identifications $x_j(t) = u_j(t)$ and $k_j = s_j$, $j = 1, \ldots, N - 1$, we see that the system (5.56) has precisely the same form as the difference approximation system (5.33).

The continual reduction in amplitude of a free vibration of a physical system is caused by energy dissipation in the system. For the beam problem discussed above it would take the form of internal structural damping. Perhaps the most fundamental mathematical representation of such a phenomenon is that which characterizes a resisting force due to fluid friction in an ideal dashpot. Thus, for linear systems the viscous damping force is taken as a function of the velocity $(-c\dot{r})$. The minus sign indicates that this force (usually) opposes the systems motion.

If we now introduce material damping into the system of Figure 5.2 the new mathematical model is as shown in Figure 5.4.

The reader will gain experience with the mathematical model of Figure 5.4 in the problem set which follows.

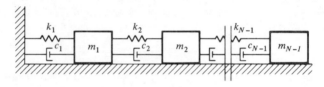

FIGURE 5.4

PROBLEMS

1. In obtaining the system of Equations (5.56) much care was taken to write the specific equations for $j = 1$ and $j = N - 1$. Was this necessary? Can these equations be considered an integral part of the second of (5.56)?
2. Discuss the relative merits of the method used in Section 5.4 to obtain the discrete equations of motion as opposed to that used in this section.
3. Suppose that a uniform rod fixed at one end and free at the other is acted upon by a longitudinal forcing function $f(t)$ at its free end. Considering the rod to be undamped use the method of this section to derive the equation(s) of motion for the system.
4. Derive the equation(s) of motion for the system of Figure 5.4.
5. Suppose that the system of Figure 5.4 is acted upon by a longitudinal forcing function $f(t)$ at its fixed end. Derive the equations of motion for the system. Give a physical interpretation to this system.
6. In Problem 5 determine the general solution for steady-state motion of a general mass m_i when $f(t) = F_0 e^{i\omega t}$. Find an expression for the

phase shift at any position along the beam. Give a physical interpretation of the phase angle. (See Problem 25, Section 1.7.)
7. Given the massless beam shown in Figure 5.5 with the concentrated mass m at its end being acted upon by the forcing function $F_0 e^{i\omega t}$. Derive the discrete equation(s) of motion for the system.

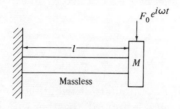

FIGURE 5.5

8. Determine the general solution for the steady-state motion of a generic point on the beam of Figure 5.5. Find an expression for the phase shift at any position along the beam. Give a physical interpretation of the phase angle.
9. Make a comparison of the solutions to Problems 6 and 8.

5.7 THE INFLUENCE FUNCTION

In this section we shall formulate the problem of Section 5.1 by a method quite different from that used previously. It will turn out that the formulation will be rather general and hence applicable to a variety of problems. The method consists of working directly with the Green's functions, or influence functions.

Suppose that we have a particular physical object B which is oriented with respect to an axis such that it is identified by the closed interval $[0,l]$, l being the length of the object. For example, B might be a string or a beam of length l.

Definition 5.1

$K(x,s)$ equals the deflection of the point x of B in a specified direction when a unit concentrated force is applied at the point s of B in the same specified direction. The function $K(x,s)$ is called the *influence function* or the Green's function of the physical system.

For the case of a string or of the transverse vibrations of a beam $K(x,s)$ will represent the vertical deflection of B when the unit force is applied at

s transverse to the orientation of the object. For the case of the longitudinal vibrations of a bar $K(x,s)$ will represent the horizontal displacement resulting from a horizontal force.

We assume that if B is subjected to n forces of magnitude $f_1, f_2, \ldots, f_n$ transverse to the x axis at the points $s_1, s_2, \ldots, s_n$ respectively, the resulting deflection $y(x)$ at x is obtained by linear superposition. Therefore we may write

$$y(x) = \sum_{j=1}^{n} K(x,s_i) f_i. \tag{5.57}$$

More generally, if a continuous distribution of forces, transverse to the x-axis, with intensity $g(s)$ is applied, together with the concentrated forces, we shall have

$$y(x) = \int_a^b K(x,s) g(s)\, ds + \sum_{j=1}^{n} K(x,s_j) f_j, \tag{5.58}$$

where $[a,b]$ $(0 \leq a < b \leq l)$ is the interval on which f and g are different from zero.

A point s of B is called *movable* if $K(s,s) > 0$. Thus, for a bar with the end $x = 0$ fixed and the end $x = l$ free, we shall have $K(0,0) = 0$ and $K(l,l) > 0$ so that the end $x = l$ is movable and clearly the end $x = 0$ is not movable. We assume that if M is the set of all movable points, then

$$K(x,s) > 0 \quad \text{for } s \text{ in } M \quad \text{and all} \quad 0 < x < l. \tag{5.59}$$

Before proceeding further with the development let us illustrate some influence functions with the aid of examples.

(i) Let a string of length l be stretched and denote its tension by T. If a unit force is applied transverse to the string at the point s the string will have the form illustrated in Figure 5.6. The equations for the two portions of this deflected string will be

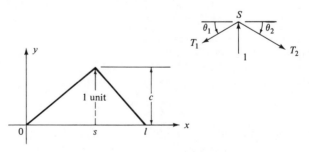

FIGURE 5.6

$$y_1 = \frac{c}{s} x, \quad 0 \leq x \leq s,$$

and

$$y_2 = \frac{c}{l-s}(l-x), \quad s \leq x \leq l.$$

With the aid of the free-body diagram of point s let us formulate the condition of static equilibrium (the sum of the forces equals zero) in the y-direction. For small displacements of the string we may regard the tension as a constant. Hence

$$-T \sin \theta_1 - T \sin \theta_2 + 1 = 0$$

$$= -T[\sin \theta_1 + \sin \theta_2] + 1 \sim -T\left[\frac{dy_1}{dx} + \frac{dy_2}{dx}\right] + 1 = 0.$$

Employing the foregoing geometric relations we obtain

$$\frac{cT}{s} - \frac{cT}{l-s} = 1,$$

or

$$c = \frac{s(l-s)}{lT}.$$

It follows from (5.57) that

$$K(x,s) = \begin{cases} \dfrac{x(l-s)}{lT}, & 0 \leq x \leq s, \\ \dfrac{s(l-x)}{lT}, & s \leq x \leq l. \end{cases} \tag{5.60}$$

(ii) As a second example, consider a hanging string with density σ per unit length and having length l. A unit force applied in the horizontal direction at a distance s from the bottom deflects the string into the position illustrated in Figure 5.7. For small displacements the horizontal component of the tension at the distance s from the lower end is (note the free-body diagram insert).

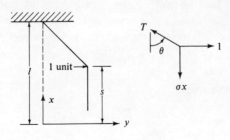

FIGURE 5.7

$$1 = T \sin \theta \sim T \frac{dy}{dx},$$

and the corresponding vertical component of the tension is

$$\sigma x = T \cos \theta \sim T.$$

Thus by eliminating T between the two equations we obtain

$$\frac{dx}{dy} = \sigma x, \quad \text{for} \quad 0 < \theta \ll 1. \tag{5.61}$$

To find K we must integrate (5.61). Thus

$$dy = \frac{dx}{\sigma x}$$

so that

$$y(x) = \int_x^l \frac{1}{\sigma u} \, du,$$

since $y(l) = 0$. For $x < s$, we have

$$y(x) = \int_s^l \frac{du}{\sigma u} = \frac{1}{\sigma} \log u \Big|_s^l = \frac{1}{\sigma} [\log l - \log s]$$
$$= -\frac{1}{\sigma} \log \frac{s}{l} = \text{const.}$$

For $x > s$, we have

$$y(x) = \frac{1}{\sigma} \int_x^l \frac{1}{u} \, du = -\frac{1}{\sigma} \log \frac{x}{l}.$$

It follows that

$$K(x,s) = \begin{cases} -\dfrac{1}{\sigma} \log \dfrac{s}{l}, & \text{if } 0 \leq x \leq s, \\ -\dfrac{1}{\sigma} \log \dfrac{x}{l}, & \text{if } s \leq x \leq l. \end{cases} \tag{5.62}$$

Note that both of the influence functions (5.60) and (5.62) satisfy the condition (5.59). In the case of the string fixed at both ends we have $K(0,0) = K(l,l) = 0$ since these are not movable points. In the case of the hanging string $K(0,0)$ is undefined, in fact we have

$$\lim_{s \to 0} K(0,s) = +\infty,$$

from which it follows that

$$\lim_{(x,s) \to (0;0)} K(x,s) = +\infty.$$

We can also interpret this unbounded behavior as implying that the point

0 is movable. Observe also that we have
$$K(x,s) = K(s,x) \tag{5.63}$$
for both of these functions.

We will show now that this equation is always satisfied for a conservative system. In fact, if the system is conservative the work required to bring object B into the position (5.57) is completely determined by the position itself and consequently depends only upon the quantities f_i, $i = 1, \ldots, n$. Mathematically, this means that if y_i is the deflection due to the load f_i at s_i, we have
$$dV = \sum_{i=1}^{n} f_i\, dy_i, \tag{5.64}$$
where V is the potential energy of the system. The reader who is familiar with structural analysis will recognize (5.64) as a form of *Castigliano's theorem*. From (5.57) we have
$$dy_i(s_i) = \sum_{j=1}^{n} K(s_i,s_j)\, df_j, \quad i = 1, \ldots, n.$$
Substituting this result into (5.64) yields
$$dV = \sum_{i,j=1}^{n} f_i K(s_i,s_j)\, df_j,$$
from which we obtain the result that
$$\frac{\partial V}{\partial f_j} = \sum_{i=1}^{n} K(s_i,s_j) f_i, \quad j = 1, \ldots, n,$$
and hence that
$$\frac{\partial^2 V}{\partial f_j \partial f_i} = K(s_i,s_j).$$
Interchanging the roles of i and j we also have
$$\frac{\partial^2 V}{\partial f_i \partial f_j} = K(s_j,s_i).$$
Consequently the hypothesis that V is twice continuously differentiable as a function of the forces gives us the result
$$\frac{\partial^2 V}{\partial f_i \partial f_j} = \frac{\partial^2 V}{\partial f_j \partial f_i},$$
or
$$K(s_i,s_j) = K(s_j,s_i). \tag{5.65}$$

Finally the fact that the points s_i, $i = 1, \ldots, n$, are arbitrary points of B implies that (5.65) is equivalent to (5.63). Again, those who are familiar with structural mechanics will recognize this formula as the *Maxwell reciprocity theorem*. We shall, however, usually express it by saying that $K(x,s)$ is symmetric in its arguments, or simply symmetric.

Let us observe that if a continuous distribution of forces with magnitude $f(s)$ and discrete concentrated forces $f_1, \ldots, f_n$ act on B, the potential energy is given by

$$2V = \left\{ \int_0^l \int_0^l K(x,s) f(x) f(x)\, dx\, ds + 2 \int_0^l \sum_{j=1}^n K(x,s_j) f_j f(x)\, dx \right. $$
$$ \left. + \sum_{i,j=1}^n K(s_i,s_j) f_i f_j \right\}. \quad (5.66)$$

If we consider forces applied to B which depend upon t, so that we must look upon y as a function $y = y(x,t)$, we have

$$f(s)\, ds = -\frac{\partial^2 y}{\partial t^2}(s,t) \rho(s)\, ds$$

where $\rho(s)$ is the mass distribution along B. Assuming that no discrete concentrated forces act upon B, we obtain the integro-differential equation

$$y(x,t) = -\int_0^l K(x,s) \frac{\partial^2 y}{\partial t^2}(s,t) \rho(s)\, ds. \quad (5.67)$$

This integro-differential equation represents an alternative formulation of the boundary-value problem of Section 5.1. It must be solved, of course, subject to the initial conditions

$$\begin{aligned} y(x,0) &= f(x), \quad 0 < x < l, \\ y_t(x,0) &= g(x), \quad 0 < x < l. \end{aligned} \quad (5.68)$$

More will be done with this formulation in the following section.

5.8 REMARKS ON THE INTEGRO-DIFFERENTIAL EQUATION

In this section we wish to deduce some facts about Equation (5.67). For simplicity let us take the equation in the form

$$y(x,t) + \int_0^l K(x,s) \rho(s) y_{tt}(s,t)\, ds = 0. \quad (5.69)$$

We first apply the energy method to Equation (5.69). Multiply the equation by $\rho(x) y_t(x,t)$ to obtain

$$\rho(x) y_t(x,t) y(x,t) + \int_0^l K(x,s) \rho(x) \rho(s) y_t(x,t) y_{tt}(s,t)\, ds = 0.$$

Now integrate with respect to x from 0 to l to obtain

$$\frac{1}{2}\int_0^l \rho(x)\frac{\partial}{\partial t}(y(x,t))\,dx + \int_0^l\int_0^l K(x,s)\rho(x)\rho(s)y_t(x,t)y_{tt}(s,t)\,ds\,dx = 0. \tag{5.70}$$

Observe that if the roles of x and s are reversed in (5.69) it will read

$$y(s,t) + \int_0^l K(s,x)\rho(x)y_{tt}(x,t)\,dx = 0.$$

Multiplying this equation by $\rho(s)y_t(s,t)$ and integrating with respect to s from 0 to l gives

$$\frac{1}{2}\int_0^l \rho(s)\frac{\partial}{\partial t}(y(s,t))\,ds + \int_0^l\int_0^l K(x,s)\rho(x)\rho(s)y_t(s,t)y_{tt}(x,t)\,dx\,ds = 0 \tag{5.71}$$

because of the symmetry of $K(x,s)$. Adding (5.70) and (5.71), replacing the dummy variable s by x in the single integral, and interchanging the order of integration in one of the double integrals results in the equation

$$\frac{\partial}{\partial t}\left\{\int_0^l \rho(x)y^2(x,t)\,dx + \int_0^l\int_0^l K(x,s)\rho(x)\rho(s)y_t(s,t)y_t(x,t)\,ds\,dx\right\} = 0. \tag{5.72}$$

Let us now observe that the potential and kinetic energies of the system are

$$2V = \int_0^l \rho(x)y^2(x,t)\,dx,$$

and

$$2T = \int_0^l\int_0^l K(x,s)\rho(x)\rho(s)y_t(x,t)y_t(s,t)\,ds\,dx. \tag{5.73}$$

Equation (5.72) is then seen to be precisely the principle of the conservation of energy for the system. That is,

$$T + V = \text{const.} \tag{5.74}$$

Next we observe that the method of separation of variables can also be applied to (5.69). Let us apply this method using the general form

$$y(x,t) = \varphi(x)\sin(\lambda t + \delta). \tag{5.75}$$

Then

$$y_{tt} = -\lambda^2\varphi(x)\sin(\lambda t + \delta)$$

and substitution into (5.69) yields

$$\varphi(x)\sin(\lambda t + \delta) - \lambda^2\int_0^l K(x,s)\rho(x)\varphi(s)\sin(\lambda t + \delta)\,ds = 0.$$

On setting $\omega = \lambda^2$,

$$\sin(\lambda t + \delta)\left\{\varphi(x) - \omega \int_0^l K(x,s)\rho(s)\varphi(s)\,ds\right\} = 0.$$

Since $\sin(\lambda t + \delta) \not\equiv 0$ for all choices of λ and δ, we require that the function φ satisfy the integral equation

$$\varphi(x) = \omega \int_0^l K(x,s)\rho(s)\varphi(s)\,ds. \tag{5.76}$$

Equation (5.76) is called a *Fredholm integral equation*. It has nonzero solutions only for certain discrete values of the parameter ω. The values of ω for which nonzero solutions φ exist are called the *eigenvalues of the integral equation* and the corresponding functions φ are called *eigenfunctions*. The theory of such integral equations is sketched in the rest of this chapter. For the present let us point out two facts of considerable importance. The first concerns the symmetrizability of the integral equation (5.76). Keep in mind that the function ρ is always positive since it is a physical property, density. Accordingly the function $\rho^{1/2}$ is well-defined. We multiply (5.76) by $\rho^{1/2}(x)$ to obtain

$$\rho^{1/2}(x)\varphi(x) = \omega \int_0^l K(x,s)[\rho(x)\rho(s)]^{1/2}\rho^{1/2}(s)\varphi(s)\,ds.$$

Now defining the functions

$$\begin{aligned}\psi(x) &= \rho^{1/2}(x)\varphi(x), \\ \tilde{K}(x,s) &= K(x,s)[\rho(x)\rho(s)]^{1/2},\end{aligned} \tag{5.77}$$

we then have

$$\psi(x) = \omega \int_0^l \tilde{K}(x,s)\psi(s)\,ds. \tag{5.78}$$

The integral equation (5.78) is called a *symmetric integral equation* because the function $\tilde{K}$, called the *kernel* of the integral equation, is symmetric. That is,

$$\tilde{K}(x,s) = \tilde{K}(s,x). \tag{5.79}$$

The theory of symmetric integral equations is much simpler than the theory of general integral equations. For this reason the fact that (5.76) is symmetrizable will prove to be important to us in our work in Sections 5.9 and 5.10.

The second fact of importance to us establishes the close connection of the integral equation (5.76) with the eigenvalue problem of Section 5.3 defined by equations having the form

$$(\rho(x)X')' + \lambda r(x)X = 0, \tag{5.80}$$

where

$$X(0) = X'(l) = 0. \tag{5.81}$$

We can prove that there exists a function $G(x,s)$ satisfying the symmetry condition
$$G(x,s) = G(s,x),$$
such that this problem can be transformed into the problem of solving the integral equation
$$X(x) = \lambda \int_0^l G(x,s)r(s)X(s)\,ds. \tag{5.82}$$
The reader should recognize the function $G(x,s)$ as the Green's function for the boundary-value problem (5.80), (5.81). The function $r(x)$ is positive by virtue of its physical significance so that (5.82) can also be symmetrized. Moreover, the function $G(x,s)$ is equal to the influence function $K(x,s)$ as was pointed out in Section 5.7. Consequently we see that the problem of solving the integral equation (5.78) of the present section is equivalent to that of solving the Sturm-Liouville problem of Section 5.3.

PROBLEMS

1. Consider the differential equation
$$u_{tt} - a^2 u_{xx} = 0, \quad 0 < x < l, \quad 0 < t,$$
with boundary conditions
$$u(0,t) = u(l,t) = 0, \quad 0 < t.$$
The Green's function for this system is obtained by writing the solution of the boundary-value problem defined by the initial conditions
$$u(x,0) = 0, \quad u_t(x,0) = f(x), \quad 0 < x < l,$$
in the form
$$u(x,t) = \int_0^l f(y)G(x,y,t)\,dy,$$
where function $G(x,y,t)$ is the Green's function. Find this function for the above system.

2. Show that if the system of Problem 1 is solved by separation of variables there results the boundary-value problem
$$X'' + \lambda^2 X = 0, \quad X(0) = X(l) = 0.$$

3. The influence function associated with the system of Problem 1 is that function $K(x,\xi)$ which satisfies
   (i) $$X'' = 0$$
   except when $x = \xi$,

(ii) $$K(0,\xi) = 0 = K(l,\xi),$$

and

(iii) $K(x,s)$ is continuous for $0 \leq x \leq l$, $0 \leq \xi \leq l$,

and $\partial K/\partial x$ has a jump discontinuity of magnitude 1 along the line $x = s$. Prove that

$$K(x,\xi) = \begin{cases} \dfrac{x(l-\xi)}{l}, & 0 \leq x \leq \xi, \\ \dfrac{\xi(l-x)}{l}, & \xi \leq x \leq l. \end{cases}$$

4. Find the Green's function in the sense of Problem 1 for the wave equation

$$u_{tt} - a^2 u_{xx} = 0, \quad 0 < x < l, \quad 0 < t,$$

with boundary conditions

$$u_x(0,t) = u(l,t) = 0.$$

5. Find the influence function for the differential equation

$$X'' = 0$$

with boundary conditions

$$X'(0) = X(l) = 0.$$

6. Find the Green's function for the wave equation with boundary conditions

$$u(0,t) = 0, \quad \alpha u(l,t) + u_x(l,t) = 0.$$

7. Find the influence function for $X'' = 0$ with boundary conditions

$$X(0) = 0, \quad \alpha X(l) + X'(l) = 0.$$

8. Show that the influence function for $X'' + X = 0$ with boundary conditions $X(0) = X(l) = 0$, is given by

$$K(x,\xi) = \begin{cases} \sin x \sin (l-\xi)/\sin l, & 0 \leq x \leq \xi, \\ \sin (l-x) \sin \xi/\sin l, & \xi \leq x \leq l. \end{cases}$$

9. Prove that the influence function of Problem 3 satisfies the inequalities

$$\begin{vmatrix} K(x_1,\xi_1) & K(x_1,\xi_2) \\ K(x_2,\xi_1) & K(x_2,\xi_2) \end{vmatrix} > 0,$$

and

$$\begin{vmatrix} K(x_1,\xi_1) & K(x_1,\xi_2) & K(x_1,\xi_3) \\ K(x_2,\xi_1) & K(x_2,\xi_2) & K(x_2,\xi_3) \\ K(x_3,\xi_1) & K(x_3,\xi_2) & K(x_3,\xi_3) \end{vmatrix} > 0,$$

where $0 < x_1 < x_2 < x_3 < l$ and $0 < \xi_1 < \xi_2 < \xi_3 < l$. What happens if some of the variables coincide?

10. Show that the inequalities of Problem 9 are also valid for the influence function of Problem 8.

11. Find the Green's function for the equation
$$u_{tt} - a^2 u_{xx} + hu = 0,$$
where h is a constant, with the boundary conditions
$$u(0,t) = u(l,t) = 0.$$

12. Find the Green's function for the heat equation
$$u_t - a^2 u_{xx} = 0$$
with boundary conditions
$$u_x(0,t) = u(l,t) = 0.$$

13. Show that the Green's function for the boundary-value problem
$$u''(x) + F(x) = 0$$
$$u(0) = u(1) = 0,$$
where
$$u(x) = \int_0^1 G(x,s) F(s)\, ds$$
is
$$G(x,s) = \begin{cases} x(1-s), & 0 \le x \le s \le 1, \\ s(1-x), & 0 \le s \le x \le 1. \end{cases}$$

14. Show that the function
$$u(x) = \int_0^1 G(x,s) s\, ds$$
satisfies the boundary-value problem
$$u''(x) + x = 0,$$
$$u(0) = u(1) = 0.$$

15. Formulate the integro-differential equation for the longitudinal vibrations of a uniform bar with density ρ per unit length if the bar is fixed at one end and free at the other end.

16. Formulate the integro-differential equation for the bar of Problem 15 if it is fixed at one end and elastically supported at the other end. That is, $\alpha u(l,t) + u_x(l,t) = 0$.

17. Solve the integro-differential equation of Problem 15 by the method of separation of variables.

18. Show that, if the physical continuum B is loaded by masses m_1, m_2, ..., m_n at the points $\xi_1 < \xi_2 < \cdots < \xi_n$ and the mass of the continuum itself is neglected, the displacements of the points $\xi_1, \ldots, \xi_n$ satisfy the system of ordinary differential equations

$$y_i = -\sum_{k=1}^{n} K(\xi_i,\xi_k) m_k \ddot{y}_k.$$

19. Explain how the result of Problem 18 can be looked upon as a discretization of the problem of vibration of the continuum B.

20. Prove that there exists a function $G(x,s)$ satisfying the symmetry condition

$$G(x,s) = G(s,x),$$

such that this problem can be transformed into the problem of solving the integral equation (5.82).

21. Assume that in addition to the continuous distribution of time dependent forces acting upon B there are also discrete concentrated forces $f_1(t), \ldots, f_n(t)$ acting upon B. Write the integro-differential equation analogous to (5.67) for this case.

5.9 SOME PROPERTIES OF THE SYMMETRIC INTEGRAL EQUATION

In Section 5.8 we reduced the problem of solving the integro-differential equation

$$y(x,t) + \int_0^l K(x,s)\rho(s)y_{tt}(s,t)\,ds = 0 \tag{5.83}$$

to that of solving an integral equation by the method of separation of variables. Let us write this integral equation (5.78) in the form

$$\psi(x) = \omega \int_0^l k(x,s)\psi(s)\,ds \tag{5.84}$$

where $k(x,s) = K(x,s)[\rho(x)\rho(s)]^{1/2}$, and k satisfies the symmetry condition

$$k(x,s) = k(s,x).$$

To be more specific, the kernel k is a real-valued function of two variables defined over the square $0 \leq x,\, s \leq l$. This function generates an integral transformation defined by

$$K\psi = \int_0^l k(x,s)\psi(s)\,ds, \tag{5.85}$$

in the sense that a real-valued function ψ defined on $0 \leq x \leq l$ is trans-

formed by (5.85) into a real-valued function
$$z(x) = K\psi$$
also defined on $[0,l]$.

Now suppose we set
$$\mu = \frac{1}{\omega}, \tag{5.86}$$
then (5.84) appears in the symbolic form
$$K\psi = \mu\psi. \tag{5.87}$$
Our aim is to discuss the properties of solutions of Equation (5.87). To do this we shall require some further hypotheses concerning the nature of the kernel k.

Definition 5.2

The kernel k is called a *Hilbert-Schmidt kernel* and the transformation K generated by k and defined by (5.85) is called a *Hilbert-Schmidt integral operator* if
$$\int_0^l \int_0^l k^2(x,s) \, dx \, ds < \infty. \tag{5.88}$$

Recall that in Definition 3.2 we first introduced the concept of a square integrable function of one variable and made considerable use of that concept thereafter. In terms of this fundamental notion we can characterize a Hilbert-Schmidt kernel (to be called hereafter an H-S kernel) as a square integrable function of two variables.

We shall assume henceforth that k is a symmetric H-S kernel. Let us pause to give some examples.

EXAMPLE 1

Let
$$k(x,s) = \begin{cases} \dfrac{x(l-s)}{lT}, & 0 \leq x \leq s, \\ \dfrac{s(l-x)}{lT}, & s \leq x \leq l. \end{cases}$$

Then
$$\int_0^l \int_0^l k^2(x,s) \, dx \, ds = \frac{1}{l^2 T^2} \left\{ \int_0^l \int_0^s x^2(l-s)^2 \, dx \, ds \right. $$
$$\left. + \int_0^l \int_0^l s^2(l-x^2) \, dx \, ds \right\}.$$

Now on $[0,l]$ we have $s^2 \leq l^2$, $x^2 \leq l^2$, $(l-s)^2 \leq l^2$, and $(l-x)^2 \leq l^2$, hence

$$\int_0^l \int_0^l k^2(x,s)\, dx\, ds \leq \frac{l^2}{T^2} \int_0^l \int_0^l dx\, ds = \frac{l^4}{T^2} < \infty.$$

Thus k is an H-S kernel.

EXAMPLE 2

The kernel of Example 1 is bounded on the square $0 \leq x,s \leq l$. Our next example is

$$k(x,s) = \begin{cases} -\dfrac{1}{\sigma} \log \dfrac{s}{l}, & 0 \leq x \leq s, \\ -\dfrac{1}{\sigma} \log \dfrac{x}{l}, & s \leq x \leq l, \end{cases}$$

a kernel which is unbounded when $x = 0$, $s = 0$, or when both $x = 0$ and $s = 0$. Nevertheless k is an H-S kernel. In fact we have again

$$\int_0^l \int_0^l k^2(x,s)\, dx\, ds = \frac{1}{\sigma^2} \left\{ \int_0^l \int_0^s \log^2 \frac{s}{l}\, dx\, ds + \int_0^l \int_s^l \log^2 \frac{x}{l}\, dx\, ds \right\}.$$

In the second integral on the right we can interchange the order of integration to obtain

$$\int_0^l \int_s^l \log^2 \frac{x}{l}\, dx\, ds = \int_0^l x \log^2 \frac{x}{l}\, dx \leq l \int_0^l \log^2 \frac{x}{l}\, dx.$$

Similarly, the first integral is simply equal to

$$\int_0^l s \log^2 \frac{s}{l}\, ds \leq l \int_0^l \log^2 \frac{s}{l}\, ds.$$

Therefore by introducing the change of variable $s = ly$ in the first integral and $x = ly$ in the second, we have

$$\int_0^l \int_0^l k^2(x,s)\, dx\, ds \leq \frac{2l^2}{\sigma^2} \int_0^l \log^2 y\, dy.$$

But $\int_0^l \log^2 y\, dy = [y(\log^2 y - 2\log y + 2)]_0^1 = 2$ (in order to evaluate $y \log^2 y$ and $y \log y$ at zero L'Hospital's rule must be used). Consequently

$$\int_0^l \int_0^l k^2(x,s)\, dx\, ds \leq \left(\frac{2l}{2}\right)^2 < \infty,$$

showing that k is indeed an H-S kernel.

EXAMPLE 3

The two previous examples were both of the form

$$k(x,s) = \begin{cases} f(x)g(s), & 0 \le x \le s, \\ g(x)f(s), & s \le x \le l. \end{cases} \quad (5.89)$$

For any such kernel we have

$$\int_0^l \int_0^l k^2(x,s)\, dx\, ds = \int_0^l \int_0^s f^2(x)g^2(s)\, dx\, ds + \int_0^l \int_s^l g^2(x)f^2(s)\, dx\, ds.$$

In the second integral we interchange orders of integration and obtain

$$\int_0^l \int_0^l k^2(x,s)\, dx\, ds = 2\int_0^l g^2(s)\left(\int_0^s f^2(x)\, dx\right) ds. \quad (5.90)$$

With this formula at hand, let us prove

Lemma 5.2

Let k be a kernel having the form (5.89) where f and g are square integrable on $[0,l]$, then k is an H-S kernel.

PROOF. Let

$$\int_0^l f^2(y)\, dy = F < \infty, \quad \int_0^l g^2(y)\, dy = G < \infty,$$

then

$$\int_0^s f^2(x)\, dx \le \int_0^l f^2(x)\, dx = F.$$

Consequently

$$\int_0^l g^2(s)\left(\int_0^s f^2(x)\, dx\right) ds \le \left(\int_0^l g^2(s)\, ds\right)\left(\int_0^l f^2(x)\, dx\right) = FG.$$

Hence

$$\int_0^l \int_0^l k^2(x,s)\, dx\, ds \le 2FG < \infty,$$

proving the lemma.

Here are two examples of H-S kernels which are not of the form given in Examples 1, 2, and 3.

EXAMPLE 4

$$k(x,s) = |x - s|^{-\alpha}, \quad 0 \le \alpha < \tfrac{1}{2}.$$

EXAMPLE 5

$$k(x,s) = \begin{cases} \dfrac{x(s-l)}{6} (x-s)^2, & 0 \leq x \leq s, \\ \dfrac{s(x-l)}{6} (x-s)^2, & s \leq x \leq l. \end{cases}$$

The fact that an H-S kernel is not necessarily bounded as illustrated by Examples 2 and 4 above leads us to place an additional condition upon the kernel k.

Definition 5.3

Let k be an H-S kernel. Then the function

$$k_1(x,s) = \int_0^l k(x,\xi)k(\xi,s)\,d\xi$$

is called the *first iterated kernel*. We say k is a *regular kernel* if k_1 is a bounded and continuous function in the square $0 \leq x,s \leq l$.

We shall assume henceforth that k is a regular kernel. This assumption does not rule out any of the examples above. It is, of course, clear that if k is itself bounded and continuous then k_1 is also. We shall call k a smooth kernel if it is bounded and continuous.

Let us show now that the unbounded kernel of Example 2 is a regular kernel. For this purpose we will make use of the fact (see Problem 7 below) that the symmetry of k implies that of k_1. Thus we need only show k_1 is bounded and continuous for $0 \leq x \leq s \leq l$. With this inequality valid we have

$$k_1(x,s) = \frac{1}{\sigma^2}\left\{x\log\frac{s}{l}\log\frac{x}{l} + \log\frac{s}{l}\int_x^s \log\frac{\xi}{l}\,d\xi + \int_0^l \log^2\frac{\xi}{l}\,d\xi\right\}$$

$$= \frac{1}{\sigma^2}\left\{x\log\frac{s}{l}\log\frac{x}{l} + l\log\frac{s}{l}\int_{x/l}^{s/l} \log y\,dy + l\int_{s/l}^1 \log^2 y\,dy\right\}$$

$$= \frac{1}{\sigma^2}\left\{x\log\frac{s}{l}\log\frac{x}{l} + l\log\frac{s}{l}[y\log y - y]_{x/l}^{s/l} \right.$$

$$\left. + l[y\log^2 y - 2y\log y + 2y]_{s/l}^1\right\}.$$

Thus we obtain

$$k_1(x,s) = \frac{1}{\sigma^2}\left[(x+s)\log\frac{s}{l} + 2(l-s)\right] \quad \text{for} \quad 0 \leq x \leq s \leq l.$$

This function is obviously bounded and continuous except when $s = 0$ where it is undefined by the formula. But

$$\lim_{s \to 0} k_1(x,s) = 2l/\sigma^2,$$

since $x \leq s$. Thus we can complete the definition of k_1 by defining it to have the value $2l/\sigma^2$ when $s = 0$ and it will be bounded and continuous on $0 \leq x \leq s \leq l$. Finally by symmetry we set $k_1(0,s) = 2l/\sigma^2$ and k_1 is then bounded and continuous on the entire square $0 \leq x,s \leq l$.

We shall now place one final restriction on the kernel k. To understand the purpose of this restriction let us write (5.87) in the form

$$\mu\psi(x) = \int_0^l k(x,s)\psi(s)\,ds$$

and multiply both sides by $\psi(x)$ to obtain

$$\mu\psi^2(x) = \int_0^l k(x,s)\psi(s)\psi(x)\,ds.$$

Now integrate over $[0,l]$. This gives

$$\mu \int_0^l \psi^2(x)\,dx = \int_0^l \int_0^l k(x,s)\psi(s)\psi(x)\,ds\,dx.$$

In the notation of Section 3.3 this equation reads

$$\mu\|\psi\|^2 = (K\psi,\psi).$$

Now suppose μ is an eigenvalue of (5.87) and ψ with $\|\psi\| \neq 0$ a corresponding eigenfunction, then we have

$$\mu = \frac{(K\psi,\psi)}{\|\psi\|^2}. \tag{5.91}$$

Definition 5.4

The kernel k is said to be positive definite and the corresponding operator K is said to be positive if there exists a constant $k_0 > 0$ such that

$$(Ku,u) \geq k_0\|u\|^2$$

for every function u such that $\|u\| \neq 0$.

It is obvious that if K is a positive operator every eigenvalue is positive from (5.91) and the converse is also true, although this is not quite so obvious.

We shall assume henceforth that k generates a positive operator. This hypothesis has considerable significance for our original integro-differential equation. To see this is so we return to formula (5.73) for the kinetic energy, namely,

$$2T = \int_0^l \int_0^l K(x,s)\rho(x)\rho(s)y_t(x,t)y_t(s,t)\,ds\,dx$$
$$= \int_0^l \int_0^l k(x,s)\rho^{1/2}(x)y_t(x,t)\rho^{1/2}(s)y_t(s,t)\,ds\,dx$$
$$= (K\rho^{1/2}y_t, \rho^{1/2}y_t).$$

On the basis of the above formula we can state

Theorem 5.2

Let $y(x,t)$ be a solution of the integro-differential equation (5.69) with the property that

$$\|\rho^{1/2}y_t\| = \int_0^l \rho(x)y_t^2(x,t)\,dx^{1/2} \neq 0$$

for a fixed value of t and let k be a positive definite kernel. Then the kinetic energy in the system at time t is positive. In any case we have

$$2T \geq k_0 \|\rho^{1/2}y_t\|^2 \geq 0$$

for all t.

In other words, the hypothesis that k generates a positive operator guarantees that the kinetic energy is positive when the solution is nonzero and this fact corresponds with our physical intuition concerning the underlying problem.

PROBLEMS

1. Show that the kernel of Example 4 is an H-S kernel.
2. Show that the kernel of Example 5 is an H-S kernel.
3. Let k be a kernel having the form

$$k(x,s) = \begin{cases} f_1(x)g_1(s) + f_2(x)g_2(s), & 0 \leq x \leq s, \\ f_1(s)g_1(x) + f_2(s)g_2(x), & s \leq x \leq l \end{cases}$$

where f_1, f_2, g_1, and g_2 are square integrable on $[0,l]$. Show that k is an H-S kernel.

4. Let k be defined by

$$k(x,s) = \begin{cases} \dfrac{x(s-l)}{6}(x-s)^2, & 0 \leq x \leq s, \\ \dfrac{s(x-l)}{6}(x-s)^2, & s \leq x \leq l. \end{cases}$$

Show that k is an H-S kernel.

5. Define k by

$$k(x,s) = 1 - \frac{2}{l}|x - s|.$$

Show that k is an H-S kernel.

6. Let $p_1, \ldots, p_n$ be linearly independent functions defined on $[0,l]$ and define k by the formula

$$k(x,s) = \sum_{j=1}^{n} p_j(x)p_j(s).$$

Determine the conditions the functions $p_1, \ldots, p_n$ must satisfy in order that k be an H-S kernel. Discuss the solution of the integral equation with k as kernel. [*Hint:* For $\mu \neq 0$, ψ must have the form $\sum_{j=1}^{n} c_j p_j(x)$. What happens if $\mu = 0$?]

7. Prove that if k is a symmetric kernel, then k_1 is also.
8. Show that the kernel of Example 4 is a regular kernel.
9. Suppose the kernel k is defined as in Example 3 and satisfies the conditions of Lemma 5.2. Using the inequality

$$\int_0^l |f(x)g(x)|\, dx \leq \left(\int_0^l f^2(x)\, dx\right)^{1/2} \left(\int_0^l g^2(x)\, dx\right)^{1/2} \quad (5.92)$$

valid for any pair of square-integrable functions, show that k_1 is an H-S kernel.

10. If $k(x,s) = xs$ solve the integral equation.

5.10 SOLUTIONS OF THE INTEGRAL EQUATION

In Section 5.8 we called the values of ω for which (5.83) has a nontrivial solution eigenvalues of the integral equation. Let us now agree to call the reciprocals of the eigenvalues the characteristic values. In most of the modern literature the basic theorems are formulated so that they apply directly to the characteristic values. We will follow this procedure here. We will start by quoting without proof a number of basic results. Proofs can be found in a variety of texts bearing the title of integral equations or boundary-value problems or functional analysis, but they involve methods generally beyond the scope of this text.

Theorem 5.3

Let k be a symmetric, positive definite, H-S kernel. Then the integral equation (5.87) has an infinite sequence of positive characteristic values

$$\mu_1 \geq \mu_2 \geq \cdots$$

with
$$\lim_{n\to\infty} \mu_n = 0. \tag{5.93}$$

The largest characteristic value μ_1 satisfies
$$\max_u \frac{(Ku,u)}{\|u\|^2} = \mu_1, \tag{5.94}$$

and the maximum is achieved when $u = c\psi_1$ where c is an arbitrary nonzero constant and ψ_1 is any eigenfunction corresponding to μ_1. In general, the $(n+1)$th characteristic value satisfies
$$\max_u \frac{(Ku,u)}{\|u\|^2} = \mu_{n+1} \tag{5.95}$$

where the functions u are subject to the constraints
$$(u,\psi_1) = 0, \ldots, (u,\psi_n) = 0 \tag{5.96}$$

where $\psi_1, \ldots, \psi_n$ are eigenfunctions belonging to the characteristic values $\mu_1, \ldots, \mu_n$. The maximum is achieved when $u = c\psi_{n+1}$ where c is an arbitrary nonzero constant and ψ_{n+1} is any eigenfunction corresponding to μ_{n+1}.

In order to understand this theorem properly we must make several remarks. In (5.94) the maximum is to be computed over all square-integrable functions. Observe that if the function $u = \psi_1$ satisfies
$$\frac{(K\psi_1,\psi_1)}{\|\psi_1\|^2} = \mu_1$$

and if $c \neq 0$ is a constant, then
$$(Kc\psi_1,c\psi_1) = c^2(K\psi_1,\psi_1)$$

and
$$\|c\psi_1\|^2 = (c\psi_1,c\psi_1) = c^2(\psi_1,\psi_1) = c^2\|\psi_1\|^2,$$

hence
$$\frac{(Kc\psi_1,c\psi_1)}{\|c\psi_1\|^2} = \mu_1.$$

In particular we may choose c such that $\|\psi_1\| = 1$. Such an eigenfunction will be called a *normalized eigenfunction* corresponding to μ_1. Now given μ_1 and a corresponding eigenfunction ψ_1, formula (5.95) states that μ_2 is the maximum of the quotient $(K\psi,\psi)/\|\psi\|^2$ where the maximum is taken over all square-integrable functions u satisfying $(\psi_1,u) = 0$, that is, over all u orthogonal to ψ_1. Since this class of functions is smaller than the class of all square-integrable functions, it must be true that

$$\mu_2 = \max_{\substack{u \\ \text{with } (u,\varphi_1)=0}} \frac{(Ku,u)}{\|u\|} \leq \max_u \frac{(Ku,u)}{\|u^2\|} = \mu_1.$$

Similarly we must have $\mu_{n+1} \leq \mu_n$. Under the hypothesis of the theorem there is no guarantee that strict inequality will hold for a given value of n.

In view of these remarks it makes sense to speak of the distinct characteristic values as the distinct positive real numbers appearing in the sequence $\{\mu_1, \mu_2, \ldots |\}$. We may also speak of the multiplicity of a characteristic value; defining it to be the number of times it appears in the above sequence. The multiplicity is also the number of linearly independent eigenfunctions which belong to the given characteristic value. One might guess that this is so from the fact that the successive eigenfunctions are orthogonal to all previous eigenfunctions.

Theorem 5.3 actually implies that every characteristic value of (5.87) has finite multiplicity. We state this formally in:

Corollary 5.1

Every characteristic value of (5.87) has finite multiplicity.

Theorem 5.3 also implies that the eigenfunctions belonging to the characteristic values form an orthogonal sequence. We have also pointed out that they can be normalized. We formalize these facts in:

Corollary 5.2

The sequence of eigenfunctions satisfies

$$(\psi_n, \psi_m) = 0, \quad \text{if} \quad n \neq m \quad (\psi_n, \psi_n) = 1, \quad \text{for all } n. \tag{5.97}$$

It should be observed that Theorem 5.3 and its two corollaries does not use the hypothesis that k is regular. When this is adjoined we obtain additional important consequences. In order to formulate these effectively we must use the concept of uniform convergence.

Definition 5.5

We say the infinite series $\sum_{n=1}^{\infty} u_n(x)$ converges uniformly on $[0,l]$ to the function $u(x)$ if given $\epsilon > 0$ there exists an integer M depending only upon ϵ (and not upon x in $[0,l]$) such that for all $m > M$

$$|u(x) - s_m(x)| < \epsilon$$

where $s_m(x)$ is the mth partial sum

$$s_m(x) = \sum_{n=1}^{m} u_n(x).$$

In words, the partial sum must be close to u for sufficiently large m on the entire interval.

It is, of course, true that if a series converges uniformly on $[0,l]$ to u, then for each fixed x_0, in $[0,l]$ the ordinary numerical series $\sum_{n=1}^{\infty} u_n(x_0)$ converges to $u(x_0)$.

Theorem 5.4

Let k be a symmetric, positive definite, and regular H-S kernel. Then

(i) every eigenfunction is continuous on $[0,l]$, and
(ii) every function of the form Kf where f is square-integrable can be written in the form

$$Kf = \sum_{n=1}^{\infty} \mu_n(Kf,\psi_n)\psi_n, \tag{5.98}$$

and the series converges uniformly on $[0,l]$.

Theorem 5.5

Let k satisfy the conditions of Theorem 5.3, then the sequence of eigenfunctions is complete.

We recall from Section 3.3 that completeness means that given any square-integrable function u we have

$$\left\| u - \sum_{n=1}^{N} (u,\psi_j)\psi_j \right\| \to 0 \quad \text{as} \quad N \to \infty.$$

This in turn implies that we have the unique representation

$$u = \sum_{n=1}^{\infty} (u,\psi_j)\psi_j, \tag{5.99}$$

where the series is understood to converge in the sense just indicated above.

With these facts at hand let us work our way back to the integro-differential equation of Sections 5.7 and 5.8. From Theorem 5.3 and the relation (5.86) we deduce that the eigenvalues ω form an infinite sequence

SOLUTIONS OF THE INTEGRAL EQUATION

$$0 < \omega_1 \leq \omega_2 \leq \cdots \leq \omega_n \leq \cdots,$$

with

$$\lim_{n \to \infty} \omega_n = +\infty.$$

In order to make the most effective use of the theorems quoted above we will multiply Equation (5.69) by $\rho^{1/2}(x)$ and write it in the form

$$z(x,t) + \int_0^l k(x,s) z_{tt}(s,t)\, ds = 0,$$
$$z(x,t) = \rho^{1/2}(x) y(x,t). \tag{5.100}$$

Since ρ is positive in the sense that $0 < \rho_0 \leq \rho(x) \leq \rho_1$ on $[0,l]$, we may solve for z and easily recover y.

Now if we seek solutions in the form

$$z(x,t) = \psi(x) \sin(\sqrt{\omega}\, t + \delta)$$

we find that

$$\sin(\sqrt{\omega}\, t + \delta) \left\{ \psi(x) - \omega \int_0^l k(x,s) \psi(s)\, ds \right\} = 0.$$

In other words, ψ must satisfy Equation (5.84). Therefore in view of Theorem 5.3 we have an infinite sequence of solutions

$$z_n(x,t) = \psi_n(x) \sin(\sqrt{\omega_n}\, t + \delta_n)$$

corresponding to the eigenvalues ω_n, $n = 1, 2, \ldots$.

By (5.68) we have

$$y(x,0) = f(x), \quad y_t(x,0) = g(x),$$

hence

$$z(x,0) = \rho^{1/2}(x) f(x) = F(x), \quad z_t(x,0) = \rho^{1/2}(x) g(x) = G(x). \tag{5.101}$$

In order to find a solution of (5.99) also satisfying (5.100) we use the formal series

$$z(x,t) = \sum_{n=1}^{\infty} a_n \psi_n(x) \sin(\sqrt{\omega_n}\, t + \delta_n), \tag{5.102}$$

in which the constants a_n and δ_n are to be determined. From the first of (5.100) we derive

$$F(x) = z(x,0) = \sum_{n=1}^{\infty} a_n \sin \delta_n \psi_n(x).$$

But according to (5.98) we must have

$$F = \sum_{n=1}^{\infty} (F,\psi_n)\psi_n$$

and this representation is unique. Comparing the two series expansions we deduce that

$$a_n \sin \delta_n = (F,\psi_n), \quad n = 1, 2, \ldots . \qquad (5.103)$$

From the second of (5.100) we derive

$$G(x) = z_t(x,0) = \sum_{n=1}^{\infty} a_n \sqrt{\omega_n} \cos \delta_n \psi_n(x)$$

and from (5.99)

$$G = \sum_{n=1}^{\infty} (G,\psi_n)\psi_n,$$

the representation being unique. Therefore

$$a_n \cos \delta_n = \frac{1}{\sqrt{\omega_n}} (G,\psi_n), \quad n = 1, 2, \ldots . \qquad (5.104)$$

Working out (5.103) and (5.104) we obtain finally the formal representation

$$z(x,t) = \sum_{n=1}^{\infty} \left[(F,\psi_n)^2 + \frac{1}{\omega_n} (G,\psi_n)^2 \right]^{1/2} \psi_n(x) \sin (\sqrt{\omega_n}\, t + \delta_n),$$
$$\delta_n = \tan^{-1} \frac{\sqrt{\omega_n}\, (F,\psi_n)}{(G,\psi_n)}. \qquad (5.105)$$

PROBLEMS

1. Compute the constants a_n and δ_n when $F(x) \equiv 0$ on $[0,l]$.
2. Solve Problem 1 if $G(x) \equiv 0$ on $[0,l]$.
3. Suppose $F(x) \equiv G(x)$ on $[0,l]$. Show that in this case $\lim_{n \to \infty} \delta_n = \pi/2$. What value does the limit have if $F = -G$ on $[0,l]$?

The next problems refer to the nonhomogeneous equation

$$K\psi = \mu\psi + f, \qquad (5.106)$$

where f is a given square-integrable function defined on $[0,l]$.

4. Let $\sum_{n=1}^{\infty} a_n \varphi_n$ and $\sum_{n=1}^{\infty} b_n \varphi_n$ be the Fourier series of ψ and f respectively,

where $\{\varphi_k\}$ are the eigenfunctions and $\{\mu_k\}$ the eigenvalues of the homogeneous equation. Show that

$$a_n(\mu_n - \mu) = b_n, \quad n = 1, 2, \ldots, n, \ldots,$$

if ψ is a solution of (5.106).

5. Assume $\mu \neq 0$ and $\mu \neq \mu_k$ for all k. Show that in this case

$$\psi = -\frac{f}{\mu} + \sum_{n=1}^{\infty} \frac{\mu_n b_n}{\mu(\mu_n - \mu)} \varphi_n$$

is a solution of (5.106).

6. Assume $\mu \neq 0$ and $\mu = \mu_m$ for some fixed integer m. Then if μ_m is a simple eigenvalue;
   (a) if $b_m \neq 0$, the integral equation (5.106) has no solution and
   (b) if $b_m = 0$, the equation can have infinitely many solutions. Give the form of these solutions in case (b). In particular, show that such solutions exist only if $(f, \varphi_m) = 0$.

7. Suppose μ satisfies the conditions of the previous problem and that $b_m = 0$ and μ_m possess the eigenfunctions $\varphi_m, \varphi_{m+1}, \ldots, \varphi_{m+p}$, that is, μ_m has multiplicity $p + 1$. Show that the integral equation (5.106) has a solution only if $(f, \varphi_m) = (f, \varphi_{m+1}) = \cdots = (f, \varphi_{m+p}) = 0$ are all satisfied. Give the form of all possible solutions in this case.

Index

Absolute convergence, 143
Adjoint, 228
Algebraic complement, 89
Angular velocity, 38, 100
Approximation, of a derivative, 246
   of an integral, 247
Asymptotic behavior of solutions, 24, 106, 162

Beam, Bernoulli-Euler, 210, 231
   cantilever, 212
   Rayleigh (Problem 2), 235
   Timoshenko, 231
Bessel's identity, 136
Bessel's inequality, 137
Boundary conditions, 119
Boundary-value problem, 119, 128 (Problems 2–16)
   solution to, 119
Bounded real-valued function, 24
Buckling load, 222

Castigliano's theorem, 263
Cauchy heat-equation problem, 121
Cauchy wave-equation problem, 183, 187
Central difference, 246

Chain rule, 187, 188
Characteristic, equation, 5, 219
   lines, 191
   number (*see* Eigenvalue)
   triangle, 196
Characteristic value, 277
   multiplicity of, 279
Closed interval, 28
Commutative, 68
Complementary minor, 89
Complete sequence of functions, 137 (Definition 3.4), 280 (Theorem 5.5)
Complex eigenvalues, 7, 102
Complex number, 4
   product with a vector, 72
Complex solution, ODE, 7
   PDE, 124
   system of ODE (Proposition 2.1), 57
Complex vector, 76
Conduction, equation of, 120
Conservation of energy, 32 (Problem 7), 108, 198, 214 (Definition 4.1)
Conservative system (*see* Conservation of energy)

Continuous dependence (Corollary 3.2), 151
Convergence, absolute, 143
  conditional, 143
  of Fourier series (Theorem 3.6), 146
  uniform (Definition 5.5), 279
Cramer's rule, 6
Critical angular velocity, 100

D'Alembert's formula, 188
Damped system (Problem 9), 61
Damping matrix, 101
Degree of freedom, 58
Dependent variable, 4
Derivative, one-sided, 145
Determinant, of a matrix, 89
  Wronski, 12
Diagonal matrix (Problem 4), 78
Difference, approximation, 246
  quotient, 145, 247
Differential equation, 3, 115
  homogeneous, 21
  linear homogeneous, 122
  nonlinear, 171
  partial, 118
  system of, 81
  variable coefficients, 24ff., 237
Diffusion problem, 139
Directional derivative, 195
Discrete system, 256–258
Domain of dependence, 196
Double pendulum, 52
Duffing's equation (Problem 19), 49

Eigenfunctions, of a boundary-value problem, 131
  of an integral equation, 266
Eigenvalues, of an integral equation, 266
  of a matrix, 70
  of an ODE, 5
  of a system of ODE, 82, 100, 252
Eigenvectors, of a matrix, 70, 101
  of a system of ODE, 82
Energy, conservation of, 32 (Problem 7), 108, 198, 214 (Definition 4.1)
  dissipation of, 26, 108
  identity, 30, 108
Energy method, 21
  nonlinear, 172
  ODE, system of, 107
    with variable coefficients, 24
  PDE, 162, 193, 212
    with variable coefficients, 239
Equilibrium position, 43
Euler, buckling problem, 226
  critical load, 227
  formula for a complex number, 5
Existence of a solution (Theorem 4.1), 189
Exponential form of a solution, 4

Finite differences, 246, 256
First-order system, cascaded loops, 65
  general, 96
Follower problem, 215, 224
Forcing function, 1, 50 (Problem 25), 154
Fourier method, 131
Fourier series, 127, 143
  coefficients of, 126, 144
  convergence of (Theorem 3.6), 146
  eigenfunctions of, 131
Fourier, sine series, 127
Fredholm integral equation, 266
Free-body diagram, 257
Functions, Green's (*see* Green's function)
  infinite sequence of, 131
  linear independence of (Definition 3.1), 131
  nondecreasing, 28
  norm of, 133
  normalized, 134
  piecewise continuous, 144
  quasi-differentiable (Definition 3.6), 145
  real-valued, 12
  strictly increasing, 28

Generalized coordinates, 52
  (*See also* Degree of freedom)
Green's function, for ODE, 45
  for PDE, 127, 156, 207

INDEX

Half-open interval, 28
Heat, density of source, 116
    flow in a rod, 115
Heat equation, 118
    Cauchy problem, 121
    nonhomogeneous equation, 154
    nonlinear equation, 171
    properties of, 122
    uniqueness of solution, 149
Hilbert-Schmidt, integral operator, 271
    kernel, 271
Hooke's Law, 182
Homogeneous, differential equation (Problem 12), 21, 122
    material, 115
    problem, ODE, 21
    PDE, 119

Identity matrix, 77 (Problem 3), 89
Independent variable, 4
Infinite sequence of functions, 131
Influence function, 259
Initial condition, 3
Initial-value problem, 3
    well-posed, 54
Inner product (*see* Scalar product)
Integro-differential equation, 2, 264
Interior point, 12
Inverse of matrix (*see* Matrix)

Kernel, 266
    positive, 275
    regular kernel, 274
Kinetic energy, 42, 52, 265
Kirchhoff's voltage law, 1
Kroneker, delta, 90

Lagrange's equations of motion, for conservative systems, 44
    for nonconservative systems (Problem 13), 48
Laplace, stability in the sense of, 24
Leakage problem, 4
Liebnitz' rule (Problem 11), 19
Limit, from left (Definition 3.5), 144
    from right (Definition 3.5), 144

Linear combination, 73
Linear superposition, ODE (Proposition 1.1), 6
    PDE, 125
    system of ODE (Proposition 2.1), 57
    vectors (Proposition 2.2), 72
Linearization, 53
Linearly dependent functions (Definition 1.3), 12
Linearly independent, functions (ODE) (Definition 1.3), 12
    functions (PDE) (Definition 3.1), 131
    orthonormal sequence (Theorem 3.4), 134
    solutions (PDE), 124
    vectors, 74, 86

Mathematical model, 256, 257
Matrix, 67
    column vector, 67
    determinant of, 89
    diagonal (Problem 4), 78
    identity, 77 (Problem 3), 89
    inverse of, 90
    minor of, 89
    nonsingular (Definition 2.2), 90
    product with a matrix, 68
    product with a vector, 68
    row vector, 67
Maximum-value principle (Theorem 3.7), 147
Maxwell reciprocity theorem, 264
Mean-value theorem, 40
    of differential calculus (Theorem 3.1), 117
    of integral calculus (Theorem 3.2), 118
Movable point, 260

Newton's second law, 37
Nonconservative problem, 215, 223 (Problem 13)
    circuit (Problem 12), 32
Nondecreasing function, 28

288 INDEX

Nonlinear systems, Duffing's equation (Problem 19), 49
  heat equation, 171
  pendulum, 38
Nonself-adjoint, 228
Nonsingular matrix (Definition 2.2), 90
Nontrivial solution, ODE, 12
  PDE, 124
Norm of a function, 133
Normalized, eigenfunction, 135, 278
  function (Definition 3.3), 134

One-sided derivative, 145
Open interval, 12
Order of differential equation, 4
Ordinary differential equations, 1, 51
  nonconstant coefficients, 24, 98 (Example 3)
  solution to (Definition 1.1), 3
  (*See also* Differential equation)
Orthogonal, functions (Definition 3.3), 134
  set (Definition 3.3), 134
Orthonormal sequence (Definition 3.3), 134

Partial differential equations, 115, 181
  nonconstant coefficients, 182, 225, 237
  solution to, 119
  (*See also* Differential equation)
Pendulum, double, 52
  linear, 44
  nonlinear, 36
Periodic extension, 145
Phase angle, 47, 50 (Problem 25)
Phase plane, 39
  closed trajectories, 41
  nonsingular trajectories, 40
  separatrices, 41
  singular solution, 39
  singular trajectories, 39
Piecewise continuous function, 144
Potential energy, 42, 53, 265
Principle of superposition (*see* Linear superposition)

Quasi-differentiable function, 145

Real-valued function, 12
Regular kernel (Definition 5.3), 274
RLC circuit, 1
  capacitance (c), 1
  charge (q), 1
  electromotive source $(e(t))$, 1
  general solution, 15
  inductance (L), 1
  leakage (homogeneous) problem, 4, 23
  resistance (R), 1
Rotating shaft, 100, 106

Scalar product, 67, 133
  of real-valued functions (Definition 3.2), 133
Schwarz's inequality (Problem 10), 153
Second-order differential equation, 4
Self-adjoint, 228
Separation of variables, 201, 216, 240
  beam equation, 216
  variable coefficient, 240
  wave equation, 201
Simple harmonic motion, 46
Simple pendulum, 36
Solution to, ODE (Definition 1.1), 3
  PDE, 119
Stable, asymptotically, 24
  differential equation, 24
  in the sense of Laplace, 24
Stiffness matrix, 101
Strictly increasing function, 28
String, vibrations of, 184 (Problems 1, 2), 238
Sturm-Liouville problem, 240
Superposition of solutions (*see* Linear superposition)
Symmetric integral equation, 266
  properties of, 270
Systems, 51

Taylor expansion with remainder, 44, 246

Thermal conductivity, 116
Trivial solution, 11

Uniform convergence (Definition 5.5), 279
Uniqueness of, ODE (Theorem 1.3), 33, 35 (Theorem 1.4)
 heat-conduction problem, 149
 leakage problem, 20
 wave equation, 193, 199 (Theorem 4.4)

Variable coefficients, ODE, 24
 PDE, 237
Variation of parameters, 8, 16, 87
Vector, complex, 75
 components, 66
 linear independence, 74 (Definition 2.1), 86
 product with a scalar, 66
 scalar product, 67
 sum, 66
Vector-valued function, 69

Vibrations, longitudinal of a rod, 181, 237
 string (*see* String vibrations)
 torsion of a rod (Problem 4), 184 (Problem 15), 186
 transverse of a rod, 210, 231
 (*See also* Beam, Pendulum, *RLC* circuit)

Wave equation, 189
 homogeneous, 183, 187
 nonhomogeneous, 182
 traveling waves, 189
 variable coefficients, 237
Wave front, 191
Well-posed problem, 54, 118, 152
Whirling motion of a rotating system, 106
Wronski determinate, 12
Wronskian (Definition 1.4), 13

Zero, of an equation, 71
 vector, 66